Molecular Basis of Oncology

MOLECULAR BASIS OF CLINICAL MEDICINE

General Series Editor

C. Thomas Caskey, MD

Director, Institute for Molecular Genetics

Investigator, Howard Hughes Medical Institute

Baylor College of Medicine

Houston, Texas

Other titles in the series:

P. Michael Conneally: Molecular Basis of Neurology

R. Roberts: Molecular Basis of Cardiology

Molecular Basis of Oncology

EDITED BY

Emil J Freireich, MD, DSc (Hon)

Division of Hematology
University of Texas
M.D. Anderson Cancer Center
Houston, Texas

Sanford A. Stass, MD

Division of Hematopathology Program
Division of Laboratory Medicine
University of Texas
M.D. Anderson Cancer Center
Houston, Texas

Blackwell
Science

Blackwell Science
EDITORIAL OFFICES:
238 Main Street, Cambridge, Massachusetts 02142, USA
Osney Mead, Oxford OX2 OEL, England
25 John Street, London WC1N 2BL, England
23 Ainslie Place, Edinburgh EH3 6AJ, Scotland
54 University Street, Carlton, Victoria 3053, Australia
Arnette Blackwell SA, 1 rue de Lille, 75007 Paris, France
Blackwell Wissenschafts-Verlag GmbH, Kurfürstendamm 57, 10707 Berlin, Germany
Blackwell MZV, Feldgasse 13, A-1238 Vienna, Austria

DISTRIBUTORS:
North America
Blackwell Science, Inc.
238 Main Street
Cambridge, Massachusetts 02142
(Telephone orders: 800-215-1000 or 617-876-7000)
Australia
Blackwell Science Pty Ltd
54 University Street
Carlton, Victoria 3053
(Telephone orders: 03-347-5552)
Outside North America and Australia
Blackwell Science, Ltd.
c/o Marston Book Services, Ltd.
P.O. Box 87
Oxford OX2 0DT
England
(Telephone orders: 44-865-791155)

Acquisitions: Victoria Reeders
Development: Coleen Traynor
Production: Michelle Choate
Manufacturing: Kathleen Grimes
Typeset by Best-set Typesetter, Hong Kong
Printed and bound by Braun-Brumfield, Ann Arbor, MI

Printed in the United States of America
95 96 97 98 5 4 3 2 1

Library of Congress Cataloging in Publication Data
Molecular basis of oncology / [edited by] Emil J Freireich, Sanford A. Stass.
p. cm.—(Molecular basis of clinical medicine)
Includes bibliographical references and index.
ISBN 0-86542-254-0
1. Tumors—Genetic aspects. 2. Tumors—Molecular aspects. I. Freireich, Emil J, 1927– . II. Stass, Sanford A., 1943– III. Series.
[DNLM: 1. Neoplasms—genetics. 2. Molecular Biology. QZ 200 M718316 1995]
RC268.4.M635 1995
616.99'4042—dc20
DNLM/DLC
for Library of Congress 94-43841
CIP

Contents

Contributors

David Berger, MD
Department of Tumor Biology, University of Texas, M.D. Anderson Cancer Center, Houston, Texas

C. Thomas Caskey, MD
Department of Molecular and Human Genetics, Baylor College of Medicine, Houston, Texas

Kun-Sang Chang, PhD
Division of Laboratory Medicine, University of Texas, M.D. Anderson Cancer Center, Houston, Texas

Leland W.K. Chung, PhD
Urology Research Laboratory, University of Texas, M.D. Anderson Cancer Center, Houston, Texas

Jeff Deyo
Department of Tumor Biology, University of Texas, M.D. Anderson Cancer Center, Houston, Texas

Zeev Estrov, MD
Department of Clinical Investigation, University of Texas, M.D. Anderson Cancer Center, Houston, Texas

Marsha L. Frazier, PhD
University of Texas, M.D. Anderson Cancer Center, Houston, Texas

Emil J Freireich, MD
Department of Hematology, University of Texas, M.D. Anderson Cancer Center, Houston, Texas

Marc F. Hansen, PhD
Department of Molecular Genetics, University of Texas, M.D. Anderson Cancer Center, Houston, Texas

Cheryl F. Hirsch-Ginsberg, MD
Division of Laboratory Medicine, University of Texas, M.D. Anderson Cancer Center, Houston, Texas

Mien-Chie Hung, PhD
Department of Tumor Biology, University of Texas, M.D. Anderson Cancer Center, Houston, Texas

William B. Isaacs, MD
Department of Urology, Johns Hopkins Hospital, Baltimore, Maryland

Geoffrey R. Kitchingham, PhD
Associate Member, Department of Virology and Molecular Biology, St. Jude Children's Research Hospital, Memphis, Tennessee

Razelle Kurzrock, MD
Department of Clinical Investigation, University of Texas, M.D. Anderson Cancer Center, Houston, Texas

Richard S. Morrison, MD
Department of Neurological Surgery, University of Washington Medical Center, Seattle, Washington

Shawn Panzer, MD
Department of Molecular and Human Genetics, Johns Hopkins University School of Medicine, Baltimore, Maryland

Mark A. Pershouse, PhD
Department of Neuro-Oncology, The Brain Tumor Center, University of Texas, M.D. Anderson Cancer Center, Houston, Texas

Jack A. Roth, MD
Department of Thoracic Surgery, University of Texas, M.D. Anderson Cancer Center, Houston, Texas

E. Cristy Ruteshouser, PhD
Department of Molecular Genetics, University of Texas, M.D. Anderson Cancer Center, Houston, Texas

Hideyuki Saya, MD, PhD
Department of Neuro-Oncology, The Brain Tumor Center, University of Texas, M.D. Anderson Cancer Center, Houston, Texas

David S. Schrump, MD
University of Texas, M.D. Anderson Cancer Center, Houston, Texas

Michael J. Siciliano, PhD
Department of Molecular Genetics, University of Texas, M.D. Anderson Cancer Center, Houston, Texas

Sanford A. Stass, MD
Division of Laboratory Medicine, University of Texas, M.D. Anderson Cancer Center, Houston, Texas

Peter A. Steck, PhD
Department of Neuro-Oncology, The Brain Tumor Center, University of Texas, M.D. Anderson Cancer Center, Houston, Texas

Michael A. Tainsky, PhD
Department of Tumor Biology, University of Texas, M.D. Anderson Cancer Center, Houston, Texas

Moshe Talpaz, MD
Department of Clinical Investigation, University of Texas, M.D. Anderson Cancer Center, Houston, Texas

Jan Trapman, MD
Department of Pathology, Erasmus University, Rotterdam, The Netherlands

Meir Wetzler, MD
Department of Clinical Investigation, University of Texas, M.D. Anderson Cancer Center, Houston, Texas

Alan M. Yahanda, MD
Department of Neurosurgery and Tumor Biology, and Surgical Oncology University of Texas, M.D. Anderson Cancer Center, Houston, Texas

Dihua Yu, MD, PhD
Department of Tumor Biology and Surgical Oncology, University of Texas, M.D. Anderson Cancer Center, Houston, Texas

Introduction

All physicians and health care professionals are aware of the revolution in biology which is a consequence of the new knowledge of molecular genetics. Although for many areas of fundamental knowledge the application to clinical medicine is gradual and deliberate, in the case of molecular genetics, the impact of clinical medicine has been immediate and quite dramatic. Perhaps for no other area of clinical medicine has molecular genetics had a more profound impact than on the field of oncology. Cancer is fundamentally a genetic disease. The primary change in the genetic makeup of a host cell capable of providing a survival advantage for that population results in unregulated accumulation of the abnormal cell population. The survival advantage includes not only a failure to respond to regulatory signals, but also the ability to metastasize, that is, to spread to other organs, and to compromise their function. The molecular genetic changes in the clonal population of cells that ultimately results in malignancy are not only fundamental to the process but also enormously heterogeneous. Virtually every malignancy is associated with multiple and extremely heterogeneous changes. For some tumors, specific subsets that identify a small fraction of all patients with the same histologic diagnosis are associated with unique clinical behavior, unique requirements for treatment, and unique prognostic implications. The editors of this publication felt that an introductory volume of the *Molecular Basis of Oncology* was important to provide insight into the genetics of cancer at the molecular level. Although this field has existed for less than 30 years, it is already enormously complex and encompasses a literature that is much too extensive to be covered in a single useful volume. Therefore, the editors have attempted to identify fundamental mechanisms that are shared by many different forms of malignancy and to augment that information with examples of specific malignancies in which molecular genetic information has had a profound influence on not only diagnosis but also early detection, prevention, treatment and prognosis.

The first chapter focuses on some of the principles of molecular biology as it relates to malignant diseases. We then turn to genetic events that increase susceptibility to malignancy and an understanding of how they interact with the acquired changes. We then move to the prototypic oncogene, the RAS family of oncogenes that promote growth and follow that with a lucid discussion of the retinoblastoma gene and other tumor suppressors. The important information about the HER-2/*neu* gene, particulary important in breast cancer and ovarian cancer, both for prognosis and as a target for treatment is followed by a discussion of the exciting new field of tumor angiogenesis and other proteins that regulate tumor growth and metastasis. The seventh chapter deals with chronic myelogenous leukemia, a disease for which detailed knowledge of molecular genetic changes is available. This is followed by discussions of the acute leukemias describing how cytogenetic and molecular genetic information has become fundamental to the diagnosis and management of patients. The final four chapters deal with the common malignancies of lung cancer, colon cancer, prostate cancer, and finally central nervous system tumors. These tumors are associated with molecular events that are important in understanding the biology and may also represent therapeutic targets. Thus, the rapidly developing molecular knowledge about these tumors required special presentation.

The editors hope that this volume will serve to move the reader quickly to the level of basic knowledge about molecular genetics in clinical oncology. The ability to maintain fluency in this rapidly growing field will augment the physician–scientist's ability to deal with future issues in oncologic diseases in the most effective way for the benefit of the patient.

Notice

The indications and dosages of all drugs in this book have been recommended in the medical literature and conform to the practices of the general medical community. The medications described do not necessarily have specific approval by the Food and Drug Administration for use in the diseases and dosages for which they are recommended. The package insert for each drug should be consulted for use and dosage as approved by the FDA. Because standards of usage change, it is advisable to keep abreast of revised recommendations, particularly those concerning new drugs.

Molecular Basis of Oncology

CHAPTER 1

Principles of Molecular Oncology

Cheryl F. Hirsch-Ginsberg
Sanford A. Stass
Emil J Freireich

Cancer is a group of diseases characterized by an autonomous proliferation of neoplastic cells that are genetically dysfunctional. All cellular functions are controlled by proteins, and because these proteins are encoded by DNA organized into genes, molecular studies have shown that cancer is a paradigm of acquired genetic disease. In 1953 James Watson and Francis Crick discovered the structure of DNA. Since that time, the development of tools and techniques to explore the structure and function of the genes has fostered an evolving understanding of how the various genetic alterations affect protein structure, thus inducing oncogenesis. In order to discuss the specific known oncogenic alterations, it is important to have a basic understanding of some key concepts of gene structure, as well as the historical background on which these concepts are built.

The basic concepts in genetics—that genes reside on chromosomes, which are composed of proteins and DNA, and that each protein or enzyme is encoded by a particular gene—began with Mendel's pea plant breeding experiments in the 1860s, and involved work by such noted scientists as Nettie Stevens, Edmund Wilson, Archibald Garrod (author of the one protein–one gene hypothesis), and Linus Pauling, who in 1949 published a paper on sickle cell anemia and it's origin from mutant hemoglobin. In the 1920s, Phoelous demonstrated that DNA was composed of deoxyribose, along with either of two purine bases (adenine and guanine) and two pyrimidine bases (thymine and cytosine), while Alexander Todd in the early 1950s showed that a DNA chain is formed by phosphodiester bonds connecting the 5′ carbon atom of one deoxyribose to the 3′ carbon of the successive nucleotide. In 1952, building on works of Oswald Avery, Colin MacLeod, Maclyn McCarty, and others, Alfred Hershey and Martha

Chase established, once and for all, that the chromosomal DNA, and not the chromosomal protein, was responsible for genetic specificity. The final key element preceding Watson and Crick's seminal paper was the work of Edwin Chargraff, who showed that although the amounts of the individual bases in any given piece of DNA could vary widely, the ratios of adenosine and guanine to thymine and cytosine, respectively, were always equal. At the same time the crystallography experiments of M. H. F. Wilkins and Rosalind Franklin demonstrated 1) the helical nature of DNA, 2) that DNA was composed of more than one chain, and 3) that the 3.4 Å thick purine and pyrimidine bases are stacked perpendicular to the helical axis. The final answer to DNA structure, solved by Watson and Crick in 1953, was a double-stranded helical polymer composed of a core consisting of stacked hydrogen-bonded complementary base pairs (A-T and G-C), with the attached deoxyribose chains as an outer backbone, with each strand running in opposite (antiparallel) directions (5′ to 3′ on one chain, and 3′ to 5′ on the other) (Fig. 1.1).

Concurrent with the discoveries of the intricacies of DNA structure were the discoveries of another single stranded nucleic acid polymer, RNA. The only differences in this molecule are that the pyrimidine uracil takes the place of thymine, and that ribose, rather than deoxyribose, forms the sugar backbone (still connected by 5′-3′ phosphodiester linkages). With the observation that most RNA is found in the cytoplasm, where protein synthesis was thought to take place, Francis Crick in 1956 proposed his central dogma for the flow of genetic information from chromosome to protein:

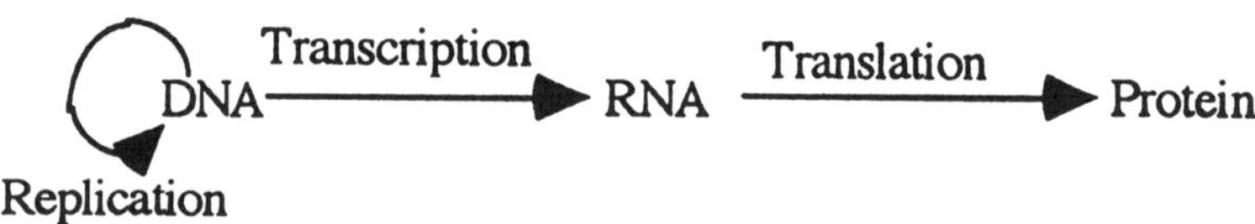

The circular arrow stands for DNA strands, which because of their complementary structure are able to act as templates on which another strand is produced. These strands also act as a template upon which RNA is produced (transcribed). This RNA molecule is then transported to the cytoplasm, where it is translated in protein (Fig. 1.2).

Previous knowledge that there are 20 amino acids involved as building blocks of proteins, and simple mathematical calculations as to the minimum number of nucleotides that would have to be grouped together to provide the 20 discrete choices, led to the conclusion that three nucleotides grouped together would be necessary to code for all of the specific amino acids. This universal genetic code, which provides the key for the translation of groups of three nucleotides (codons) into proteins,

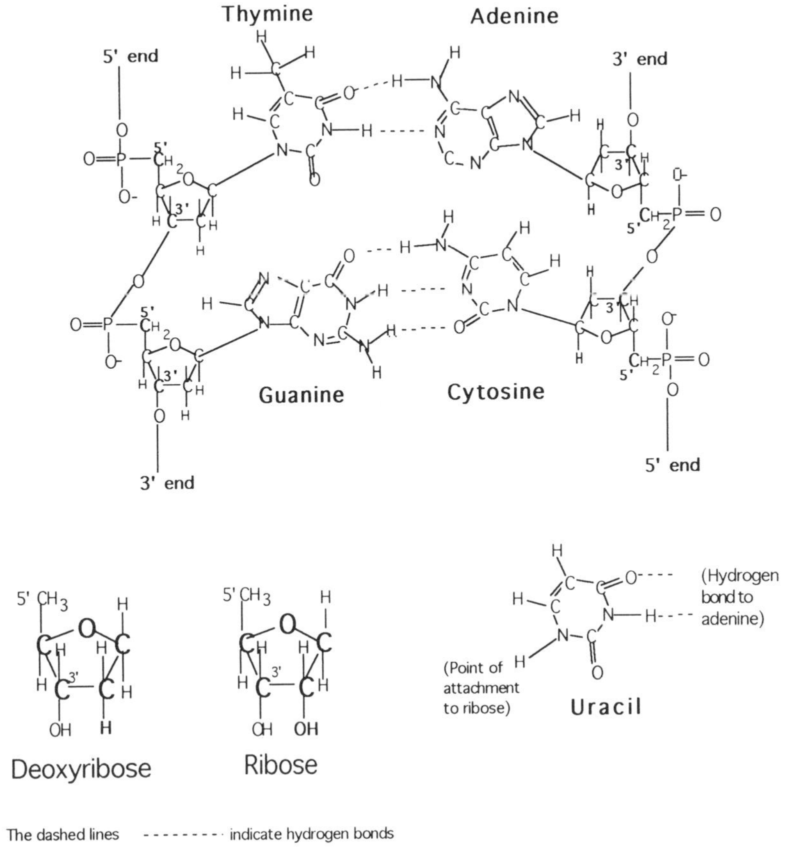

The dashed lines -------- indicate hydrogen bonds

Fig. 1.1: Schematic diagram of the molecular structure of DNA including the hydrogen-bonded nucleotides (dashed lines) as well as the configuration of phosphodiester bonds in the two deoxyribose chains.

was finalized in 1966 and involved the work of many, including C. Yanofsky, A. Brenner, M. W. Nirenberg, J. H. Matthaei, F. Crick, and H. G. Khorana. Their works revealed that 61 of 64 ($4 \times 4 \times 4$) possible permutations coded for amino acids, with most amino acids being encoded by more than one triplet (degenerate), while the three additional possibilities serve as "stop" (translation) signals. Other investigations revealed that translation always starts with the triplet codon ATG, which corresponds to the amino acid methionine (Met) (Fig. 1.3).

The elucidation of the processes of transcription and translation involved the discovery of many enzymes, including DNA and RNA poly-

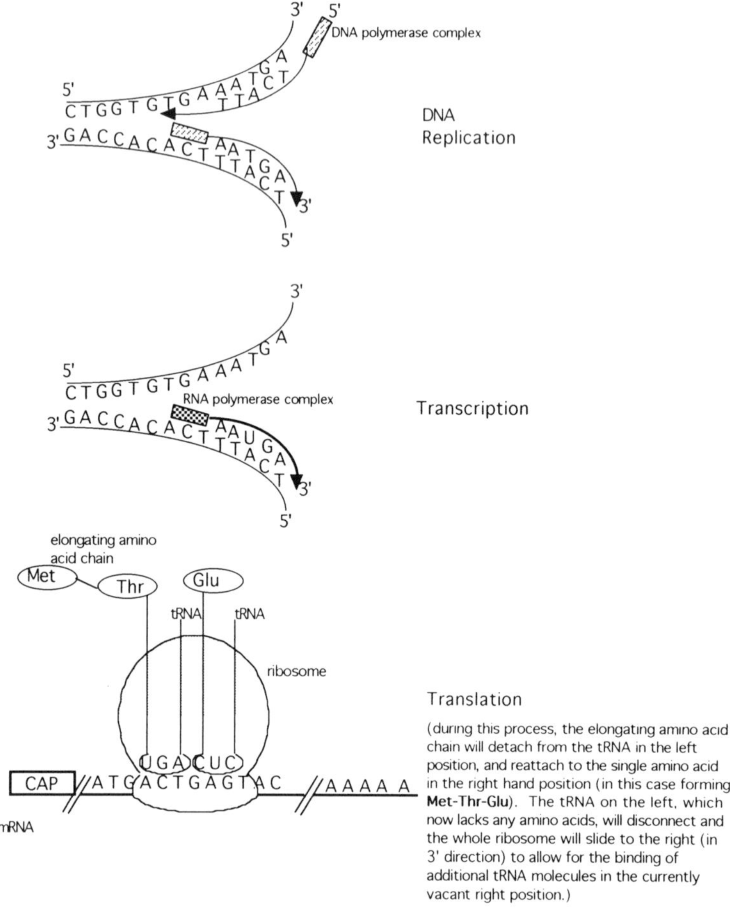

Fig. 1.2: Graphic representation of DNA replication, RNA transcription, and messenger RNA translation.

merases, which use the triphosphate forms of the nucleotide bases to elongate the DNA and RNA chains, respectively. Other essential discoveries involve the different classes of RNA, including messenger RNA (mRNA), which serves as the actual template for protein production (translation); ribosomal RNA (rRNA), which together with proteins forms the ribosomes responsible for RNA-protein translation; and transfer RNA (tRNA), to which the various amino acids are attached. In the ribosomes, which are either free in the cytoplasm or attached to the endoplasmic reticulum (forming rough endoplasmic reticulum), the anticodons of the

Genetic Code

	U	C	A	G	
U	UUU =**Phe**nylalanine UUC =" UUA =**Leu**cine UUG ="	UCU = **Ser**ine UCC = " UCA = " UCG = "	UAU =**Tyr**osine UAC = " UAA =***Stop*** UAG =***Stop***	UGU = **Cys**teine UGC = " UGA =***Stop*** UGG =**Trp**tophan	U C A G
C	CUU =**Leu**cine CUC = " CUA = " CUG = "	CCU =**Pro**line CCC = " CCA = " CCG = "	CAU =**His**tidine CAC = " CAA =**Gl**utami**n**e CAG = "	CGU =**Arg**inine CGC = " CGA = " CGG = "	U C A G
A	AUU =**Isoleu**cine AUC = " AUA = " AUG =**Met**hionine	ACU =**Thr**eonine ACC = " ACA = " ACG = "	AAU =**As**para**g**ine AAC = " AAA =**Lys**ine AAG = "	AGU =**Ser**ine AGC = " AGA =**Arg**inine AGG = "	U C A G
G	GUU =**Val**ine GUC = " GUA = " GUG = "	GCU =**Ala**nine GCC = " GCA = " GCG = "	GAU =**Asp**artic acid GAC = " GAA =**Glu**tamic acid GAG = "	GGU =**Gly**cine GGC = " GGA = " GGG = "	U C A G

The **first column** contains the **first nucleotide** of the triplet codon.
The **top row** contains the **second nucleotide** of the triplet codon.
The **last column** contains the **third nucleotide** of the triplet codon.
In the names of the amino acids, the letters in **bold type** constitute the three letter abbreviation denoting the amino acid.

Fig. 1.3: Tabular representation of the genetic code, with the three-letter abbreviations of the amino acids highlighted in boldface. The first column contains the first nucleotide of the triplet codon. The top row contains the second nucleotide of the triplet codon. The last column contains the third nucleotide of the triplet codon.

tRNA molecules bind to complementary coding segments (codons) on the mRNA chain being translated, and deliver their attached amino acid to the elongating protein chain. Recently, another class of small nuclear RNA (snRNA) has been discovered that, with or without additional proteins, is responsible for other RNA processing.

The discovery of enzymes and techniques with which portions of DNA can be manipulated (cut, separated, sequenced, manufactured, duplicated, and recombined with other chromosomal and extrachromosomal DNA) led concurrently to the elucidation of individual genes, as well as to an evolving understanding of the general structure and regulation of genes.

Both prokaryotic and eukaryotic genes contain portions of DNA that do not code for protein. The nucleotide sequences of these noncoding segments of the genes influence their regulation. For instance, upstream (5′) of the coding portion of the gene are transcriptional regulatory elements (promoters), which may be on either DNA chain and influence the binding of RNA polymerase II (polymerase responsible for mRNA production). In addition, upstream (5′), within, or downstream (3′) of the coding regions, DNA contains sequences (enhancer or silencer elements) that bind regulatory proteins that affect the amount, rate, and timing of transcription from a given promoter region. These enhancer or silencer elements function preferentially, or exclusively, in certain cell types and are most likely responsible for tissue-specific expression and developmental regulation. In addition, one gene may have more than one promoter or enhancer element that is independently regulated, and these could give rise to different, tissue-specific 5′ ends.

In 1993 Richard Roberts and Philip Sharp won the Nobel prize for their 1977 discoveries that in eukaryotes, genes exist on the chromosomes in discontinuous sequences, with regions included in the mature mRNA (exons) separated by regions not included in the mRNA (introns). In some genes 5′ exons or 3′ exons or both contain important nontranslated (?regulatory) sequences, and in general, the separate translated exons encode different functional domains of the protein. During the process of transcription, the first molecule produced is pre-mRNA, which retains all of the exon and intron sequences in linear order. The introns, which contain conserved sequences at the 5′ (GT) and 3′ (AG) ends, are "spliced out" (excised) by an snRNA-containing "spliceosome," to generate the open reading frame (start translation codon–stop translation codon), which is subsequently translated into the protein. Owing to this discontinuous structure, nuclear pre-mRNAs can generate different final mRNAs by a process known as "alternative splicing," in which certain exons may be excised from the final mRNA product. This splicing process and its regulation, which is specific to both tissue and developmental stage, are not completely understood, although it is clear that alternative splicing explains how different, but related proteins can be produced from a single gene. Following transcription, the mRNA is polyadenylated (addition of adenine residues at the 3′ end) to increase its stability and effect its transport from the nucleus to the cytoplasm. Also to increase the efficiency of translation of the mRNA to protein, the 5′ end is "capped" with an additional GTP residue, attached by a 5′-to-5′ link (rather than a 5′-to-3′ link).

DNA itself can also undergo excision and splicing within genes, on the same or different chromosomes. The immunoglobulin and T cell receptor genes consist of variable (V), diverse (D) (present only in the immunoglobulin heavy chain gene and T cell β and δ receptor genes), joining (J), and constant (C) elements organized in discrete regions along the chromosome. In lymphocytes stimulated by contact with antigens, the elements recombine (rearrange) to create a different gene structure that brings a given V region together with a given J region (or previously rearranged D-J region). This combinatorial joining is an important source of diversity in the immune system, and has been exploited in the analysis of lymphoid malignancies (Figs. 1.4 and 1.5). Interchromosomal splicing

Tβ Gene

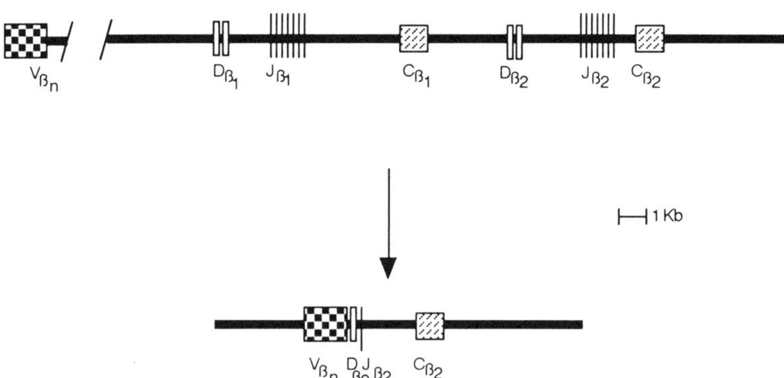

Fig. 1.4: Schematic representation of the T cell receptor β-chain gene, demonstrating the variable (V) region and two sets of diverse (D), joining (J), and constant (C) regions of the gene. The upper drawing represents the germline configuration of the gene, while the lower drawing demonstrates one of the many possible rearrangements. (*Adapted with permission from Hirsch-Ginsberg CF, Huh Yo, Kagan J, Liang JC, Stass SA. Advances in the diagnosis of acute leukemia. Hematol Oncol Clin North Am 1993;7:1–46.*)

occurs during chromosomal translocations either between homologous chromosomes, as during sister chromatid exchange associated with "crossing over," or between different chromosomes, as in the translocations associated with malignancy (many of which will be discussed below, or in subsequent chapters).

WHATEVER CAN GO WRONG WILL

As can be seen from the preceding paragraphs, the process of protein production involves a cascade of several different steps, each with its attendant enzymes, which are also encoded by DNA and regulated by other proteins. Most steps in the process can be affected, eventually leading to an alteration in the amount or structure of protein, which in turn affects cellular function. However, whereas cellular function may be altered by disturbance of one gene, malignant transformation is thought to require two or more abnormalities occurring in the same cell.

Although there are mechanisms responsible for DNA maintenance and repair, the basic structure of DNA and the order of the nucleotide

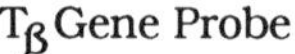

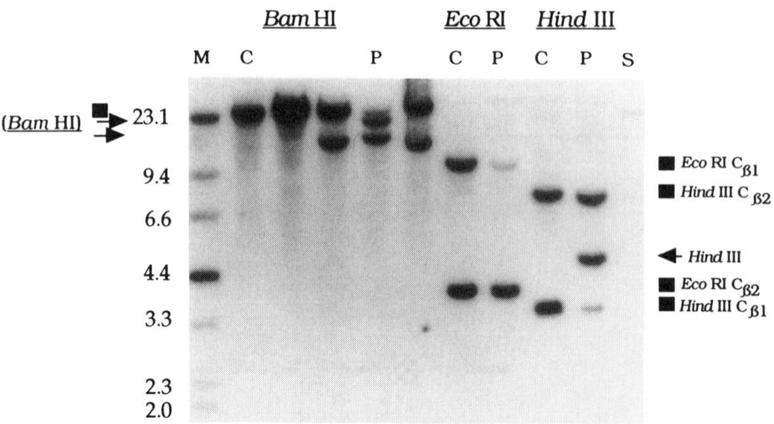

Fig. 1.5: Autoradiograph of the Southern blot probed with the T cell receptor β-chain probe. The patient's DNA (P) demonstrated rearrangements in this gene, as indicated by the arrow. The control DNA (C) is from a healthy volunteer (germline bands indicated by ■), while the molecular weight marker (M) (measured in the kilobase units to the left side of the autoradiograph) is composed of λ phage DNA digested with the HindIII and plasmid pBR322 digested with Pstl and BamHI. The 5% sensitivity lane (S) is composed of 0.5 μg of control DNA digested with a combination of HindIII and BamHI. (BamHI, EcoRI, HindIII, and PstI are restriction endonucleases whose action is described in the Methods section.) (*Reproduced by permission from Hirsch-Ginsberg CF, Huh Yo, Kagan J, Liang JC, Stass SA. Advances in the diagnosis of acute leukemia. Hematol Oncol Clin North Am 1993;7:1–46.*)

bases can be altered (mutated). These mutations can be inherited or can occur sporadically, and can be present in all cells or in only the tumor cells. Also the genes coding for proteins responsible for DNA repair can be mutated, as in xeroderma pigmentosum, which is characterized by multiple skin tumors. At the nucleotide level, these mutations can be substitutions, additions, or deletions, which can affect the protein produced 1) by causing substitution of a different amino acid (point mutations), 2) by offsetting the reading of the triplet code by the addition or deletion of a base (frameshift mutations), 3) by altering the position of a translation start or stop signal (nonsense mutations), 4) by disturbing the intron splicing signal, or 5) by affecting protein-binding domains (transcriptional domains). Several of the oncogenes discussed below, including the p53, c-fms, and Ras genes, can be activated by point mutations that lead to amino acid substitutions in critical portions of the protein. In addition, one variant of p53 has a deletion in the translation start (ATG) codon resulting in a loss of amino-terminal amino acid residues. Analysis of the retinoblastoma gene (Rb) has demonstrated splice donor and acceptor mutations and frameshift mutations. Another genetic disease produced by this class of mutations is Tay-Sachs disease, which demonstrates either an insertion of four base pairs (bp) in exon 11, or a splice junction mutation in intron 12. The β thalassemias provide other examples of nonsense, frameshift, and transcription mutations.

In addition, chromosomal translocations can lead to all or part of a gene recombining with other genes. There are two main categories of oncogenic activation by chromosomal translocation. One involves production of a chimeric fusion protein and is exemplified by Philadelphia-chromosome-positive acute and chronic leukemias (acute lymphocytic, acute myelogenous, and 95% of chronic myelogenous leukemia), in which the reciprocal t(9;22)(q34;q11) chromosome translocation results in transposition of the cellular ABL (c-ABL) gene from its location on chromosome 9 to chromosome 22. The transposed gene is recombined with the BCR gene, resulting in the expression of a hybrid (chimeric) BCR/c-ABL protein product (p190 or p210) with acquired tyrosine kinase activity.

The second category involves altered expression of an oncogene by translocation to a site regulated by different transcriptional activators or enhancers. An example of this mechanism observed in Burkitt's lymphomas involves the juxtaposition of c-MYC to the immunoglobulin heavy chain (IgH) locus by the t(8;14) translocation, resulting in the deregulation of c-MYC as a consequence of its proximity to *cis*-activating elements within the IgH locus.

The amount of a particular protein product can also be altered by amplification of the gene itself, leading to overexpression. These amplified genes often reside in double minute chromosomes or in homogeneous staining regions of a particular chromosome. Examples of overexpressed oncogenes include the neu gene, amplified in breast and ovarian carcinoma, and the N-myc gene, amplified in neuroblastoma. Other causes of overexpression (or underexpression) in the absence of DNA amplification or chromosomal translocation may be postulated to be due 1) to an alteration (by mechanisms mentioned above) in other DNA-binding proteins which either up-regulate, or fail to down-regulate transcription (activated jun, fos, rel, erbA, ets, myc, and myb oncogenes), 2) alteration of the protein binding site that effects the binding of regulatory proteins, or 3) altered DNA, RNA, or ribosomal protein structure, which enhances the life span of the mRNA by preventing degradation.

Finally, in addition to the mutation events discussed above, loss of expression (or underexpression) can occur with gene deletion (whole or partial chromosome deletion, or with chromosomal translocation, occurring either in all cells or only in tumor cells). The concomitant propensity to tumor formation due to the resulting lack of expression is seen with loss of tumor suppressor genes, such as loss of the Rb gene in retinoblastoma, or loss of the interferon-inducing IRF-1 gene in del(5q) or t5q31, observed in some myelodysplastic disorders (interferon being a potent modulator of differentiation and growth).

ONCOGENES

Category I: Protooncogenes

The discovery of tumor-producing RNA viruses (retroviruses) led to the observation that tumor production was the result of the virus introducing oncogenes into the host cell genome (Fig. 1.6). Around the same time it was observed that the DNA of various human tumors differed from nontumor tissue, and that the element responsible for malignant transformation could induce malignant transformation in other target cells. The

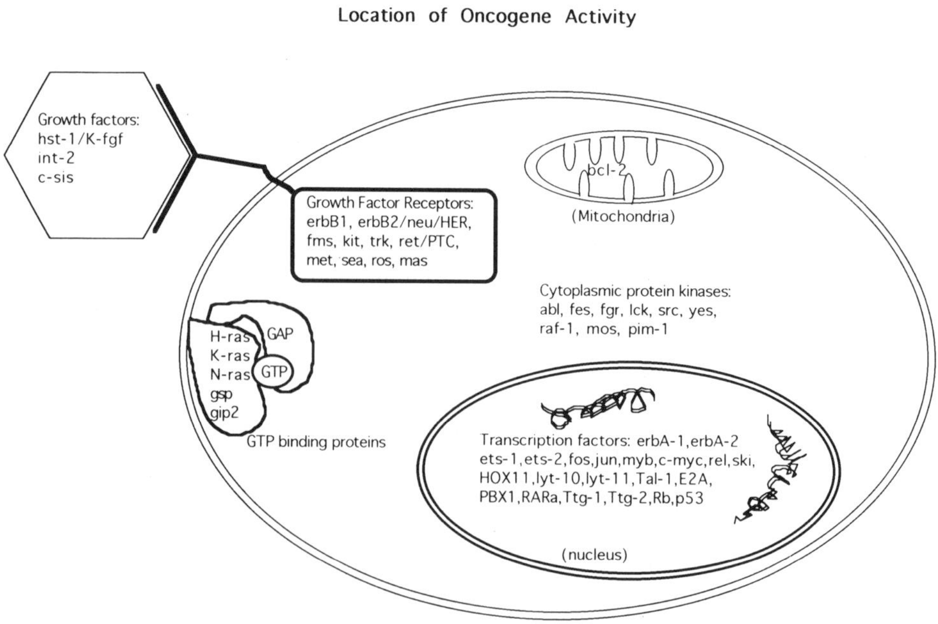

Fig. 1.6: Location of oncogene activity for the genes mentioned in the text.

homology of the viral oncogenes to the cellular oncogenes was established in 1976 by D. Stehelin, H. E. Varmus, J. M. Bishop, and P. K. Vogt with their work on the Rous sarcoma virus and the src gene, responsible for tumors in chickens. Since that time many oncogenes have been discovered. Furthermore, it has been demonstrated that the activated cellular oncogenes exist as protooncogenes, and that it is mutation or altered expression (see preceding paragraphs) that drives oncogenic transformation.

These protooncogenes can be divided according to their cellular function. The smallest group of oncogenes make up a small subset of *growth factors*, and include c-sis, producing platelet-derived growth factor (PDGF); hst-1/K-fgf, producing angiogenesis growth factor; and int-2, producing another growth factor. The hypothesized methods of oncogenesis are direct mitogenesis (PDGF) or constitutive growth stimulation through autocrine mechanisms.

Growth factor receptors form the next major class of protooncogenes, and include erb-B1, erb-B2, met, c-fms, kit, trk, ret, and sea. The structures of these genes include ligand binding, *trans*-membrane, and a cytoplasmic catalytic domains, which in the majority of the genes is a tyrosine kinase that catalyzes the transfer of a phosphate group to a target protein. Oncogenic activation leads to constitutive activation of the receptor in the absence of ligand.

As signals progress from the cell membrane to the nucleus, the next major class of oncogenes are the *signal transducers*, which are made up of *cytoplasmic protein kinases* (abl, fes, fgr, Ick, src, yes, raf-1, mos, and pim-1) and *GTP-binding proteins* (H-, K-, and N-ras, gsp, and gip2). Although some of the kinases are serine and threonine kinases, most are tyrosine kinases that share homology in their catalytic domain. Oncogenic activation appears to disrupt the function of the negative regulatory domains which allow these enzymes to constitutively phosphorylate their substrates.

The GTP-binding oncogenes form a small subset of guanine-binding proteins (G proteins) that are responsible for transmitting signals from cell surface ligands, including growth factors, hormones, and neurotransmitters, to eventual effectors such as adenylate cyclase or phospholipase C. This transduction, mediated by conversion of GTP to GDP, involves binding a GTPase-activating protein (GAP). G proteins such as the Ras family of oncogenes are activated through amplification, or point mutations that alter GTP binding, or GTPase activity, leading to prolonged stimulation of the effector enzymes.

The largest group of protooncogenes consists of *transcriptional regulators* (erbA-1, erbA-2, ets-1, ets-2, fos, jun, myb, c-myc, L-myc, and

N-myc, rel, ski, HOX11, lyt-10, lyl-1, Tal-1, E2A, PBX1 RARα, rhombotin/Ttg-1, and rhom-2/Ttg-2). These genes contain different functional domains that mediate DNA binding and protein-protein interactions. Some of the proteins encoded by these genes interact with other proteins in this group to form complex heteroduplexes, which together control transcription of a target gene. The expression of these genes is usually precisely regulated to respond rapidly to proliferation and differentiation signals. In some instances (i.e., c-myc) activation occurs through deregulated expression, which prevents the cell from terminating its proliferative phase and entering its differentiation phase.

Category II: Tumor Suppressor Genes

The theoretical basis for, and evidence of, tumor suppressor genes originated from the work of A. G. Knudson and others in their investigation of familial and sporadic retinoblastoma. Initial cytogenetic analyses of families with retinoblastoma demonstrated deletion of chromosome 13q14 in every patient cell. Further investigation, including molecular analysis of the tumor cells themselves, demonstrated loss of function, by deletion or mutation, of the remaining (second) allele of the newly discovered Rb gene on chromosome 13q14 (referred to as the second hit in the "two hit" hypothesis of oncogenesis). In the sporadic cases of retinoblastoma both the first and second Rb gene losses are confined to the tumor cells. Loss of Rb gene function has been seen also in several other cancers including breast, prostate, small-cell lung carcinoma, and some hematopoietic malignancies. The protein encoded by this gene is postulated to be a transcriptional regulator, which is also thought to be the target for inactivation during oncogenesis by the protein products of the DNA tumor viruses: papillomavirus (causative agent of cervical carcinoma), simian virus 40, and adenovirus E1A.

The gene p53 is a second tumor suppressor gene, the loss of function of which is implicated in colon, breast, lung, and brain tumors. Loss of p53 activity has also recently been implicated in the Li-Fraumeni familial carcinoma syndrome, occurring in a similar manner to familial retinoblastoma (loss of function of one allele in all cells, followed by loss of a second allele in tumor cells). Inactivation of this DNA-binding protein can occur through chromosomal loss or mutation (including single point mutations). However, occasionally these mutations can lead to increased tumor production in a fashion analogous to protooncogenes.

Category III: Regulators of Programmed Cell Death

To date, the Bcl-2 gene is the only known gene in this class of oncogenes. The bcl-2 protein functions on the inner cell membrane of the mitochondria, prolonging the life span of the individual cell by preventing apoptosis (programmed cell death). In lymphocytes, this action of the bcl-2 protein is thought to be responsible for maintaining the long life span of the memory B cells and plasma cells. This oncogene, located on chromosome 18, becomes activated by deregulated expression resulting from translocation into the IgH locus (14;18). Although increased survival of the affected cells may not confer malignant transformation, it is postulated that the extended life span of the cell provides opportunities for protooncogene activation, or loss of tumor suppressor gene function.

METHODS

No introduction to molecular oncology would be complete without a discussion of some of the major methods used to detect the oncogenes and their alterations. Southern blot analysis, first described by E. M. Southern in 1975, has been used to isolate and identify both normal and abnormal gene products. In this assay high molecular weight DNA is extracted from cells in the sample of interest (blood, bone marrow, surgical biopsy specimen, or fine needle aspirate) using a combination of proteases, detergents, and organic solvents. The DNA is then digested into smaller fragments with restriction endonucleases, which recognize palindromic base pair sequences as points for DNA cleavage. These restriction fragments are separated by size during electrophoresis through agarose gel. Although this standard procedure is used to separate fragments from several hundred base pairs to 30,000 base pairs, larger fragments generated by restriction enzymes that cut less frequently can be separated by pulsed field gel electrophoresis, in which the direction of the electric current is alternated between several different sets of electrodes. Once separated, the DNA fragments are denatured in the gel (the strands are separated from each other), then transferred to a membrane support by capillary action (blotting), vacuum suction, or electrophoretic transfer. The membranes (blots), with affixed DNA, are incubated with DNA or RNA probes labeled for radioactive (usually with phosphorus-32), chemical, or fluorescent detection, then washed to remove the excess probe. Both the incubation and washing steps are carried out under conditions that maximize specific hybridization of the probe to the DNA segments of interest, while minimizing nonspecific hybridization. If the probes were labeled for a chemical reaction, the membranes are further processed in order to produce a color reaction. If fluorescent or radioactive detection systems are used, the blots are then exposed to x-ray film, or are scanned with devices that detect fluorescence or β particle emissions. Often Southern blotting is used to detect size differences in the fragments produced by the restriction endonucleases (restriction fragment length polymorphisms), which may be 1) normal variations between individuals, 2) produced by clonal (from a population of cells from a single progenitor) gene rearrangements, as in the immunoglobulin or T cell receptor genes, or 3) produced from pathological changes in the genes of interest (as occurring during chromosomal translocation) (Fig. 1.7).

A modification of Southern blot analysis has been applied to RNA. This technique, referred to as Northern blot analysis (a pun on Dr. Southern's name), detects the presence of, or alterations in, the amount or size of transcribed RNA. Although the reagents used in working with

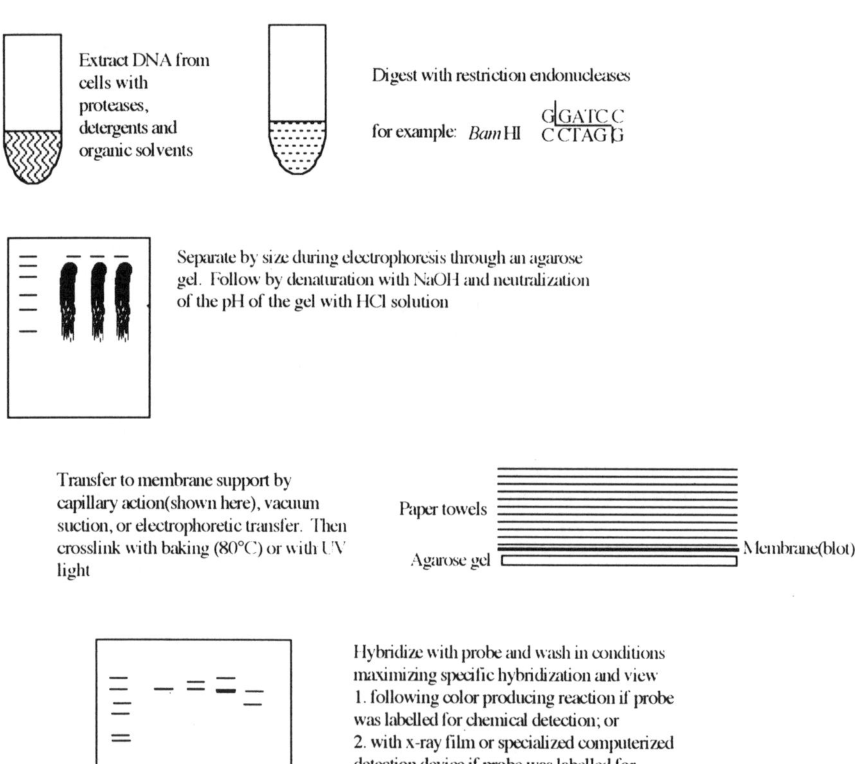

Fig. 1.7: Steps required for Southern blot analysis described in the text. (*Reproduced by permission from Hirsch-Ginsberg CF, Kagan J, Liang JC, Stass SA. Recent advances in the diagnosis of acute leukemia. The Cancer Bulletin, Department of Educational Publishing, University of Texas M. D. Anderson Cancer Center, 1993.*)

RNA differ from the DNA reagents, the basic steps are similar and involve extraction of RNA, size fractionation through a gel, transfer to a membrane support, and hybridization with a labeled probe. Alternatively, extracted RNA may be blotted directly to a membrane support and exposed to labeled probe. Although this technique yields less information about the RNA transcript (i.e., it gives no information on size or number of related transcripts), it is quicker and can be used even when the RNA is degraded.

The use of analysis based on polymerase chain reaction (PCR) is much more sensitive than traditional Southern blotting (up to 10^4-fold). The method, originated by Kary Mullis, for which he received the 1993 Nobel prize in chemistry, amplifies small target segments (up to 1000 bp) of DNA by multiple repetitions of three successive steps: 1) denaturation of DNA at a high temperature (90°–95°C); 2) annealing of primers complementary to both strands of DNA flanking the target segment (42°–72°C); and 3) chain elongation by a heat-stable DNA polymerase, in most cases Taq (72°C). The exact temperatures and times of each step vary and are optimized for different uses. This technique has been used to amplify segments of genes to detect point mutations, as well as to amplify chromosomal breakpoints that are clustered in relatively small regions (up to approximately 1000 bp). An extension of the technique, which has been used to amplify RNA, includes a reverse transcriptase step, to produce cDNA (DNA complementary to the mRNA) for use as a substrate in the DNA amplification procedure. This modification has allowed for chromosomal breakpoint detection in cases in which a chimeric fusion mRNA transcript is produced and the lack of a clustering of the breakpoint precludes DNA amplification. In addition, both DNA and RNA PCR have become the mainstay in generating quantities of homogeneous DNA for cloning, sequencing, and other complex analyses (Fig. 1.8).

CONCLUSION

Although the years since the discovery of the structure of DNA have witnessed an explosion in the understanding of the molecular basis of oncogenesis, they have also uncovered unexpected complexity in the genetic machinery. The plethora of transcriptional regulators, as well as other classes of oncogenes, the understanding of their complex interrelations, and the knowledge of their role in oncogenesis are continually evolving. The chapters that follow give a glimpse of the current state of knowledge of some individual oncogenes and their contribution to carcinogenesis in the various organ systems.

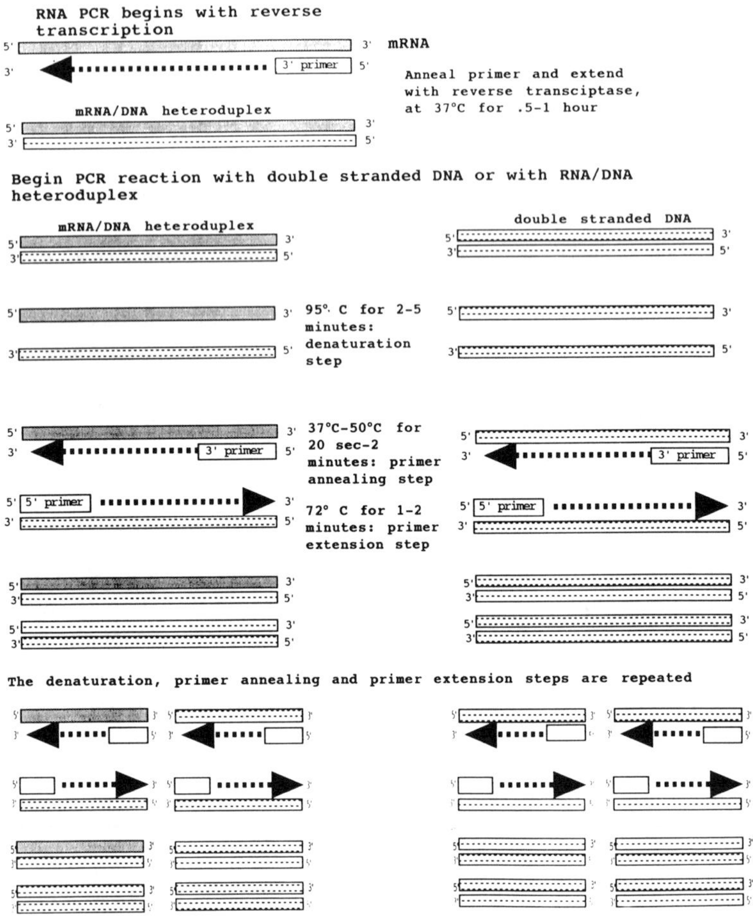

Fig. 1.8: Schematic representation of the steps required for RNA and DNA polymerase chain reaction.

GLOSSARY

adenine (adenosine, deoxyadenosine): (abbreviation A) purine base, which becomes the ***nucleoside adenosine*** or ***deoxyadenosine*** with the addition of ribose or deoxyribose (indicated by "d" when abbreviated), respectively. It is converted to a nucleotide with the addition of one or more phosphate groups at the 5′ carbon (AMP, ADP, ATP, dATP) (see Fig. 1.1).

alternative splicing: the process for generating different final mRNA products from a single gene, based on the exclusion or inclusion of different exons (as in the rat calcitonin gene, which produces calcitonin in the thyroid C cells and calcitonin-gene-related peptide in the brain).

anticodon: trinucleotide portion of the tRNA that binds to the codon of the mRNA. The first base of the anticodon (last base of the codon) may contain the purine inosine, which can hydrogen bond to cytosine, uracil, or adenine. Also in this position guanine can bind with uracil. These different hydrogen-bonding possibilities allow for the degeneracy of the genetic code (more than one codon per amino acid).

antiparallel: refers to the reverse orientation of the sugar-phosphate backbone of two hydrogen-bonded nucleotide chains.

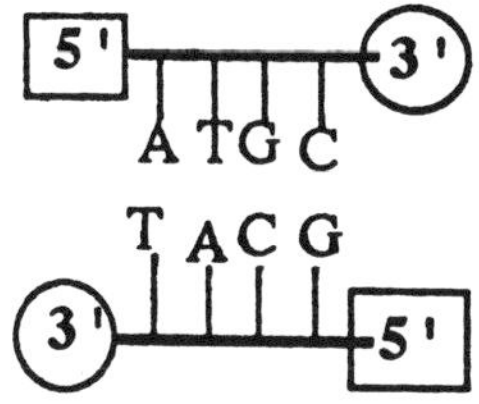

antisense: nucleotide chain complementary to the chain containing the actual protein coding sequences.

apoptosis: programmed cell death, part of the natural limitations of a cell's life span. Apoptosis has been implicated as an important part of the process eliminating the tissue webs between the digits in the developing fetus, while the delay of apoptosis is responsible for the longevity of memory lymphocytes.

autocrine: autostimulation through the production and secretion of growth factors by a cell, which also has receptors for the growth factor.

cDNA: DNA complementary to mRNA and therefore containing only the exon sequences for a particular gene. The first strand is produced using reverse transcriptase with the mRNA as the template, and the second strand is synthesized using a DNA polymerase with the previously synthesized DNA strand as a template. This second strand has the same nucleotide sequence as the mRNA (with the exception of the U-to-T substitutions).

***cis*-acting factors**, or ***cis*-activators**: regions of the gene (same chromatin [DNA] strand), either upstream, within, or downstream, that are binding sites for transcriptional regulating proteins (general term for enhancer or silencer elements).

codon: triplet nucleotide sequence coding for a particular amino acid.

complementary: nucleotide chain capable of hydrogen binding with, or acting as template for, another nucleotide chain by exploiting the pairing of A-T (or A-U if RNA) and G-C. The chain in the box below is *complementary* to the chain in the oval.

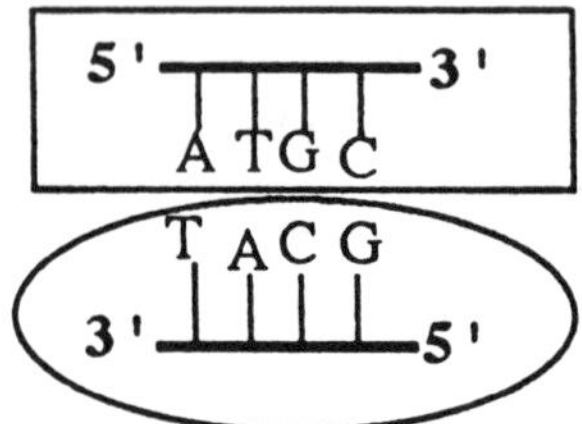

constitutive: automatic and unregulated activity.

cytosine (cytidine, deoxycytidine): (abbreviation C) pyrimidine base, which becomes the *nucleoside cytidine* or *deoxycytidine* with the addition of ribose or deoxyribose (indicated by "d" when abbreviated), respectively. It is converted to a nucleotide with the addition of one or more phosphate groups at the 5′ carbon (CTP, dCTP) (see Fig. 1.1).

denature: to pull apart complementary DNA or DNA-RNA chains by breaking the hydrogen bonds between the base pairs. This is usually accomplished by heating (>94°C) or by using a strong base such as sodium hydroxide.

enhancer: *cis*-acting element that binds regulator proteins to increase transcription of the gene.

eukaryotic: organisms with DNA residing in the nuclei.

exon: segment of DNA transcribed into RNA.

guanine (guanosine, deoxyguanosine): (abbreviation G) purine base, which becomes the *nucleoside guanosine* or *deoxyguanosine* with the addition of ribose or deoxyribose (indicated by "d" when abbreviated), respectively. It is converted to a nucleotide with the addition of one or more phosphate groups at the 5′ carbon (GMP, GDP, GTP, dGTP) (see Fig. 1.1).

homologous (homology): variable parts of the coding sequences of genes are similar, and presumably produce similar protein products.

Quantitation of the amount of homology is an indicator of the relatedness of a gene or protein.

intron: segment of DNA excised from the final mRNA.

mitogen(esis): substance (process) that induces mitosis.

nucleoside: purine or pyrimidine base with attached sugar (ribose or deoxyribose).

nucleotide: nucleoside with one to three phosphate groups attached to the 5′ carbon.

open reading frame: segments of DNA or RNA containing ribosome binding site sequences and the "start translation" signal (ATG or AUG, for DNA or RNA, respectively) uninterrupted by chain-terminating codons (TAA, TAG, TGA or UAA, UAG, UGA, for DNA or RNA, respectively).

phosphorylation: the process of introducing the PO_3^- group into an organic molecule, catalyzed by a kinase (which is named for the acceptor molecule).

polymerase: class of enzymes responsible for nucleotide chain synthesis from a template DNA chain (nucleotidyltransferase).

prokaryotic: organisms lacking nuclei.

promoter: segments of DNA indicating a transcriptional "start," which provide for the binding of RNA polymerase.

reverse transcriptase: retrovirus-derived DNA polymerase that uses RNA as the template.

reverse transcription: the process of synthesizing a DNA chain from an RNA template using the enzyme reverse transcriptase.

sense strand: nucleotide strand with the coding sequences that are subsequently translated into protein. As opposed to the antisense strand, which is complementary to the coding strand.

template: pattern or mold. In genetics a strand of DNA serves as a template for the synthesis of a complementary DNA or RNA strand. The codons of the mRNA strand serve as a template for protein production.

thymine (deoxythymidine): (abbreviated T) pyrimidine base, which becomes the *nucleoside deoxythymidine* with the addition of deoxyribose (indicated by "d" when abbreviated). It is converted to a nucleotide with the addition of one or more phosphate groups at the 5′ carbon (dTTP) (see Fig. 1.1).

transcription: process indicating the synthesis of RNA from DNA.

translation: process indicating the synthesis of protein from mRNA.

uracil (uridine): (abbreviated U) pyrimidine base, which becomes the *nucleoside uridine* with the addition of ribose. It is converted to a nucleotide with the addition of one or more phosphate groups at the 5′ carbon (UTP) (see Fig. 1.1).

BIBLIOGRAPHY

Boxer LM, Cleary ML, Willman CL. Transcription factors: clinical significance and biologic function. In: Schrier SL, McArthur JR, eds. The education program of the American Society of Hematology. St. Louis: American Society of Hematology, 1993:128–142.

Fenoglio-Preiser CM. Oncogenes and tumor suppressor genes in solid tumors: general considerations. In: Fenoglio-Preiser CM, Wilman CL, eds. Molecular diagnostics in pathology (Kaufman N, series ed. Techniques in diagnostic pathology). Baltimore: Williams and Wilkins, 1991:149–165.

Gaidano G, Dalla-Favera R. Protooncogenes and tumor suppressor genes. In: Knowles DM, ed. Neoplastic hematopathology. Baltimore: Williams and Wilkins, 1992:245–261.

Hirsch-Ginsberg CF, Huh YO, Kagan J, Liang JC, Stass SA. Advances in the diagnosis of acute leukemia. Hematol Oncol Clin North Am 1993;7:1–46.

Kaushansky K. Section 7. Molecular biology glossary: glossary of molecular biology terminology. In: Schrier SL, McArthur JR, eds. The education program of the American Society of Hematology. St. Louis: American Society of Hematology, 1993:183–197.

Moffat AS. Nobel prizes: in Stockholm, a clean sweep for North America. Science 1993;262:506–507.

Watson JD, Hopkins NH, Roberts JW, Steitz JA, Weiner AM. Molecular biology of the gene, 4th ed. Menlo Park: Benjamin/Cummings Publishing Company, 1987.

Watson JD, Tooze J, Kurtz DT. Recombinant DNA: a short course. New York: Scientific American Books, 1983.

Willman CL. Molecular diagnostics: scientific fundamentals and technical approaches. In: Fenoglio-Preiser CM, Willman CL, eds. Molecular diagnostics in pathology (Kaufman N, series ed. Techniques in diagnostic pathology). Baltimore: Williams and Wilkins, 1991:1–20.

Wivel NA, Walters L. Germ-line gene modification and disease prevention: some medical and ethical perspectives. Science 1993;262:533–538.

CHAPTER 2

DNA Repair Defects Associated with Neoplasia

Shawn Panzer
C. Thomas Caskey

Numerous diseases in advanced states carry a worse prognosis than cancer, but there is probably no other diagnosis that has as great an impact on a patient's psyche. Almost everyone knows someone who has died of cancer or suffered the morbidity of various modalities of cancer treatment, and numerous others live with the fear of developing a cancer like one that killed a relative. Although great strides have been made in secondary and tertiary prevention of cancer, primary prevention is still the most cost-effective method of dealing with this type of disease. Various methods and schedules of screening have been developed or proposed by numerous organizations. Although there is debate over whether screening tests such as evaluating stool for blood to screen for colon cancer decrease overall mortality, these tests almost certainly allow the discovery of cancers at earlier stages. However, most screening tests rely on the actual physical presence of the cancer or expression of the malignant phenotype through measurable markers. The ideal screening test would be one that could reliably predict or provide a probability of the eventual development of a malignancy. This would allow the efficient but vigilant application of tests such as colonoscopy for colon cancer or elective surgery for persons concerned about highly penetrant familial cancer syndromes. It is this kind of rationale that drives the search for the genetic basis of the malignant phenotype.

Life would not be possible without the accurate replication of DNA. It is the reliability of the system directing DNA replication, messenger RNA transcription, and protein translation that allows the propagation of single-celled and multicellular organisms. When this delicate interrelation is perturbed, a vital function is disrupted and a disease—or death—may follow. On the DNA level, numerous mutations such as deletions, transitions, transversions, and others caused by external insults have been described and related to various types of cancer. Of probable equal importance is the organism's ability to repair the disruptive effect of such insults to its homeostasis. There is little doubt that with better understanding of the DNA repair of specific genes, the details of coordinated efforts of subcellular and cellular repair systems will soon follow, giving a coherent picture of a given pathological condition, such as cancer (1).

MOLECULAR DEFENSES AGAINST MUTAGENS

The number of mutagens identified is immense, and the fact that the development of neoplastic cells is the exception rather than the rule speaks to the importance of the DNA repair processes in the maintenance of cellular proliferation control. Damage of genomic DNA occurs as a result of agents that may be physical (e.g., ultraviolet or x-irradiation, heat), chemical (e.g., known antigens, chemotherapeutics), or biological (e.g., viruses) (2). These insults can inflict direct damage, or may initiate a cascade of intracellular responses or biochemical reactions that lead to DNA damage. It has been estimated that almost 50% of all chemicals could act as carcinogens and that carcinogenic potential could be accentuated by high-level exposure because of cellular responses such as inflammation and cellular proliferation (2). Because of the specialized nature of differentiated tissues, agents may demonstrate organ or tissue specificity in their damage and subsequent carcinogenesis. Examples of this might be the association between chronic hepatitis B infection and hepatocellular carcinoma, or between the previous use of alkylating agents for the treatment of lymphoma and the subsequent development of acute myelogenous leukemia.

The type of mutagen and its specific structural effect on the target DNA dictates the type of repair required. Ultraviolet (UV) radiation causes cyclobutane pyrimidine dimers and (6-4) pyrimidine-pyrimidones; ionizing radiation generates single-strand and double-strand breaks and single-base damage; electrophilic agents like aflatoxin B_1 induce nucleotide adducts; oxidating agents can modify bases; cytotoxic drugs can cause DNA-DNA crosslinks, DNA-protein crosslinks, and strand breaks; and there can be spontaneous apurinic and apyrimidinic sites (3). Organisms have developed systems to repair these defects, and current knowledge suggests that DNA repair genes have been conserved throughout evolution. Examples of this are the similarities between the mammalian long-patch repair system and the equivalent repair system in *Escherichia coli*, and the mammalian short-patch repair system and its equivalent very-short-patch repair system in *E. coli* (4). An accurate estimate of the

Table 2.1: Human DNA Repair Genes

Gene	*Disorder*	*Chromosome*	*Predicted Function*
ERCC1	Unknown	19q13	Unknown
ERCC2	XP-D,CS,TTD	19q13	DNA helicase
ERCC3	XP-B,CS-C	2q21	DNA helicase
ERCC5	Unknown	13q	Unknown
ERCC6	CS-B	10q11	DNA helicase
XPAC	XP-A	9q34	Zinc-finger protein
XPCC	XP-C	Unknown	Unknown
FACC	FA-C	9q22	Unknown
XRCC1	Unknown	19q13	Unknown
DNA lig1	BS-like	19q13	DNA ligase
APE	Unknown	14q11.2–12	AP endonuclease
[ADP-rib]n	Unknown	1q42	Polymerase
MGMT	Unknown	Unknown	06-Methylquanine DNA methyltransferase
ANPG	Unknown	Unknown	Alkyl-*N*-purine DNA glycosylase
UNG15	Unknown	12	Uracil DNA glycosylase
HHR6A	Unknown	Xq24	Ubiquitin-conjugating enzymes
HHR6B	Unknown	5q23	Ubiquitin-conjugating enzymes

Adapted from Wevrick R, Buchwald M. Mammalian DNA-repair genes. Curr Opin Genet Dev 1993;3:470–474.

number of genes involved in DNA repair in mammalian cells is not yet available, but there are reasons to expect that the number will be at least as large as it is in organisms such as *E. coli* and *Saccharomyces cerevisiae* (5).

The human damage and repair genes that have been found thus far (Table 2.1) are heterogeneous and homologous to helicases, ligases, endonucleases, polymerases, methyltransferases, and glycosylases. The presence of helicases on this list is unique to eukaryotes and suggests that the unwinding of DNA is important in the eukaryotic DNA repair process (6). There surely are numerous other proteins that play important roles in the DNA repair process. An example is the human single-stranded DNA-binding protein (HSSB/RPA). HSSB seems to be involved in protein-protein interactions and stimulates DNA polymerase α. It may also work in concert with DNA helicase to displace the incision proteins associated with the area of DNA damage, protect the gapped region and termini from being degraded, and interact with other proteins to mediate DNA repair synthesis (7). The cellular pathways by which these families of proteins effect DNA repair are summarized in Table 2.2. These pathways seem to be activated according to the type of insult. Pyrimidine dimers

Table 2.2: Cellular Pathways for the Repair and Tolerance of DNA Lesions

Process	*Substrate*	*Repair Enzymes*	*DNA Modification/Product*
Direct Repair			
1. Reversal	Pyrimidine dimer	Photolyase	Light-dependent splitting
2. Removal	Adduct	Alkyltransferase	Methyl group removal
Excision Repair			
1. Base removal	Modified base	Spontaneous; glycosylase	AP/APy site
2. Strand cleavage	Distorted helix or AP/APy site	Specific endonuclease or spontaneous	Lesion excision or generation of substrate for repair replication
3. Repair replication	Strand break	Exonuclease; polymerase and ligase	Lesion removal and insertion of repair patch
Tolerance			
1. Translesion synthesis	Persistent damage	Polymerase complex	Error-prone bypass of lesion and replicative dilution
2. Post-replication repair	Gap in daughter strand	Recombination and repair replication enzymes	Intact daughter strand

Reproduced by permission from Smith PJ. Carcinogenesis: molecular defences against carcinogens. Br Med Bull 1991;47:3–20.

seem to activate the long-patch repair system, while the lesions induced by ionizing radiation are substrates for repair enzymes including DNA exonucleases and endonucleases and DNA glycosylases which remove oxidatively damaged portions of the DNA molecule, thereby initiating excision-repair (3). Other mammalian genes are being sought, and the approaches used include gene transfer and library screening of transformant DNA, chromosome transfer with deletion mapping and gene localization, complementary DNA cloning and transfection or microinjection, retroviral inactivation and repair gene disruption, evolutionary walking between species searching for homology, and working backward from a protein to the corresponding gene using oligonucleotide probes or antibodies (5).

Considering that there are approximately 100,000 genes of varying size in the human genome, it seems likely that at least cell survival would be jeopardized if there were no priority in the repair process. There are solid data supporting the heterogeneity of DNA repair, as reflected in the preferential repair of active genes (8). This has been demonstrated by Bohr and associates (9) in normal Chinese hamster ovary cells (CHO), in which it may take up to 24 hours to repair the entire genome, but much less for the dihydrofolate reductase (DHFR) gene (10,11). This differential repair may be related more to the transcriptional activity of the gene than to accessibility of the DNA. Madhani and colleagues (12) examined the repair of two UV-irradiated protooncogenes, c-*mos* (transcriptionally silent) and c-*abl* (transcriptionally active), in Swiss mouse 3T3 fibroblasts. A dose of UV light was applied to confluent cell cultures that were either lysed immediately or incubated for 24 hours to allow repair before lysis. The appropriately restricted DNA samples were either treated or mock treated with T_4 endonuclease V before electrophoresis, Southern transfer, and hybridization with a c-*abl* or c-*mos* probe (13). The investigators initially found similar dimer frequencies in both genes, but c-*abl* was preferentially repaired compared with c-*mos*. It seems that the methylation status also dictates preferential repair. Methylation of CpG islands is an important method by which cells inactivate genes, and the methylation of selected areas in the genome may be coordinated with preferential DNA repair activity (14). It has also been shown that within a gene, portions that are more prone to mutation are preferentially repaired. Bohr has shown increased repair efficiency in the 5′-flanking region of the c-*myc* gene in an area prone to rearrangements (15). Finally, certain types of damage, such as interstrand crosslinks, are repaired more efficiently than other lesions, such as psoralen photomonoadducts (16).

CLINICAL IMPLICATIONS OF DNA REPAIR GENE FUNCTION

The amount of damage sustained by the DNA in a human being is mediated by numerous exogenous factors including diet, alcohol consumption, smoking history, socioeconomic status, and geographical location. Thus, any estimate of the number of DNA-damaging events per cell per day is likely to be low (17). The estimated numbers of DNA-damaging events per mammalian cell per day are 10,000 depurinations, 500 depyrimidinations, 100 to 300 deaminations, 10,000 base damages, and 20,000 to 40,000 single-strand breaks (17). It is obvious that dysfunction of one or more components of an organism's DNA repair systems could have dire consequences for cell function, including proliferation control, and ultimately for the organism itself. There are five genetic instability syndromes: xeroderma pigmentosum, Bloom's syndrome, Cockayne's syndrome, Fanconi's anemia, and ataxia telangiectasia. These are autosomal recessive diseases in which single gene defects lead to genetic instability, increased mutation rates, and eventually cancer.

Xeroderma pigmentosum (XP) is a rare autosomal disorder manifested by excessive solar damage to the skin due to deficient excision of UV-induced DNA photoadducts (18,19). The increased frequency of various cancers in those with XP makes it a paradigm of a pathological condition, revealing that carcinogenesis involves DNA damage, and that DNA repair plays a major role in maintaining cell homeostasis (1). The incidence is approximately one in 250,000, and some patients also may have the other features of neurologic defects, growth retardation, and hypogonadism (20).

Candidate DNA repair genes are associated with recognized clinical syndromes through complementation studies, in which the gene is delivered into cultured cells from persons with the disease and shown to correct the repair deficiency. Xeroderma pigmentosum consists of seven complementation groups and a further XP-variant group, in which a different basic defect is postulated and whose patients show a different clinical picture—including a lower incidence of cancer (21). As mentioned above, patients with XP have defective nucleotide excision repair of UV lesions (18), and it is felt that the various complementation groups have defective genes coding for separate proteins essential to the early steps of excision repair. These groups also demonstrate variable sensitivity to UV irradiation, with complementation group XP-A demonstrating less than 5% of the normal capacity to perform excision repair (20). Group XP-E, by comparison, exhibits 25% to 70% of normal DNA repair capacity. One study demonstrated this difference by application of x-irradiation during

G_2 phase to normal skin fibroblasts and stimulated blood lymphocytes (22). Cytosine arabinoside (ara-C) was then applied at fixed intervals to inhibit DNA repair synthesis. By measuring the accumulation of chromatid breaks and gaps as a measure of DNA incision activity, the authors showed a negligible increase in breaks or gaps in XP-A and XP-D and a higher but less than normal rise in XP-C cells.

Another study examined the repair of cisplatin-induced DNA lesions (23). Cisplatin is thought to cause the formation of DNA intrastrand adducts between adjacent bases in GpG and ApG sequences, and DNA interstrand crosslinks. The study compared the repair of lesions in the DHFR and ribosomal RNA genes in normal and XP-A cell lines. Normal human fibroblast cells repaired 84% of the interstrand crosslinks in the DHFR gene after 24 hours compared with 32% in XP-A, while 69% of the intrastrand adducts in the DHFR gene were repaired in normal cell lines versus 22% for XP-A cell lines. The pattern was similar for the ribosomal RNA gene. The majority of the detected mutations are G:C-to-A:T transitions, most (76% to 91%) occurring at positions where the 5′ neighbor of the cytosine residue was another cytosine or a thymine (24,25).

There has been success in cloning the XP complementation group genes including XP-A (26,27), XP-B (28,29), XP-C (30), XP-D (31,32), and XP-G (33). An increased incidence in a variety of neoplasms has been noted in XP patients including head and neck squamous cell carcinoma, thyroid adenoma, infiltrating carcinoma of the breast, acute lymphocytic and myelogenous leukemia, and brain tumors including medulloblastoma and glioblastoma multiforme with sarcomatous component (34). The presence of malignancies in internal organs speaks to the importance of the DNA excision repair system in neutralizing the effects of mutagens other than UV irradiation and endogenous reactive metabolites of normal cellular function. No direct causal relation has been drawn between the malignancies in XP and specific genetic changes. One study did observe an association between the persistence of chromatid breaks and gaps after G_2 phase irradiation with x-rays or near-UV visible light in XP skin fibroblasts (increased cancer risk) and the lack of such findings in Cockayne's syndrome (no increase in sunlight-induced cancer) (35). Several activated oncogenes have been found in the DNA of tumors from XP patients, including codon 61 mutations of N-*ras*, amplification and overexpression of c-*myc*, and rearrangement and amplification of H-*ras*. Several oncogenes can be altered within the same tumor (36). The same findings were reported in a study that examined mutation frequency in the *ras* family in skin tumors from XP patients (37). Using tissue taken from excised skin tumors from XP patients and tissue taken from excised skin tumors of normal individuals, the authors found a 50% mutation frequency in the *ras* genes in XP patients versus 22% in control tumors. The mutations were opposite pyrimidine-pyrimidine sequences, which are hot spots for UV-induced mutations. It is likely that unrepaired mutations are also frequent in tumor suppressor genes, and that these changes are associated with a higher incidence of various neoplasms.

Bloom's syndrome (BS) is manifested in low birth weight, stunted growth, a normally proportioned dwarfed body, photosensitivity, butterfly telangiectasia of the face, and high susceptibility to infection resulting from moderate to severe immunodeficiency (38). Cells show increased occurrence of chromosomal rearrangements and breakages, an elevated spontaneous mutation rate, and hypermutability by DNA-damaging agents (39,40). A reduced activity of DNA ligase I has been demonstrated in BS cell lines. In several cases this is associated with heat lability of the enzyme, and some cells contain a dimeric form of DNA ligase I (41), although sequencing of the gene and analysis of mRNA levels in BS cells have shown normal results (42). The defect may therefore be related to post-translational modification of DNA ligase I (43). Another described defect is somatic recombination causing doubled expression of one allele of glycophorin A (GPA), a red blood cell surface antigen. This process normally occurs at a frequency of 1×10^{-5} to 8×10^{-5} but rises to 1×10^{-3} to 3×10^{-3} in BS patients (44,45). When examined with restriction fragment length polymorphisms (RFLPs), BS cells showed loss of heterozygosity due to somatic crossing over for 5 to 7 cM of chromosome 3q. A tumor suppressor gene could easily be lost in such a process, leading to malignancy. An alteration in uracil DNA glycosylase has been reported in BS cells (46). An increased incidence of leukemia (lymphoid and nonlymphoid), lymphoma, and adenocarcinoma has been associated with BS (34).

Cockayne's syndrome (CS) is an autosomal recessive disease manifested in growth and developmental arrest leading to dwarfism, photosensitivity, deafness, intracranial calcifications, mental deficiency, sunken eyes, disproportionately long arms and legs, and a general appearance of premature aging (38). It has three complementation groups (CS-A to CS-C). Lymphoid cell lines from CS-A and CS-B show normal DNA repair synthesis in plasmids with UV photoproducts (47). This corresponds to results of in vivo studies showing normal overall repair capacity. CS-C and XP-B share the same phenotype (29) and are caused by mutations in the *XPBC/ERCC3* gene, and CS and XP overall are closely related. Mutations in excision repair genes may lead to either XP or CS, or a combination of both phenotypes. The defect in CS groups A and B seems to be an inability to perform preferential repair of active genes (48). Patients with CS do not demonstrate a higher incidence of cancer than the normal population. This is unusual in that the phenotypic features overlap with those of XP, yet only XP patients are susceptible to higher rates of malignancy. One clue to the potential answer was provided in a study showing that CS fibroblasts repaired DNA damage inflicted during G_2 phase by either x-rays or fluorescent light, but were deficient in the repair of damage inflicted during G_1 and S phases of the cell cycle (35). As mentioned above, XP-C cells were deficient in G_2 repair, and these patients are prone to malignancies. Unrepaired strand breaks could lead to deletions of acentric fragments during subsequent mitoses, and unrepaired "open" chromatid breaks which persist into a subsequent G_1 phase could lead to gene rearrangements (35).

Fanconi's anemia (FA) is an autosomal recessive disorder characterized by congenital and developmental disorders such as mental retardation, microcephaly, skeletal malformations, renal and ocular malformations; café au lait spots; insulin-dependent diabetes mellitus; and variable progressive pancytopenia (49). There are four complementation groups (FA-A to FA-D). The gene that is mutated in FA-C has been cloned but is not homologous to any known gene (50). The FA cell lines do demonstrate increased sensitivity to mutagens. One study examined the effects of psoralen at the HPRT locus in FA-B cells (51). Exposure of normal cells to psoralen led to point mutations (or deletions of less than 100 bp) in 100% of the monoadduct lesions and 90% of the crosslink lesions. This occurred one-half as often in the FA-B cells, but the frequency of class 2 deletions (defined as loss of individual exons or clusters of exons in the HPRT gene) was much higher (46%) in FA-B cells than in normal cells (5%). It has been postulated that FA cells demonstrate reduced interstrand crosslink removal. Zhen and associates showed that normal fibroblasts repaired 84% of cisplatin-induced interstrand crosslinks in the DHFR gene at 24 hours versus 50% in the FA-A cell lines, and 69% of the intrastrand adducts versus 24% in FA-A cell lines (23). The sensitivity of FA cells to the blastogenic effects of crosslinking agents such as mitomycin C, cisplatin, psoralens with exposure to UV-A, and other agents, compared with that of normal cells, is the basis of prenatal testing (52,53). Persons with FA might have a defect in an error-prone pathway operating during replication (21). They have a 15,000-fold increased incidence of acute myelogenous leukemia (54). There may also be an increased incidence of other types of acute leukemias, squamous cell carcinoma, and hepatocellular carcinoma (34). No direct association has been drawn between the aforementioned genetic defects and the neoplasms seen in FA, and no activated oncogenes have been reported (21).

Ataxia telangiectasia (AT) is an autosomal recessive disorder with an incidence of one in 40,000 live births (38). Features include oculocutaneous telangiectasia, progeric changes of the hair and skin, freckling, and hyperpigmented macules; mental retardation; progressive cerebellar ataxia, choreoathetosis, oculomotor defects, dysarthric speech, and dystonic posturing of fingers; and mild diabetes mellitus (41). There are four complementation groups (AT-A to AT-D), and the genes for AT-A, AT-B, and AT-C have been localized to chromosome 11q22.3 (43). The AT-D gene has been mapped to the same region (11q22.3–23.1). The AT cells are hypersensitive to ionizing radiation and other DNA-damaging agents, but no DNA repair defect has been identified (20). There is also a stable, clonal translocation involving chromosomes 7 and 14 (55). An unusual feature is supernormal resistance of DNA synthesis in AT cells to x-irradiation (56), as well as radioresistant replication initiation and chain elongation (38). Patients with AT have an elevated overall incidence of cancer of approximately 10% (57). These include lymphomas and lymphoid leukemia; stomach, ovarian, breast, and skin cancer; and brain tumors (34). As up to 2% of the population are heterozygous carriers of a defective AT gene allele (20) and demonstrate an approximately fourfold increase in cancer incidence (58), identification of the genes could lead to more vigilant cancer screening. This might be particularly true for breast cancer screening, as it has been estimated that 9% to 18% of breast cancer patients may be AT heterozygotes (59).

Two skin conditions, *hereditary dysplastic nevus syndrome* (DNS) and *nevoid basal cell carcinoma syndrome* (NBCCS), are consistent with deficiency in DNA repair. DNS manifests as multiple dysplastic nevi (atypical pigmented skin lesions) (60). Cells from these patients are hypermutable when exposed to UV irradiation and 4-nitroquinoline-1-oxide, but have normal DNA excision repair (47). The incidence of malignant melanoma is increased 400-fold compared with that in normal individuals (20). Patients with NBCCS have multiple basal cell carcinoma lesions, with up to 75% of affected persons developing malignant changes in the lesions (8). There is a 25% reduction in maximal DNA repair as reflected in unscheduled DNA synthesis (1), and there may be delayed removal of UV-induced pyrimidine dimers (61).

The previous discussion demonstrated the importance of intact DNA repair pathways in both development and control of cellular proliferation. However, these cancer-prone syndromes are rare relative to the incidence of sporadic lung, breast, and colon carcinoma. It seems likely that defective DNA repair will be further implicated in spontaneous carcinogenesis and its clinical behavior. Tremendous variation in the expression of three human DNA repair genes (*ERCC1*, *ERCC2*, and *ERCC6*) has been demonstrated in nonmalignant bone marrow specimens from cancer patients (62). Defective repair systems could provide the substrate that predisposes certain persons to tumor suppressor gene loss or protooncogene activation, particularly in cancer-prone families. A review of lung cancer, and the myriad associated genetic changes seen in genes such as *ras*, *myc*, *neu*, and the tumor suppressor genes *rb* and p53 (63), suggests the presence of some other underlying defect that allows such critical defects to go unrepaired. As discussed below, this association has been demonstrated in hereditary nonpolyposis colon cancer. It is also likely that within a tumor, heterogeneity of expression of DNA repair genes between cells could result in the selection of a subpopulation of cells with enhanced ability to survive and clinical progression of the malignancy. The same process would be of tremendous significance in the tumor's response to anticancer modalities such as radiation therapy and chemotherapy, in which DNA-damaging agents are delivered to tumor tissue (and incidentally to normal tissue) with anticipated lower DNA repair capacity (possibly secondary to the dedifferentiated state) (64). One study demonstrated that patients with ovarian carcinoma with a diminished clinical response to treatment protocols based on cisplatin or carboplatin had a 2.6-fold higher expression of *ERCC1* than did patients who responded to that therapy (65). Thus, DNA repair genes may help effect carcinogenesis through underexpression and subsequent mutation persistence, or tumor progression through overexpression and subsequent survival advantage. A defect in the DNA repair systems may also be important in short tandem repeat expansion. Short tandem repeats are stretches of DNA with continuous, repeated nucleotide combinations such as $(CA)_n$, $(CAG)_n$, and $(CTAT)_n$. In six diseases (fragile X-A syndrome, fragile X-E syndrome, spinobulbar muscular atrophy, myotonic dystrophy, Huntington's disease, and spinocerebellar ataxia type 1), an expansion of a triplet repeat results in a profound change in phenotype, usually within a single generation. Although no cancer predisposition has been found in these diseases, any mutational process that can produce from phenotypically normal parents an offspring with multisystem defects has to be considered potentially carcinogenic.

SHORT TANDEM REPEATS AND THE NORMAL HUMAN GENOME

The normal human genome has a variety of repetitive elements, but short tandem repeats have generated intense interest because of their usefulness in personnel identification and their recent association with heritable diseases. A short tandem repeat (STR) is said to be polymorphic if the number of repeated units varies within the normal population at an incidence of more than 3%. The polymorphism is easily examined using polymerase chain reaction (PCR) amplification (66) and sequence gel

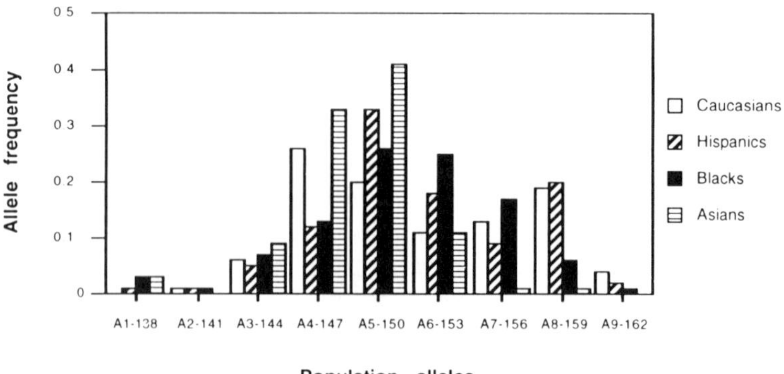

Fig. 2.1: Relative frequencies in four populations of different-sized alleles of a triplet short tandem repeat (AAT) located on 6q21-qter. Each allele has a different number of repeat units in the PCR product.

allele separation. Trimeric and tetrameric repeats are frequent within the human genome, occurring every 150 to 500 kilobases (kb) (67). The variability can be quantified by the term heterozygosity, with most useful repeats having a heterozygosity of more than 0.7. A number of STR loci can be examined in a single reaction (multiplexing) for use in personnel identification, correction of sample mislabeling, and identification of casualties (68). Figure 2.1 shows the allele distribution, in four populations, of an AAT triplet repeat located on chromosome 6q.

The exact mechanism responsible for the formation of STRs (and other repeats) is uncertain (Table 2.3). The high tendency to variability suggests that there is intrinsic instability in addition to any extrinsic influence. The repeat may impair transcriptional fidelity or efficiency (69), leading to the addition of multiple units of the repeat. This may be particularly important in the GC-rich repeats because of the strong hydrogen bonding. One model postulates asymmetric difficulty of replication of GC-rich sequences leading to multiple incomplete single strands of complementary, triplet, reinitiated sequences (70). This might be accompanied by strand switching between leading and lagging strands during replication and repeat expansion. Other ideas postulate interference with nucleosome packaging, and DNA polymerase slippage (71) in which the primer and template strand transiently dissociate during replication of the repeat tract and then reassociate in a misaligned configuration. This could lead to a deletion or expansion depending on whether the unpaired bases were in the template or primer strand. The latter hypothesis was supported in a recent study of mutations in three yeast genes involved in DNA-

Table 2.3: Possible Mechanisms of Triplet Repeat Expansions

Impairment of transcriptional fidelity or efficiency
Production of multiple incomplete single strands of complementary, triplet, reinitiated sequences
Strand switching between leading and lagging strands during replication
Interference with nucleosome packaging
DNA polymerase slippage with dissociation and reassociation in misaligned configuration
Defective DNA repair genes

mismatch repair (72). These investigators showed that mutations in three yeast genes involved in DNA-mismatch repair (*PMS1*, *MLH1*, and *MSH2*) led to 100- to 700-fold increases in poly(GT) tract instability, while mutations that eliminated the proofreading function of DNA polymerases had little effect. The evolutionary preservation of these variable repeats can be ascribed teleologically to their presence largely in introns. Here, the variability has no impact on the structure of a vital protein, and no selection pressure is involved. However, STRs do occur in transcribed and coding regions (73) and may affect various properties of the mRNA and the protein. Only trimeric repeat contraction and expansion would preserve the reading frame of the encoded protein, and one would expect these to attract less selection pressure than repeats that change the reading frame.

Diseases associated with triplet repeats are attributed to changes in the production or function of a protein caused by the genetic variability of an STR at a particular locus. Six such diseases have been described: fragile X-A syndrome, fragile X-E syndrome, spinal bulbar muscular atrophy (Kennedy's disease), myotonic dystrophy, Huntington's disease, and spinocerebellar ataxia type 1. A review of these diseases is important in any discussion of how triplet repeats may cause cancer.

TRIPLET REPEAT DISEASES

Fragile X syndrome is the most common single cause of mental retardation. The gene has been localized to the Xq27.3 fragile site. Fragile X syndrome may occur as frequently as once in 1200 males and once in 800 females. Somatic features include an increase in head circumference, large and prominent ears, hypotonia, and enlargement of the jaw and forehead. Nine of 10 affected males also develop macroorchidism after puberty (74). Other features include seizures, emotional outbursts, and findings related to defective connective tissue formation such as mitral valve prolapse, joint hyperextensibility, and others (75). Female carriers also show gonadal enlargement. The cytogenetic correlate of this disease was first described by Lubs in 1969 (76). The involved X chromosome contains a "fragile" site on the distal end of the long arm, and consists of a constriction of the chromosomal material at this point. In vitro appearance of this phenomenon is dependent on certain tissue culture conditions (77), and other fragile sites have been described on the X chromosome and autosomes with variable frequencies. The genetic mutation represents a new class of mutation that has an extremely high mutation rate. This can have a significant impact on phenotype in one meiosis. The molecular event is the alteration of a "normal" triplet repeat DNA sequence to a premutation allele (larger, but does not produce disease), and then further alteration to the full mutation allele (markedly expanded sequence). The crucial discovery in the search for the fragile X gene was the identification of proximal and distal markers near the locus in a single yeast artificial chromosome (YAC) (78). Rare restriction endonuclease recognition sites found near the fragile site identified a GC-rich region called a "CpG island," (79) often associated with the 5′ end of commonly expressed genes. The addition of a methyl group to the cytosine moiety in the CpG dinucleotide may down-regulate the expression of the gene (80). This CpG island was methylated in fragile X individuals (which might shut off the gene) and was unmethylated in normal subjects (81). The resulting diminished expression of the protein could lead to the observed phenotype. When analyzed closely, it was found that a $(CGG)_n$ STR was expanded in fragile X patients (82). The CGG repeat was present in the 5′ untranslated portion of the FMR-1 gene. This repeat was shown to be polymorphic in the normal population, with alleles ranging from 6 to 52 repeats in length. Phenotypically normal males and female carriers had expansions of 54 to 200 repeats. These larger alleles are associated with a normal phenotype but are the source of the high-risk individuals who give rise to affected offspring with markedly expanded repeats. Study of premutation alleles in mothers who bear males with fragile X syndrome reveals a mutation frequency of one when the premutation allele exceeds 80 repeats. It was this study that explained the molecular basis of "anticipation" (82). Simply stated, as the premutation allele enlarges, the likelihood of its expansion to a disease-producing mutation increases. Thus, the increasing incidence and severity of fragile X syndrome have a molecular correlation. Studies suggest that the FMR-1 protein has properties of RNA binding (83) and localizes to the cytoskeleton. Both FMR-1 mRNA and protein are diminished in fragile X individuals (84), possibly resulting from delayed replication of the gene (85). This is thought to be secondary to abnormal methylation of the CpG island in fragile X individuals (86) as described above. These 5′ sites are unmethylated in normal males (81) and methylated on the inactive X of normal females. The methylation leads to underexpression of FMR-1 in fragile X males and affected female heterozygotes. The resulting diminished expression of the protein correlates with the phenotype of retardation. Several investigators have demonstrated somatic mosaicism in the FMR-1 $(CGG)_n$ repeat (82,84). This mitotic instability has been demonstrated by both Southern blots and PCR analysis, and has been found in the majority of affected males. Also, approximately 20% of affected males have premutation alleles in addition to the larger smear of hybridization seen in Southern blot analysis (73). Examples of nontransmitting males and female carriers exhibiting mosaic premutation alleles have been demonstrated (82), showing that premutation alleles are also unstable.

The preceding discussion of the fragile X syndrome is a description of a clinical syndrome attributed to a particular fragile site, FRAXA, involving the gene FMR-1. Several families have been described in the literature with features of the fragile X syndrome but lacking mutations in the FMR-1 gene (87–89). Investigation of such families with fluorescent in situ hybridization (FISH) revealed another fragile X site, FRAXE, 500 kb distal to FRAXA (90,91). This fragile site was cloned recently and found to possess amplifications of a GCC repeat adjacent to a CpG island in the Xq28 region (92). Normal individuals had 6 to 25 copies of the GCC repeat, while mentally retarded FRAXE-positive individuals had more than 200 repeats and abnormal methylation of the CpG island. The situation is different from that of FRAXA (fragile X syndrome) in that the repeat can expand or contract and is equally unstable when passed through the male or female lines. Initial localization by FISH of a third fragile site, FRAXF, associated with mental retardation and normal FMR-1 testing has been described, 1 to 2 Mb distal to FRAXE in Xq28 (93).

Spinobulbar muscular atrophy (SBMA, or Kennedy's disease) is an X-linked form of hereditary spinal muscular atrophy. It is of insidious onset and usually progresses slowly, with signs and findings of lower motor neuron degeneration. In the original description, proximal limb weakness with bulbar muscle involvement was followed by atrophy some years later (94). Muscle cramps and tremors may precede the specific symptoms, and facial fasciculations may be present. The clinical features are diagnostic in most cases (95), and creatine kinase levels are usually high. Because gynecomastia is present in more than 50% and testicular atrophy is also a feature (96), it seemed possible that an abnormality in androgen activity caused the disease phenotype. The testosterone level in these patients is normal, suggesting an abnormality in the androgen receptor itself. Other studies showed evidence of reduced androgen binding by patient fibroblasts in vitro, reduced ability of motor neurons to bind androgens (97–99), and the linkage of the disease locus to Xq11–12, where the androgen receptor gene is located (100–104). A CAG repeat that is highly polymorphic within the normal human population had been identified previously within the androgen receptor gene (105). Careful study of this STR in patients with Kennedy's disease revealed expanded alleles not found in the normal population. Normal alleles varied from 17 to 26 repeats, with a average of 21 units. Analysis of patients with Kennedy's disease revealed lengths between 40 and 52 CAG repeat units (105). The repeat is in the first exon of the androgen receptor gene, is translated, and codes for a polyglutamine stretch with unknown function. This exon does not appear to be important in the activity or DNA-binding properties of the mature protein. It is the 3′ coding region that is most commonly associated with mutations accounting for testicular feminization (androgen receptor deficiency). Because of the polymorphic nature of the repeat, it seems that the receptor structure can tolerate some variability in the polyglutamine region without affecting activity significantly. Thus, with the Kennedy's disease mutations, one has a new phenomenon—acquisition of an unexpected deleterious central nervous system trait related to a polyglutamine tract expansion in the coding region of the gene. Because Kennedy's disease is X-linked, it is important to point out that it exhibits classic X-inactivation in females (i.e., methylation). Unlike fragile X syndrome, there are few, if any, affected female heterozygotes and the 5′ region of the androgen receptor is not methylated in males with triplet expansions. Thus, while the gene's function in health and disease is well documented, the acquisition of new unexpected features of the polyglutamine tract on CAG-containing mRNA of affected patients is unexplained.

Dystrophia myotonica (DM) is a multisystem disease that can affect the eyes, skin, heart, central nervous system, and other organs with smooth muscle (106). It is an autosomal dominant disease (107) and is the most common form of adult muscular dystrophy (prevalence one in 8000). The atrophy affects mostly the muscles of the distal limbs, face, and neck. Myotonia of the affected muscles is typical but is not exclusive of other diseases. It is thought that the primary defect must be in the sarcolemmal membrane or the T tubule of the muscle fiber itself. Specifically, the defect may affect the flux of Ca^{2+} (108) or result in abnormal sodium and chloride transport (109–111). There are both animal model and similar human disease correlates for such hypotheses (112–117). The average age of onset is 25 years, but the symptoms can appear at any stage of life (118), and the clinical expression is quite variable. This variability exists even among affected family members. However, earlier onset of symptoms generally indicates more severe expression of the phenotype (119). DM has the genetic feature of anticipation mentioned previously for fragile X syndrome, as well as in Huntington's disease (120). Anticipation refers to the earlier and more severe onset and incidence of a genetic disease in successive generations of a family. Congenital dystrophia myotonica (CDM) is the most extreme example of genetic anticipation. The form is transmitted only by the mother (121), who is usually only mildly affected or asymptomatic. In contrast, the infant with CDM has severe hypotonia, respiratory diseases, and feeding difficulties, but no muscle atrophy or myotonia. Molecular analysis of the gene has shown CDM and DM to be caused by the same alteration at the same gene locus. The expansion of the CTG STR in the MT-Pk gene is the molecular basis of anticipation in DM and CDM, with CDM on average having a greater amplification of the repeat than that in the noncongenital DM population (122). The DM locus was linked to the C3 complement factor gene on chromosome 19 in 1983 (123,124), and further work with known markers narrowed the position to 19q13.3 (125–130). Probing of YAC and cosmid clone contigs with an oligonucleotide sequence ACG identified a repeat element mapping to this region. Probing of Southern blots from DM families with these positive clones revealed a disease-specific expansion in a polymorphic EcoRI fragment. In controls, two alleles of 9.8 and 8.6 kb (with frequencies of 0.53 and 0.47, respectively) were detected by specific probes. In DM patients, the 9.8-kb allele expanded to 11 to 15 kb (131–133), and the size increase cosegregated with the disease in DM families. The clinical demonstration of anticipation in these families occurred concomitantly with an increase in the size of the expanded fragment. The cause of the expansion was elucidated through the sequencing of the 1.4-kb BamHI subclone of the EcoRI band (pDMY1) (134). The triplet repeat AGC was found in this fragment (134,135), and PCR analysis revealed considerable polymorphism in the general population. The normal alleles varied between 5 and 32 AGC repeats (136). Alleles with more than 50 CTG repeats were found in asymptomatic individuals who gave rise to affected progeny. In most families with genetic anticipation, a progressive repeat expansion in successive generations was also found (131–136). At least two reports have shown a correlation between disease severity and progressive expansion of the CTG repeat, but there is variability in the relation (122,137). As in fragile X syndrome, some patients exhibit mosaicism of the repeats when DNA from peripheral lymphocytes is analyzed. This indicates somatic mutation in addition to meiotic instability. The CTG repeat is localized in the 3′ untranslated region and is part of a transcribed unit (134), and Northern blot experiments have found high expression in the skeletal muscle, heart, and brain (136). Further analysis of the cDNA revealed strong homology to protein kinases, particularly cyclic adenosine monophosphate–dependent serine and threonine protein kinases. A defect in protein kinase activity was suggested previously by reduction of membrane protein phosphorylation in erythrocytes and myocytes of patients with DM (138,139). These data were supported by study of the recombinantly produced MT-Pk. Thus, MT-Pk is a serine-threonine protein kinase; that is, it is part of a signal transduction system. We have recently identified by in vitro methods the β-subunit of the dihydropyridine receptor (DHPR) as a protein ligand labeled by MT-Pk. This complex regulates calcium flux in muscle cells, a previously noted dysfunction in DM. The triplet repeat expansion diminishes the steady-state level of the MT-Pk mRNA and protein. How this reduction in a signal transduction kinase leads to myotonia is under investigation.

Huntington's disease (HD) is a progressive, typically adult-onset, hereditary disorder characterized by a movement disorder (usually chorea), dementia, and a personality disorder (140). The disease can appear as early as 5 years of age or as late as age 70. The prevalence is 4 in 100,000 to 9 in 100,000, and the disease is more common in whites than in other ethnic groups. Histologically, HD is characterized by loss of neurons in the cerebral cortex. The caudate nucleus and putamen are severely affected, and there is atrophy of the striatopallidal and striatonigral fibers.

Huntington's disease, which had been linked to the short arm of chromosome 4 (4p16.3), demonstrates autosomal dominant inheritance with full penetrance (141). The eventual characterization of the gene relied on information provided by a number of investigators before the discovery of a candidate gene by the Huntington's Disease Collaborative Group (142). The candidate gene, called IT15, was found to have a CAG STR near the 5′ end. The repeat is thought to be in the coding region. Of the 173 normal chromosomes tested, 98% had repeat lengths between 11 and 24 repeats with an overall heterozygosity of 80%. A smaller $(CCG)_7$ found downstream of the CAG was included in the PCR product and could have contributed to the variation. When patients were examined with the same PCR primers, HD heterozygotes were found to have one allele in the normal size range and one much larger. Of the expanded alleles, 41% were between 42 and 47 repeats and 59% were above 48 repeats in length. Family analysis revealed that the normal alleles were inherited in mendelian fashion. The abnormal alleles were not inherited in mendelian fashion and demonstrated rough correlation between earlier age of onset of HD and progressive expansion of the identified repeats. Meiotic expansion of the repeat was also confirmed in two pedigrees with possible spontaneous HD. There is no incidence of large (10- to 100-fold) expansion of triplet repeats as seen in the fragile X syndrome and DM. There would appear to be an acquisition of a dominant phenotype with mutation that is similar to Kennedy's disease. There is no evidence of methylation associated with the CAG expansion. Since the function of IT15 is unknown and measurements of mRNA and protein steady-state are not available at this time, one must be cautious in forecasting pathogenesis.

Spinocerebellar ataxia type 1 (SCA 1) is characterized by motor weakness, ataxia, and ophthalmoparesis and is known to demonstrate an autosomal dominant inheritance pattern (143–146). Once the disease starts in the third or fourth decade of life, progressive neurodegenerative changes lead to eventual death 10 to 20 years after onset. Like those with the other triplet diseases, SCA 1 families show anticipation. Recently, a highly polymorphic CAG repeat was discovered in the SCA 1 locus on chromosome 6p (147). PCR analysis of this triplet repeat in normal subjects revealed a heterozygosity of 84% and a range of 25 to 36 repeat units. Analysis of individuals affected with SCA 1 revealed one normal allele and an expanded allele in the range of 43 to 81 repeats. Also, the more severely affected juvenile patients tended to demonstrate greater repeat expansion (59 to 81 units). Of the expansion diseases studied thus far, SCA 1 shows the greatest correlation of disease severity with repeat length. The SCA 1 gene produces a 10-kb transcript.

HETEROGENEITY OF TRIPLET REPEAT EXPANSION

The previous discussion highlighted the pathogenetic mechanisms underlying the six diseases that have been linked directly to triplet repeat expansion. It is reasonable to suspect that other diseases that share similar clinical and genetic features with the fragile X syndrome (X-A and X-E), spinobulbar muscular atrophy, myotonic dystrophy, Huntington's disease, and spinocerebellar ataxia 1 will demonstrate triplet repeat expansion as an underlying cause. Although these six diseases have CAG, GCC, or CGG expansions, the wide spectrum of polymorphic triplet, tetrameric, and pentameric repeats demonstrated in the literature present more candidate STRs.

These six diseases clearly demonstrate a heterogeneity of molecular events. The involved STR shows instability (polymorphism) in normal populations, and the allele size is roughly comparable: AR, 11 to 31; FMR-1, 15 to 54; FRAXE, 6 to 25; MT-Pk, 5 to 30; HD, 11 to 24; and SCA 1, 25 to 36 repeat units (Fig. 2.2). All six diseases also demonstrate genetic anticipation, with earlier disease onset or severity existing in approximate parallel with increasing length. Also, the three involved repeats

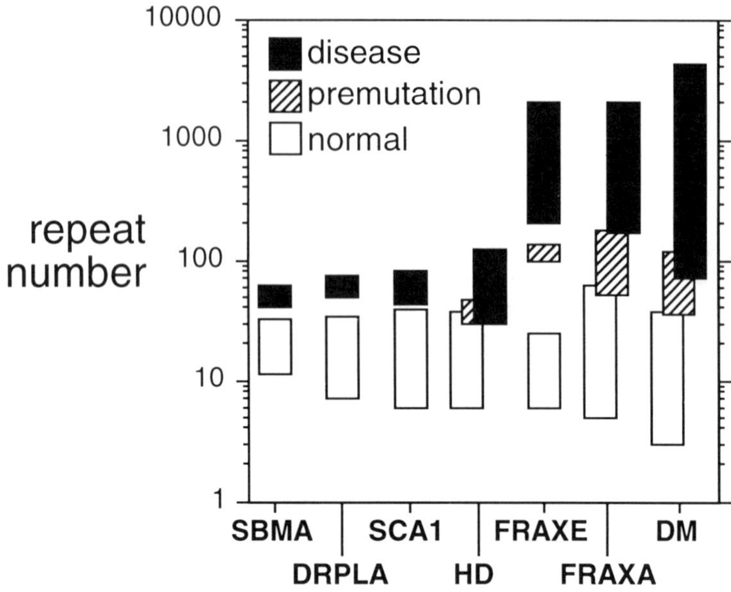

Fig. 2.2: Relation between repeat length and phenotype in the genes associated with the triplet repeat diseases based on current knowledge. Patients with a repeat length within the normal range are phenotypically normal, and their progeny have no increased risk of repeat expansion. Patients with premutation repeat lengths are either phenotypically normal or slightly affected, but their progeny have a variably high risk of repeat expansion and disease expression. Patients with repeat lengths in the disease range have phenotypic manifestations of defective gene function. Note the variability in the noted characteristics of these genes.

(CAG, GCC, and CGG) are GC-rich. There are also numerous differences. The triplet repeats are present in the coding regions of the androgen receptor gene, the Huntington's disease gene, and the SCA 1 gene; the 5′ noncoding region of the FMR-1 gene; and the 3′ untranslated portion of the MT-Pk gene. In contrast to fragile X syndrome and DM, there is no report of a doubling of allele sizes found in the normal population in the AR, HD, or SCA 1 genes. The remarkable difference between the amplification of autosomal CAG repeats of DM (large), HD, and SCA 1 remains unexplained. Furthermore, the mechanism of the AD phenotype most likely differs, because the repeats are in the coding regions for HD, SBMA, and SCA 1, but not DM.

Another significant difference between fragile X syndrome and DM is the pattern of disease allele transmission and large expansion in progeny.

Table 2.4: Comparison of the Triplet Repeat Diseases

Disease	*Repeat*	*Normal Allele Size (bp)*	*Disease Allele Size (bp)*	*Location*	*Genetic Anticipation*	*Meiotic Event*
FRAXA	CGG	15–54	200+	Xq27.3(5′nc)	Yes	Female
FRAXE	GCC	6–25	200+	Xq28(5′nc)	Yes	Female
SBMA	CAG	11–31	40–52	Xq11–12(5′c)	Yes	Female = Male
DM	CTG	5–30	?50+	19q13.3(3′nc)	Yes	Female > Male
HD	CAG	11–24	42+	4p16.3(5′c)	Yes	Unknown
SCA 1	CAG	25–36	43–81	6p(?c)	Yes	Unknown

Both diseases are characterized by the presence of premutation alleles. However, such alleles are at a variably high risk of expansion into the disease range. For example, a rapidly increasing risk of fragile X disease is found in males whose mothers have repeat alleles between 60 (17% risk) and 90 (100% risk) in the FMR-1 gene. In fragile X syndrome, the large CGG amplification associated with the disease (over 200 repeats) occurs as a female meiotic event. In contrast, the premutation alleles in DM appear to be between 50 to 100 GCT repeats, and the high-risk alleles can expand via both male and female meiosis, although more significantly in female meiosis. (The fragile X-E GCC expansion seems to be equally unstable when passed through the male or female line, and can expand or contract.) The most severe form of DM is congenital and occurs exclusively upon maternal, not paternal, inheritance of the allele. The triplet repeat diseases are summarized in Table 2.4.

Triplet repeat amplification is a heterogeneous process that can disrupt gene function in a number of ways. It is also a form of mutation that can have profound effects on phenotype and thus must be considered as a potential source of carcinogenesis.

TRIPLET REPEAT EXPANSION AND CANCER

Because cancer is currently the second leading cause of mortality in the United States, the potential etiologies leading to the malignant state have been investigated extensively, and associations varying from the geographical to molecular levels have been offered in the literature. However, it seems likely that cancer may be a malady of genes, arising from a diversity of genetic damage: recessive and dominant mutations, large rearrangements of DNA, and point mutations, all leading to distortions of

Table 2.5: Genes Containing Triplet Repeats

Gene	*Triplet Repeat (Accession No.)*
Androgen receptor	CAG (M21748)
IT15 (Huntington's)	CAG (L12392)
TATA-binding protein	CAG (M55654/M34960)
Transcription factor TFEB	CAG (M33782)
Transforming growth factor β	CAG (M60315)
Protective protein	CTG (M22960)
Placental alkaline phosphatase	CTG (M12551)
Zinc-finger protein (ETR 103)	AGC (M80583/M62829)
Natriuretic peptide	CAG (M31776)
Protooncogene pim-1	CAG (M27903)
Erythrocyte 2,3 bisphosphoglycerate mutase	CTG (M04327)
Myotonin kinase	CAG (M87313)
Proenkephalin B	CAG (X00177)
Protein 80K-H (protein kinase C substrate)	GAG (J03075)
Histidine-rich calcium binding protein	GAG (M60052)
Cyclin D	GAG (M64349)
Chromogranin A	GAG (J03483)
hCENP-B gene for centromere autoantigen B	GAG (X55039)
α_2-Adrenergic receptor	GAG (M34041)

the expression or biochemical function of genes (148). Table 2.5 presents genes known to contain triplet repeats. This diverse list includes kinases, oncogenes, growth factors, and many others. Also, new genes are being added to this list by several laboratories, including ours (Table 2.6). With this in mind, we suggest that triplet repeat expansion is a genetic mutation that could lead to the malignant phenotype.

As reviewed by Yunis (149) and Bishop (148) separately, the evidence supporting the link between genetic alterations and malignancy is diverse and compelling. Many different types of chromosomal alterations have been noted, from translocations such as the 9:22 translocation seen in more than 90% of chronic myelogeneous leukemia cases and deletions such as that seen on chromosome 3p in small cell carcinoma of the lung (150); to changes related to fragile sites such as those on chromosomes 11q, 12q, and 16q (151,152) seen in some neoplasias. It is also interesting to note the well-known association between (exogenous) cytotoxic chemotherapy exposure, secondary acute myelogenous leukemia, and alterations or loss of chromosomes 5 or 7 or both (153). Furthermore, these alterations are not particular to individual neoplasms but may be seen in a number of diverse neoplasms. For example, the duplication of 1q25q32

Table 2.6: Analysis of cDNA CAG Repeats

Name	*No. of Repeats*	*Chromosome*	*Polymorphism*	*Disease*
B1	6	Unknown	H = 30%	Novel
K5	7	2	H = 0%	PLAP-OV.CA.
A2	Complex	Unknown	H = 0%	Novel
F4	5	Unknown	H = 0%	Novel
J8	Complex	Unknown	Unknown	Novel
D11	13	10q21	H = 80%+	Novel
A11	8	Unknown	H − 0%	Novel
D5	6	Unknown	Low −3 ALL.	—*
A15	10 (Complex)	Unknown	H = 0%	Novel
A4	7	Unknown	H = 0%	—†
F13	17	22Q11	70%+	Novel

The clones were isolated from a placental cDNA library through the use of a 21-mer radiolabeled CAG oligonucleotide probe. H, heterozygosity; complex, areas of consecutive, uninterrupted repeat units separated by short sections of nonrepeat sequence; PLAP-OV CA, placental alkaline phosphatase, serum levels elevated in approximately 26% of patients with ovarian cancer.
* Glia fibrillary acidic protein
† MAR/SAR DNA-binding protein

has been noted in disorders as diverse clinically as cancer of the breast, cervix, colon, melanoma, myeloma, non-Hodgkin's lymphoma, preleukemia, and acute myelogenous leukemia (154). It seems that some cell proliferation switch is turned on by these diverse events, and that the alteration of genes or chromosome "topography" disturbs growth controls.

On a gene level, it seems that the driving force must be, among other factors, the direct activation or altered expression of genes that normally regulate cell proliferation but are quiescent in the nonmalignant phenotype (protooncogenes), or the loss of tumor suppressor gene function and the subsequent activation of the oncogene (148). A good example of the latter is the p53 gene. Its mutation has been described as the most common cancer-related genetic change known at the gene level (155). Although base substitutions have constituted the majority of mutations described in various cancers of the breast, colon, lung, esophagus, liver, and other tissues, other mutations have been described (156–161). Triplet repeat expansion could either directly activate an oncogene or inactivate a tumor suppressor gene, according to what is known about Huntington's disease, spinobulbar muscular atrophy, DM, and the fragile X syndrome. Conceptually, it seems more likely that triplet repeat expansion would cause loss of tumor suppressor function than direct oncogene activation. The activation of two of the most well-characterized oncogenes, c-*myc* and c-*abl*, seems to be related to translocations of DNA fragments containing the oncogene, which place them in proximity to inappropriate regulatory elements or remove the constraints of the original location. In Burkitt's lymphoma, the 8:14 translocation may place c-*myc* in proximity to immunoglobulin genes (162). This might drive excessive transcription of the gene, or the physical damage caused by the translocation might somehow stabilize the mRNA. In chronic myelogenous leukemia, more than 90% of patients show the 9:22 translocation that places the c-*abl* oncogene in proximity to the breakpoint cluster region (*bcr*) (163,164). The final product is a chimeric protein that is more active than the normal c-*abl* gene product (165). There is no correlate of this process in our current knowledge of the aforementioned triplet repeat diseases. Although this may change in the future, triplet repeat expansion does not seem to lead to increased DNA instability manifested as deletions or translocations. Also, the known inhibitive effect of the CGG repeat expansion on expression of the FMR-1 gene or the GTC repeat expansion on MT-Pk gene expression is diametrically opposite to what seems to occur with c-*abl* and c-*myc* oncogene activation. Rather than enhancing mRNA production or stabilizing an mRNA species, transcription or translation is disrupted, and the gene is underexpressed compared with its expression in normal subjects.

The perturbation of tumor suppressor gene function or expression seems a more plausible relation between carcinogenesis and triplet repeat expansion. As may occur in Huntington's disease, such a deletion or recombination could cause loss of function of the unaltered allele, and might also involve abnormal patterns of methylation as seen in some cancers (166-176).

A region of GC-rich expansion could form a DNA fragile site. DNA fragile sites are occasionally found near CpG islands (177), perhaps because of the high GC content, and may be more prone to breakage because methylated DNA replicates later than unmethylated DNA. Deletions could result in loss of tumor suppressor function, while rearranged oncogenes could lead to the production of an altered regulatory protein and a malignant phenotype as mentioned above. The higher GC content of an expanded area could increase the frequency of point mutations. It is known that cytosine-to-thymine transitions are responsible for a large number of normal human polymorphisms and point mutations in genetic diseases (178). Deamination of methylated cytosine generates a thymine residue, and many of the point mutations identified in the p53 tumor suppressor gene on chromosome 17p involve C-to-T transitions (179), and the CpG islands included are heavily methylated in normal DNA (178). Finally, tumor cells may have increased DNA methyltransferase activity (180,181).

Although it is possible that triplet repeat (particularly CGG) expansion leading to hypermethylation or DNA instability could lead to the malignant phenotype, it is possible that such a mechanism could cause familial cancer syndromes. It is obvious that, upon fertilization, two independent haploid genomes cooperate in directing the development of the embryo. A process known as genomic imprinting, which refers to maintenance of genetic memory of gamete origin in early development, marks certain genes for differential utilization by their passage through the maternal or paternal germlines (182). DNA methylation has been proposed as a molecular mechanism for the propagation of imprinting (183,184). The presence of genomic imprinting in humans is uncertain. However, experiments have shown that the two c-*ras* alleles in human cells can have different methylation patterns (185). Thus, the instability created by triplet repeat expansion not only may be secondary to the physical effects on the locus structure and function, but also may be attributed to an increased incidence of point mutations and deletions. Such a functionally hemizygous gene would be much more susceptible to disruption by carcinogens than would normally be true in a diploid state. The differential methylation pattern imposed on genes during their passage through paternal and maternal germlines could alter their susceptibility to spontaneously occurring mutations and thus alter the inheritance of human cancer.

Triplet repeat expansion has clearly demonstrated both mitotic and meiotic instability in myotonic dystrophy, fragile X syndrome, spinobulbar muscular atrophy, and Huntington's disease. Mitotic instability could be involved in spontaneous oncogenesis or degeneration to the malignant phenotype. The heterogeneity of tumor cell populations in myriad properties, from doubling time and percentage of S phase to oxygen requirements and antigen expression patterns, is well known. Thus, it seems likely that an unstable triplet repeat could also be heterogeneous within a given tumor population and confer different growth or drug-response characteristics to cells within a tumor. Meiotic instability of a triplet repeat could be involved in familial cancer syndromes as it is in the aforementioned diseases. As mentioned above, a possible scenario would be similar to that in retinoblastoma (186). The inactivation of one tumor suppressor allele would occur once a premutation expansion reached mutation length and passed to the progeny. Tumorigenesis would result once the second allele was inactivated, either by somatic expansion of the repeat or by some other mutation. As mentioned above, genome imprinting results in a gamete-of-origin-dependent modification of the phenotype elicited by a particular allele (187–189). In other words, the phenotype produced by a gene when that gene is maternally inherited differs from that when the gene is paternally inherited (190). This process has been demonstrated in rhabdomyosarcoma (191), osteosarcomas (192), Wilms' tumor (193–196), and bilateral retinoblastoma (197,198). The putative tumor-suppressor-carrying chromosome remaining in these cases has been overwhelmingly paternal. This is similar to what is found in the fragile X syndrome, in which the large CGG amplification and resulting phenotype (male or female) seems to be predominantly a female meiotic event. Males transmit the repeat to their progeny with small changes in repeat number. In contrast, myotonic dystrophy GCT repeat expansion can be either a male or female meiotic event, and there seems to be a greater rate of instability in male meioses than in female meioses in the affected CAG repeat region of the androgen receptor gene (spinobulbar muscular atrophy).

Another feature shared between the triple repeat diseases and some inherited forms of malignancy is dominant inheritance. Hereditary retinoblastoma (186), Wilms' tumor (199), neuroblastoma and pheochromocytoma (200), cancer of the breast (201,202), colorectal cancer (203,204), ovarian cancer (205,206), bladder cancer (207), renal cell carcinoma (208,209), lymphomas (210–212), melanoma (213,214), and others all demonstrate dominant inheritance. However, many of these cancers also have a sporadic form. It is generally accepted that both forms of a malignancy (inherited and sporadic) involve changes to the same gene (215). This has been shown with both forms of retinoblastoma (216) and with the *apc* gene in colon cancer (217). Furthermore, inherited cancer in humans has never been ascribed to the inheritance of an activated oncogene (218). This may be due to the disturbance in early embryo development that a prematurely activated oncogene might cause, leading to loss of embryo or fetus. In contrast, a mutated tumor suppressor allele would be silent and the phenotype normal as long as the other allele was normal and generating a normal product. As soon as the person suffered two to three oncogenic mutations, a premalignant lesion would develop, and the fourth would lead to malignancy (219). One would expect that a certain period of time would elapse, perhaps modulated by environmental exposures (220), before the required number of mutations accumulated. Although the function of the Huntington's disease gene is unknown, symptoms first appear in the mid-thirties in many individuals. This is not to say that the Huntington's disease gene is necessarily related to the tumor suppressor gene family but that triplet repeat expansion diseases can manifest in adult life. Figure 2.3 summarizes the proposed relation between triplet repeat expansion and neoplasia.

Although the focus of this section has been expansion of the triplet repeat, there is evidence that a triplet repeat disease can be caused by

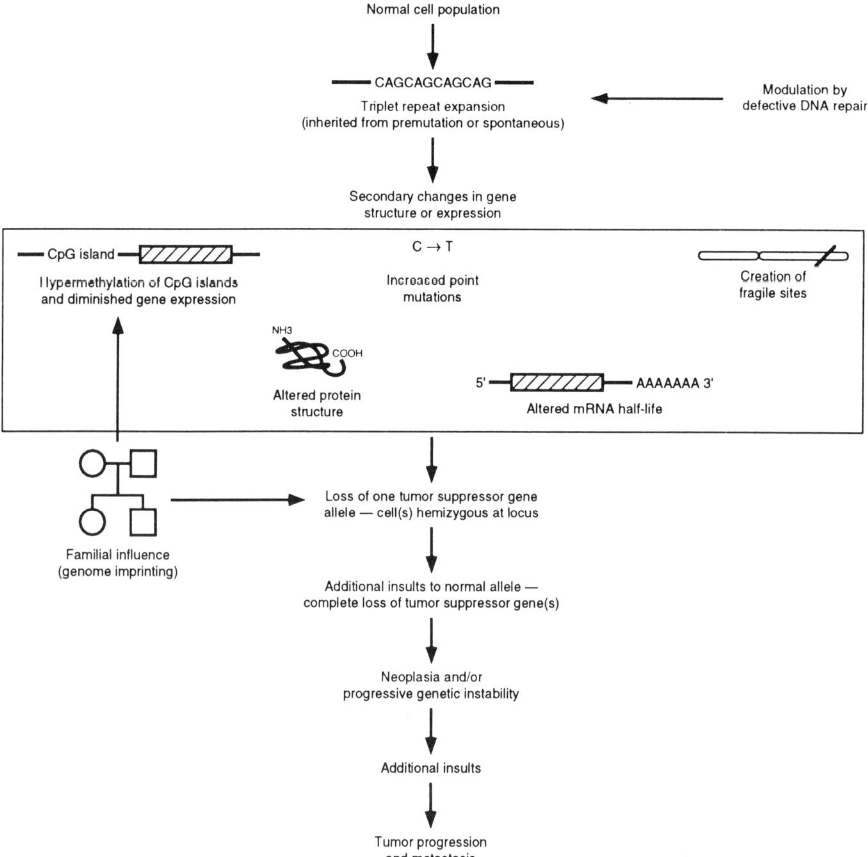

Fig. 2.3: Proposed relation between defective DNA repair, triplet repeat expansion, and neoplastic behavior.

reversion of the affected DNA fragment. This has been documented by four different groups studying myotonic dystrophy families (221–223). In one fascinating report by Brunner and associates on two myotonic dystrophy families, the affected father in one family had a DM gene with 150 to 500 CTG repeats. He appeared to pass on his DM chromosome to an unaffected son with a normal-sized allele of 19 CTG repeats. The son had no evidence of somatic or germline mosaicism. As reviewed by Brook (224), this process may represent replication slippage, unequal sister strand exchange, or direct deletion of repeat units in the genome of the parent. However, there may have also been reversal of the mutation in the early embryo. The other important point about these cases was that the reversion occurred on transmission from an affected male. This phenomenon will be further characterized in the future, but again demonstrates the importance of meiosis in the dynamics of repeat expansion and the subsequent impact on the phenotype of the progeny.

DNA REPAIR GENES AND COLON CANCER

There are data confirming the association between STR (dinucleotide) instability, DNA repair defects, and malignancy. Aaltonen and associates (225) studied large kindreds with hereditary nonpolyposis colon cancer (HNPCC) and found strong linkage at 2p15–16. Rather than finding loss of heterozygosity at this locus in tumors (compared with normal tissues from the same individuals), they found shifts in the electrophoretic mobility of $(CA)_n$ dinucleotide repeat fragments. This suggested that replication errors had occurred in these sequences during tumor development. Also, a nearby $(CAG)_n$ trinucleotide repeat also showed mobility shifts. When these changes were compared with sporadic colorectal cancer samples and corresponding normal samples, there was a much lower incidence of shifts. Further work demonstrated a human gene called *hMSH2* that is homologous to excision repair genes in *E. coli* (*MutS*) and *S. cervisiae* (*MSH2*) (226). When placed in an expression vector in *E. coli*, *hMSH2* caused a dominant mutator phenotype and disrupted bacterial mismatch repair. The gene was mapped in the vicinity of 2p21, near the proposed HNPCC locus. It so happens that *S. cervisiae MSH2* mutants exhibit increased rates of expansion and contraction of dinucleotide repeat sequences (72), which implied that *hMSH2* was the HNPCC gene. The work was supported almost simultaneously by the discovery of a gene in the HNPCC area, though mapped to 2p16, which was homologous to the *MutS* mismatch repair genes (227). The cDNA was obtained and sequenced, and somatic and germline mutations of the gene were identified in RER+ tumor cells from HNPCC kindreds. An accompanying paper by Parsons and colleagues showed that RER+ tumor cells had at least a 100-fold increase in the mutation rate of $(CA)_n$ repeats compared with RER− phenotype tumor cells. They also showed that RER+ cells have a profound defect in strand-specific mismatch repair. Another study showed that mutations in three other mismatch repair genes related to the bacterial *mutL* gene can cause HNPCC (229). One of the genes (*hMLH1*) is located on chromosome 3p21, and is within 1 cM of previously identified markers linked to cancer susceptibility in HNPCC kindreds. Mutations of *hMLH1* detected in sequence of cDNAs prepared from HNPCC kindreds included deletions, insertions, and transversions. Such mutations in the coding region of a gene were felt to be significant enough to disrupt the gene product and result in the disease. Thus, these papers relate colon cancer to defective DNA repair and repeat instability. Data like these are consistent with the idea that genetic stability is a component of the familial cancer phenotype. That such changes, or the propensity to such changes in the genome, can be passed through the genome has been demonstrated in Huntington's disease, myotonic dystrophy, fragile X syndrome (X-A and X-E), spinobulbar muscular atrophy, and now in hereditary nonpolyposis colon cancer. In the future, other associations between STR instability and disease will be elucidated. Such discoveries will contribute to the continued development of the most cost-effective health care of all—preventive medicine.

CONCLUSION

In conclusion, STRs are ubiquitous in the normal human genome and are the source of wide individual diversity. Six diseases (myotonic dystrophy, Huntington's disease, fragile X syndrome [X-A and X-E], spinobulbar muscular atrophy, and spinocerebellar ataxia 1) have been identified that are caused by triplet repeat expansion. The involved repeats have been shown to disrupt gene expression despite their location in relation to the coding portion of the gene. Moreover, these repeats can show mitotic and meiotic instability, and selectivity of effect depending on the parent of origin of the allele. The instability of STRs makes them hot spots for mutation and potential loss of cellular proliferation control. It seems that triplet repeat expansion would be seen in heritable forms of cancer or cancer associated with chromosomal fragile sites, and that they would involve loss of tumor suppressor gene function. The fact the CA dinucleotide instability on chromosome 2 has been associated with colon cancer only strengthens the probable association between STR instability and malignancy. As stated well by Richards and Sutherland, "the properties of heritable unstable elements come with their own set of rules. Finding out exactly what these rules are, and the circumstances under which these rules contribute to genetic disease and variations, represent a new and unexpected challenge in molecular genetics" (230).

REFERENCES

1. Berstam VA. Handbook of gene level diagnostics in clinical practice. Boca Raton, Florida: CRC Press, 1992.

2. Ames BN. Endogenous DNA damage as related to cancer and aging. Mutat Res 1989;214:41–46.

3. Smith PJ. Carcinogenesis: molecular defences against carcinogens. Br Med Bull 1991;47:3–20.

4. Strike P. DNA repair and mutation: recent advances. Mutat Res 1993;294:187–198.

5. Thompson LH. Properties and applications of human DNA repair genes. Mutat Res 1991;247:213–219.

6. Wevrick R, Buchwald M. Mammalian DNA repair genes. Curr Opin Genet Dev 1993;3:470–474.

7. Coverly D, Kenny MK, Lane DP, Wood RD. A role for the human single-stranded DNA binding protein HSSB/RPA in an early stage of nucleotide excision repair. Nucleic Acids Res 1992;20:3873–3880.

8. Bohr VA, Evans MK, Fornace AJ Jr. Biology of disease. DNA repair and its pathogenetic implications. Lab Invest 1989;61:143–161.

9. Bohr VA, Smith CA, Okumoto DS, Hanawalt PC. DNA repair in an active gene: rate of removal of pyrimidine dimers from the DHFR gene of CHO cells is much more efficient than the genome overall. Cell 1985;40:359–369.

10. Mellon I, Bohr VA, Smith CA, Hanawalt PC. Preferential DNA repair of an active gene in human cells. Proc Natl Acad Sci USA 1986;83:8878–8888.

11. Mellon IM, Spivak G, Hanawalt PC. Selective removal of transcription-blocking DNA damage from the transcribed strand of the mammalian DHFR gene. Cell 1987;51:241–246.

12. Madhani HD, Bohr VA, Hanawalt PC. Differential DNA repair in a transcriptionally active and inactive protooncogene: c-*abl* and c-*mos*. Cell 1986;45:417–422.

13. Hanawalt PC. Heterogeneity of DNA repair at the gene level. Mutat Res 1991;247:203–211.

14. Ho L, Bohr VA, Hanawalt PC. Demethylation enhances removal of DNA damage from the overall genome and from specific DNA sequences in CHO cells. Mol Cell Biol 1989;9:1594–1603.

15. Bohr VA. DNA repair at the level of the gene: molecular and clinical considerations. J Cancer Res Clin Oncol 1990;116:384–391.

16. Vos JM, Hanawalt PC. Processing of psoralen adducts in active gene: repair and replication of DNA containing monoadducts and interstrand crosslinks. Cell 1987;50:1789–1799.

17. Rao KS. Genomic damage and its repair in young and aging brain. Mol Neurobiol 1993;7:23–48.

18. Cleaver JE. Defective DNA replication of DNA in xeroderma pigmentosum. Nature 1968;218:652–656.

19. Setlow RB, Regan JD, German J, Carrier WL. Evidence that xeroderma pigmentosum cells do not perform the first step in the repair of ultraviolet damage to their DNA. Proc Natl Acad Sci USA 1969;64:1035–1041.

20. Hansson J. Inherited defects in DNA repair and susceptibility to DNA-damaging agents. Toxicol Lett 1992;64/65:141–148.

21. Digweed M. Human genetic instability syndromes: single gene defects with an increased risk of cancer. Toxicol Lett 1993;67:259–281.

22. Parshad R, Tarone RE, Price FM, Sanford KK. Cytogenetic evidence for

differences in DNA incision activity in xeroderma pigmentosum group A, C, and D cells after X-irradiation during G_2 phase. Mutat Res 1993;294:149–155.

23. Zhen W, Evans MK, Haggerty CM, Bohr VA. Deficient gene specific repair of cisplatin-induced lesions in xeroderma pigmentosum and Fanconi's anemia cell lines. Carcinogenesis 1993;14:919–924.

24. Yagi T, Tatsumi-Miyajima J, Sato M, Kraemer KH, Takebe H. Analysis of point mutations in an ultraviolet-irradiated shuttle vector plasmid propagated in cells from Japanese xeroderma pigmentosum patients in complementation groups A and F. Cancer Res 1991;51:3177–3182.

25. Seetharam S, Kraemer KH, Walters HL, Seidman MM. Ultraviolet mutational spectrum in a shuttle vector propagated in xeroderma pigmentosum lymphoblastoid cells and fibroblasts. Mutat Res 254:97–105.

26. Tanaka K. Molecular analysis of xeroderma pigmentosum group A gene. Jpn J Hum Genet 1993;38:1–14.

27. Tanaka K, Satokata I, Ogita Z, Uchida T, Okada Y. Molecular cloning of a mouse DNA repair gene that complements the defect of group-A xeroderma pigmentosum. Proc Natl Acad Sci USA 1989;86:5512–5516.

28. Weeda G, van Ham RCA, Masurel R, et al. Molecular cloning and biological characterization of the human excision repair gene *ERCC-3.* Mol Cell Biol 1990;10:2570–2581.

29. Weeda G, van Ham RCA, Vermeulen W, Bootsma D, van der Eb AJ, Hoeijmakers JHJ. A presumed DNA helicase encoded by *ERCC-3* is involved in the human repair disorders xeroderma pigmentosum and Cockayne's syndrome. Cell 1990;62:777–791.

30. Legerski R, Peterson C. Expression cloning of a human DNA repair gene involved in xeroderma pigmentosum group C. Nature 1992;359:70–73.

31. Flejter WL, McDaniel LD, Johns D, Friedberg C, Schultz RA. Correction of xeroderma pigmentosum complementation group D mutant cell phenotypes by chromosome and gene transfer: involvement of the human *ERCC2* DNA repair gene. Proc Natl Acad Sci USA 1992;89:261–265.

32. Weber CA, Salazar EP, Stewart SA, Thompson LH. *ERCC2*: cDNA cloning and molecular characterization of a human nucleotide excision repair gene with high homology to yeast *RAD3.* EMBO J 1990;9:1437–1447.

33. Scherly D, Nousikel T, Corlet, J, Ucla C, Bairoch A, Clarkson SG. Complementation of the DNA repair defect in xeroderma pigmentosum group G cells by a human cDNA related to yeast *RAD2.* Nature 1993;363:182–185.

34. Muller H. Recessively inherited deficiencies predisposing to cancer. Anticancer Res 1990;10:513–518.

35. Price FM, Parshad R, Tarone RE, Sanford KK. Radiation induced chromatid aberrations in Cockayne syndrome and xeroderma pigmentosum group C fibroblasts in relation to cancer predisposition. Cancer Genet Cytogenet 1991;57:1–10.

36. Suarez HG, Daya-Grosjean L, Schlaifer D, et al. Activated oncogenes in human skin tumors from a repair-deficient syndrome, xeroderma pigmentosum. Cancer Res 1989;49:1223–1228.

37. Daya-Grosjean L, Robert C, Drougard C, Suarez H, Sarasin A. High

mutation frequency in *ras* genes of skin tumors isolated from DNA repair deficient xeroderma pigmentosum patients. Cancer Res 1993;53:1625–1629.

38. Friedberg EC. DNA repair. San Francisco: W. H. Freeman, 1985.

39. Festa RS, Meadows AT, Boshes RA. Leukaemia in a black child with Bloom's syndrome. Cancer 1979;44:1507–1510.

40. Vijayalaxmi, Evans HJ, Ray JH, German J. Bloom's syndrome: evidence for an increased mutation frequency in vitro. Science 1983;221:851–853.

41. Willis AE, Weksberg R, Tomlinson S, Lindahl T. Structural alterations of DNA ligase I in Bloom syndrome. Proc Natl Acad Sci USA 1987;84: 8016–8020.

42. Petrini JHJ, Huwiler KG, Weaver DT. A wild-type DNA ligase I gene is expressed in Bloom's syndrome cells. Proc Natl Acad Sci USA 1991;88:7615–7619.

43. Lehman AR, Hoeijmakers JHJ, van Zeeland AA, et al. Workshop on DNA repair. Mutat Res 1992;273:1–28.

44. Papadopoulo D, Guillouf C, Mohrenweiser H, Moustacchi E. Hypomutability in Fanconi anemia cells is associated with increased deletion frequency at the HPRT locus. Proc Natl Acad Sci USA 1990;87:8383–8387.

45. Langlois RG, Bigbee WL, Jensen RH, German J. Evidence for increased in vivo mutation and somatic recombination in Bloom's syndrome. Proc Natl Acad Sci USA 1989;86:670–674.

46. Volberg TM, Seal G, Sirover MA. Monoclonal antibodies detect conformational abnormality of uracil DNA glycosylase in Bloom's syndrome cells. Carcinogenesis 1987;8:1725–1729.

47. Hansson J, Keyse SM, Lindahl T, Wood RD. DNA excision repair in cell extracts from human cell lines exhibiting hypersensitivity to DNA-damaging agents. Cancer Res 1991;51:3384–3390.

48. Venema J, Mullenders LHF, Natarajan AT, van Zeeland AA, Mayne LV. The genetic defect in Cockayne syndrome is associated with a defect in repair of UV-induced DNA damage in transcriptionally active DNA. Proc Natl Acad Sci USA 1990;87:4707–4711.

49. Heim RA, Lench NJ, Swift M. Heterozygous manifestations in four autosomal recessive human cancer-prone syndromes: ataxia telangiectasia, xeroderma pigmentosum, Fanconi anemia, and Bloom syndrome. Mutat Res 1992;284:25–36.

50. Strathdee CA, Gavish H, Shannon WR, Buchwald M. Cloning of cDNAs for Fanconi's anemia by functional complementation. Nature 1992;356:763–767.

51. Papadopoulo D, Laquerbe A, Guillouf C, Moustacchi E. Molecular spectrum of mutations induced at the *HPRT* locus by a cross-linking agent in human cell lines with different repair capacities. Mutat Res 1993;294:167–177.

52. Auerbach AD, Adler B, Chaganti RSK. Prenatal and postnatal diagnosis and carrier detection of Fanconi anemia by a cytogenetic method. Pediatrics 1981;67:128–135.

53. Auerbach AD, Sagai M, Adler B. Fanconi anemia: prenatal diagnosis in 30 fetuses at risk. Pediatrics 1985;76:794–800.

54. Auerbach AD, Allen RG. Leukemia and preleukemia in Fanconi anemia patients. A review of the literature and report of the International Fanconi Anemia Registry. Cancer Genet Cytogenet 1991;51:1–12.

55. McKinnon PJ. Ataxia telangiectasia: an inherited disorder of ionizing-radiation sensitivity in man. Hum Genet 1987;75:197–208.

56. Painter RB. Radioresistant DNA synthesis: an intrinsic feature of ataxia telangiectasia. Mutat Res 1981;84:183–190.

57. Spector BD. Epidemiology of cancer in ataxia-telangiectasia. In: Bridges BA, Harnden DG, eds. Ataxia-telangiectasia—a cellular and molecular link between cancer, neuropathology, and immune deficiency. New York: John Wiley and Sons, 1982.

58. Swift M, Morrell D, Massey RB, Chase CL. Incidence of cancer in 161 families affected by ataxia telangiectasia. N Engl J Med 1991;325:1831–1836.

59. Swift M, Reitnauer PJ, Morrell D, Chase CL. Breast and other cancers in families with ataxia telangiectasia. N Engl J Med 1987;316:1289–1294.

60. Greene MH, Clark WH, Tucker MA, et al. Acquired precursors of cutaneous malignant melanoma: the familial dysplastic nevus syndrome. N Engl J Med 1985;312:91–97.

61. Alcalay J, Freeman SE, Goldberg LH, Wolf JE. Excision repair of pyrimidine dimers induced by simulated solar radiation in the skin of patients with basal cell carcinoma. J Invest Dermatol 1990;95:506–509.

62. Dabholkar M, Bostick-Bruton F, Weber C, Egwuagu C, Bohr VA, Reed E. Expression of excision repair genes in non-malignant bone marrow from cancer patients. Mutat Res 1993;293:151–160.

63. Carbone DP, Minna JD. The molecular genetics of lung cancer. Adv Intern Med 1992;37:153–171.

64. Carmichael J, Hickson ID. Keynote address; mechanisms of cellular resistance to cytotoxic drugs and X-irradiation. Int J Radiat Oncol Biol Phys 1991;20:197–202.

65. Dabholkar M, Bostick-Bruton F, Weber C, Bohr VA, Egwuagu C, Reed E. *ERCC1* and *ERCC2* expression in malignant tissues from ovarian cancer patients. J Natl Cancer Inst 1992;84:1512–1517.

66. Weber JL, May PE. Abundant class of human DNA polymorphisms which can be typed using the polymerase chain reaction. Am J Hum Genet 1989;44:388–396.

67. Edwards A, Civitello A, Hammond HA, Caskey CT. DNA typing and genetic mapping with trimeric and tetrameric tandem repeats. Am J Hum Genet 1991;49:746–756.

68. Edwards A, Hammond HA, Jin L, Caskey CT, Chakraborty R. Genetic variation at five trimeric and tetrameric tandem repeat loci in four human population groups. Genomics 1992;12:241–253.

69. Mermer B, Colb M, Krontiris TG. A family of short, interspersed repeats is associated with tandemly repetitive DNA in the human genome. Proc Natl Acad Sci USA 1987;84:3320–3324.

70. Caskey CT, Pizutti A, Fu YH, Fenwick RG, Nelson DL. Triplet repeat mutations in human disease. Science 1992;256:784–789.

71. Bentley DL, Groudine M. A block to elongation is largely responsible for decreased transcription of c-*myc* in differentiated HL60 cells. Nature 1986;321:702–706.

72. Strand M, Prolla TA, Liskay RM, Petes TD. Destabilization of tracts of simple repetitive DNA in yeast by mutations affecting DNA mismatch repair. Nature 1993;356:274–279.

73. Koschinsky ML, Beisiegel U, Henne-Bruns D, Eaton DL, Lawn RM. Apolipoprotein(a) size heterogeneity is related to variable number of repeat sequences in its mRNA. Biochemistry 1990;29:640–644.

74. Sutherland GR. Fragile sites on human chromosomes: demonstration of their dependence on the type of tissue culture medium. Science 1977;197: 265–266.

75. Hagerman RJ. Physical and behavioral phenotype. In: Hagerman RJ, Silverman AC, eds. Fragile X syndrome: diagnosis, treatment and research. Baltimore: Johns Hopkins University Press, 1991:3–68.

76. Lubs HA. A marker X chromosome. Am J Hum Genet 1969;21:231–244.

77. Sutherland GR. Fragile X sites on human chromosomes: demonstration of their dependence on the type of tissue culture medium. Science 1977;197:265–266.

78. Heitz D, Rousseau F, Devys D, et al. Isolation of sequences that span the fragile X and identification of a fragile X mutation. Science 1991;251:1236–1239. Published erratum appears in Science 1991;252:494.

79. Poustka A, Dietrich A, Langenstein G, Toniolo D, Warren ST, Lerach H. Physical map of human Xq27-qter: localizing the region of the fragile X mutation. Proc Natl Acad Sci USA 1991;88:8302–8306.

80. Swain JL, Stewart TA, Leder P. Parental legacy determines methylation and expression of an autosomal transgene: a molecular mechanism for parental imprinting. Cell 1987;50:719–727.

81. Bell MV, Hirst MC, Nakahori Y, et al. Physical mapping across the fragile X: hypermethylation and clinical expression of the fragile X syndrome. Cell 1991;64:861–866.

82. Fu YH, Kuhl DP, Pizutti A, et al. Variation of the CGG repeat at the fragile X site results in genetic instability: resolution of the Sherman paradox. Cell 1991;67:1047–1058.

83. Siomi H, Siomi MC, Nussbaum RL, Dreyfuss G. The protein product of the fragile X gene, *FMR1*, has characteristics of an RNA binding protein. Cell 1993;74:291–298.

84. Pieretti M, Zhang FP, Fu YH, et al. Absence of expression of the FMR-1 gene in fragile X syndrome. Cell 1991;66:817–822.

85. Hansen RS, Canfield TK, Lamb MM, Gartler SM, Laird CD. Association of fragile X syndrome with delayed replication of the *FMR1* gene. Cell 1993;73:1403–1409.

86. Sutcliffe JS, Nelson DL, Zhang F, et al. DNA methylation represses *FMR-1* transcription in fragile X syndrome. Hum Mol Genet 1992;1:397–400.

87. Nakahori Y, Knight SJL, Holland J, et al. Molecular heterogeneity of the

fragile X syndrome. Nucleic Acids Res 1991;19:4355–4359.

88. Oberle, I, Bowe, J, Croquette M, Voelckel M, Mattei M, Mandel J. Three families with high expression of a fragile X site at Xq27.3, lack of anomalies at the *FMR-1* CpG island, and no clear phenotypic association. Am J Med Genet 1992;43:224–231.

89. Dennis N, Curtis G, MacPherson J, Jacobs P. Two families with Xq27.3 fragility, no detectable insert in the *FMR-1* gene, mild mental impairment, and absence of the Martin-Bell phenotype. Am J Med Genet 1992;43:232–236.

90. Sutherland GS, Baker E. Characterization of a new rare fragile site easily confused with the fragile X. Hum Mol Genet 1992;1:111–113.

91. Flynn G, Hirst M, Knight S, et al. Identification of the FRAXE fragile site in two families ascertained for X-linked mental retardation. J Med Genet 1993;30:97–100.

92. Knight SJ, Flannery AV, Hirst MC, et al. Trinucleotide repeat amplification and hypermethylation of a CpG island in FRAXE mental retardation. Cell 1993;74:127–134.

93. Hirst MC, Barnicoat A, Flynn G, et al. The identification of a third fragile site, FRAXF, in Xq27–q28 distal to both FRAXA and FRAXE. Hum Mol Genet 1993;2:197–200.

94. Kennedy WR, Alter M, Sung JH. Progressive proximal spinal and bulbar muscular atrophy of late onset: a sex-linked recessive trait. Neurology 1968;18:671–680.

95. Harding AE, Thomas PK, Baraitser M, Bradbury PG, Morgan-Hughes JA, Ponsford JR. X-linked recessive bulbospinal neuromyopathy: a report of ten cases. J Neurol Neurosurg Psychiatry 1982;45:1012–1019.

96. Ringel SP, Lava NS, Treihaft NM. Late onset X-linked recessive spinal and bulbar muscular atrophy. Muscle Nerve 1978;1:297–307.

97. Warner CL, Griffin JE, Wilson JD, et al. X-linked spinomuscular atrophy: a kindred with associated abnormal androgen receptor binding. Neurology 1991;41(Suppl 1):313.

98. Weiner LP. Possible role of androgen receptors in amyotrophic lateral sclerosis. Arch Neurol 1980;37:129–131.

99. Sar M, Stumpf WE. Androgen concentration in motor neurons of cranial nerves and spinal cord. Science 1977;107:77–80.

100. Fishbeck KH, Ionasescu V, Ritter AW, et al. Localization of the gene for X-linked spinal muscular atrophy. Neurology 1986;36:1595–1598.

101. Kelly TE, Lunt P, Sarfarazi M, Schnatterly P, Thomas NST, Harper PS. Evidence that X-linked Charcot-Marie-Tooth disease and X-linked bulbospinal neuronopathy loci are on opposite sides of DXYS1. Cytogenet Cell Genet 1987;46:638.

102. Lubahn DB, Joseph DR, Sullivan PM, Huntington FW, French FS, Wilson EM. Cloning of human androgen receptor and localization on the X-chromosome. Science 1988;240:327–330.

103. Chang C, Kokontis J, Liao S. Molecular cloning of human and rat cDNAs encoding androgen receptors. Science 1988;86:324–326.

104. Tilley WD, Marcelli M, Wilson JD. Characterization and expression of

a cDNA encoding the human androgen receptor. Proc Natl Acad Sci USA 1989;8:327–331.

105. La Spada A, Wilson EM, Lubahn DB, Harding AE, Fishbeck KH. Androgen receptor gene mutation in X-linked spinal and bulbar muscular atrophy. Nature 1991;352:77–79.

106. Harper PS. Myotonic dystrophy, 2nd ed. London: WB Saunders, 1989.

107. Thomasen E. Myotonia. Aarhus, Denmark: Universiteforlaget, 1948.

108. Jacobs AE, Benders AA, Oosterhof A, et al. The calcium homeostasis and the membrane potentials of cultured muscle cells from patients with myotonic dystrophy. Biochim Biophys Acta 1990;1096:14–19.

109. Rudel L, Ruppersberg JP, Spittelmeister W. Abnormalities of the fast sodium current in myotonic dystrophy, recessive generalized myotonia, and adynamia episodica. Muscle Nerve 1989;12:281–287.

110. Franke G, Hatt H, Iaizzo PA, Lehmann-Horn F. Characteristics of Na channels and Cl conductance in resealed muscle fiber segments from patients with myotonic dystrophy. J Physiol 1990;425:391–405.

111. Kuhn E, Lehmann-Horn F, Rudel R. Dystrophia myotonica (Steinert disease)—a frequently misdiagnosed disease. Nervenarzt 1990;61:323–331.

112. Steinmeyer K, Klocke R, Ortland C, Gronemeier M, Jockusch H, Grunder S, Jentsch TJ. Inactivation of muscle chloride channel by transposon insertion in myotonic mice. Nature 1992;354:304–308.

113. Rudel R, Ricker K, Lehmann-Horn F. Transient weakness and altered membrane characteristics in recessive generalised myotonia (Becker). Muscle Nerve 1990;11:202–211.

114. McClatchey AI, Van der Bergh P, Perickak-Vance MA, et al. Temperature-sensitive mutations in the III–IV cytoplasmic loop region of the skeletal muscle sodium channel gene in paramyotonia congenita. Cell 1992;68:769–774.

115. Ebers GC, George AL, Barchi RL, et al. Paramyotonia congenita and hyperkalemic periodic paralysis are linked to the adult muscle sodium channel gene. Ann Neurol 1991;30:810–816.

116. Ptacek LJ, George AL, Griggs RC, et al. Identification of a mutation in the gene causing hyperkalemic periodic paralysis. Cell 1991;67:1021–1027.

117. Rojas CV, Wang J, Schwartz LS, Hoffman EP, Powell BR, Brown RH, Jr. A methionine to valine mutation in the skeletal muscle sodium channel alpha-subunit in human hyperkalemic periodic paralysis. Nature 1991;358:387–389.

118. Harper PS. Phenotyic variation in myotonic dystrophy: causes and consequences. In: Rowland LP, ed. Pathogenesis of human muscular dystrophies. Amsterdam: Excerpta Medica, 1977.

119. Howeler CJ, Busch HFM, Geraedtis JPM, Niermeijer MF, Staal A. Anticipation in myotonic dystrophy: fact or fiction? Brain 1989;12:779–797.

120. Farrer LA, Cupples LA, Kiely DK, Conneally PM, Myers RH. Inverse relationship between age of onset of Huntington disease and paternal age suggests involvement of genetic imprinting. Am J Hum Genet 1992;50:528–535.

121. Harper PS. Congenital myotonic dystrophy in Bratain II: genetic basis. Arch Dis Child 1975;50:514–521.

122. Tsilfidis C, MacKenzie AE, Mettler G, Barcelo J, Korneluk RG. Corre-

lation between CTG trinucleotide repeat length and frequency of severe congenital myotonic dystrophy. Nature Genetics 1992;1:192–195.

123. Eiberg H, Mohr J, Nielsen L, Simonsen N. Genetic and linkage relationships of the C3 polymorphism: discovery of the C3-Se linkage and assignment of LES-C3-DM-Se-PEPD-Lu synteny to chromosome 19. Clin Genet 1983;24: 159–170.

124. Withehead AS, Solomon E, Chambers S, Bodner WF, Povey S, Fey G. Assignment of the structural gene for the third component of the human complement to chromosome 19. Proc Natl Acad Sci USA 1982;79:5021–5025.

125. Ashizawa T, Hejtmancik JF. Myotonic dystrophy—the search for the gene: progress, strategies, and prospects. Curr Neurol 1991;11:27–62.

126. Shutler G, MacKenzie AE, Brunner H, et al. Physical and genetic mapping of a novel chromosome 19 ERCC1 marker showing close linkage with myotonic dystrophy. Genomics 1991;9:500–504.

127. Johnson K, Shelbourne P, Davies J, et al. A new polymorphic probe which defines the region of chromosome 19 containing the myotonic dystrophy locus. Am J Hum Genet 1990;46:1073–1081.

128. Smeets HJM, Hermens R, Brunner HG, Ropers HH, Wieringa B. Identification of variable simple sequence motifs in 19q13.2-qter: markers for the myotonic dystrophy locus. Genomics 1991;9:257–263.

129. Harley HG, Brook JD, Floyd J, et al. Detection of linkage disequilibrium between the myotonic dystrophy locus and a new polymorphic DNA marker. Am J Hum Genet 1991;49:68–75.

130. Jansen G, de Jong PJ, Amemiya C, et al. Physical and genetic characterization of the distal segment of the myotonic dystrophy area on 19q. Genomics 1992;13:509–517.

131. Harley HG, Brook DJ, Rundle SA, et al. Expression of an unstable DNA region and phenotypic variation in myotonic dystrophy. Nature 1992;355:545–546.

132. Buxton J, Shelbourne P, Davies J, et al. Detection of an unstable fragment of DNA specific to individuals with myotonic dystrophy. Nature 1992;355:547–548.

133. Aslanidis C, Jansen G, Amemiya C, et al. Cloning of the essential myotonic dystrophy region and mapping of the putative defect. Nature 1992;355:548–551.

134. Fu YH, Pizutti A, Fenwick RG Jr, et al. An unstable triplet repeat in a gene related to myotonic muscular dystrophy. Science 1992;255:1256–1258.

135. Mahadevan M, Tsilfidis C, Sabourin L, et al. Myotonic dystrophy mutation: an unstable CTG repeat in the 3′ untranslated region of the gene. Science 1992;255:1253–1256.

136. Brook JD, McCurrach M, Harley HG, et al. Molecular basis of myotonic dystrophy: expansion of a trinucleotide repeat (CTG) at the 3′ end of a transcript encoding a protein kinase family member. Cell 1992;68:799–808.

137. Harley HG, Rundle SA, Reardon W, et al. Unstable DNA sequence in myotonic dystrophy. Lancet 1992;339:1125–1128.

138. Roses AD, Appel SH. Protein kinase activity in erythrocyte ghosts of

patients with myotonic muscular dystrophy. Proc Natl Acad Sci USA 1973;70:1855–1859.

139. Roses AD, Appel SH. Muscle membrane protein kinase in myotonic dystrophy. Nature 1974;250:245–247.

140. Fahn S. Movement disorders. In: Rowland LP, ed. Merritt's textbook of neurology, 8th ed. Philadelphia: Lea and Febiger, 1989:647.

141. Gusella JF. Huntington's disease. Adv Hum Genet 1991;20:125–151.

142. The Huntington's Disease Collaborative Research Group. A novel gene containing a trinucleotide repeat that is expanded and unstable on Huntington's disease chromosomes. Cell 1993;72:971–983.

143. Zogbhi HY, Pollack MS, Lyons LA, Ferrell RE, Daiger SP, Beaudet AL. Spinocerebellar ataxia: variable age of onset and linkage to human leukocyte antigen in a large kindred. Ann Neurol 1988;23:580–584.

144. Currier RD, Glover G, Jackson JF, Tipton AC. Spinocerebellar ataxia: study of a large kindred. Neurology 1972;22:1040–1043.

145. Nino HE, Noreen HJ, Dubey DP. A family with hereditary ataxia: HLA typing. Neurology 1980;30:12–20.

146. Schut JW. Hereditary ataxia: clinical study through six generations. Arch Neurol Psychiatry 1954;63:535–567.

147. Orr HT, Chung M-Y, Banfi S, et al. Expansion of an unstable trinucleotide CAG repeat in spinocerebellar ataxia type 1. Nature Genetics 1993;4:221–226.

148. Bishop JM. The molecular genetics of cancer. Science 1987;235: 305–311.

149. Yunis JJ. The chromosomal basis of neoplasia. Science 1983;221: 227–236.

150. Whang-Peng J, Kao-Shan CS, Lee EC. Specific chromosomal defect associated with human small-cell lung cancer: deletion 3p (14–23). Science 1982;215:181–182.

151. Yunis JJ, Bloomfield CD, Ensrud K. All patients with acute nonlymphocytic leukemia may have a chromosomal defect. N Engl J Med 1981;305:135–139.

152. Yunis JJ, Oken M, Kaplan E, Ensrud KM, Howe RR, Theologides A. Distinctive chromosomal abnormalities in histologic subtypes of non-Hodgkin's lymphoma. N Engl J Med 1982;307:1231–1236.

153. Rowley JD, Golomb HM, Vardiman JW. Nonrandom chromosome abnormalities in acute leukemia and dysmyelopoietic syndromes in patients with previously treated malignant disease. Blood 1981;58:759–766.

154. Najfeld V, Singer JV, James MC, Fialkow PJ. Trisomy of 1q in preleukemia with progression to acute leukemia. Scand J Haematol 1978; 21:24–28.

155. Hollstein M, Sidransky D, Vogelstein B, Harris CC. p53 mutations in human cancers. Science 1991;253:49–53.

156. Ahuja H, Bar-Eli M, Advani SH, Benchimol S, Cline MJ. Alterations in the p53 gene and the clonal evolution of the blast crisis of chronic myelogenous leukemia. Proc Natl Acad Sci USA 1989;86:6783–6787.

157. Wolf D, Lothar V. Minor deletions in the gene encoding the p53 tumor suppressor antigen cause lack of p53 in HL 60 cells. Proc Natl Acad Sci USA 1985;82:790–794.

158. Takahashi T, Nau M, Chiba I, et al. p53: A frequent target for genetic abnormalities in lung cancer. Science 1989;246:491–494.

159. Bressac B, Galvin KM, Liang TJ, Issellbacher KJ, Wands JR, Ozturk M. Abnormal structure and expression of p53 gene in human hepatocellular carcinoma. Proc Natl Acad Sci USA 1990;87:1973–1977.

160. Masuda H, Miller C, Koeffler HP, Battifora H, Cline MJ. Rearrangement of the p53 gene in human osteogenic sarcoma. Proc Natl Acad Sci USA 1987;84:7716–7719.

161. Miller CW, Aslo A, Tsay C, et al. Frequency and structure of p53 rearrangements in human osteosarcoma. Cancer Res 1990;50:7950.

162. Klein G. Specific chromosomal translocations and the genesis of B-cell derived tumors in mice and men. Cell 1983;32:311–315.

163. Shtivelman E, Lifshitz B, Gale RP, Canaani E. Fused transcript of *abl* and *bcr* genes in chronic myelogenous leukemia. Nature 1985;315:550–554.

164. Bartram CR, de Klein A, Hagemeijer A, et al. Translocation of c-*abl* oncogene correlates with the presence of the Philadelphia chromosome in chronic myelocytic leukemia. Nature 1983;306:277–280.

165. Konopka JB, Watanabe SM, Witte ON. An alteration of the c-*abl* protein in K562 leukemia cells unmasks associated tyrosine kinase activity. Cell 1984;37:1035–1042.

166. Feinberg AP, Vogelstein B. Hypomethylation distinguishes genes of some human cancers from their normal counterparts. Nature 1983;301:89–92.

167. Goelz SE, Vogelstein B, Hamilton SR, Feinberg AP. Hypomethylation of DNA from benign and malignant human colon neoplasms. Science 1985; 228:187–190.

168. Baylin SB, Makos M, Jianjun Wu, et al. Abnormal patterns of DNA methylation in human neoplasia: potential consequences for tumor progression. Cancer Cells 1991;3:383–390.

169. Jones PA, Buckley JD. The role of DNA methylation in cancer. Adv Cancer Res 1990;54:1–23.

170. de Bustros A, Nelkin BD, Silverman A, Ehrlich G, Poiesz B, Baylin SB. The short arm of chromosome 11 is a "hot spot" for hypermethylation in human neoplasia. Proc Natl Acad Sci USA 1988;85:5693–5697.

171. Baylin SB, Hoppener JWM, de Bustros A, Steenbergh PH, Lips CJM, Nelkin BD. DNA methylation patterns of the calcitonin gene in human lung cancers and lymphomas. Cancer Res 1986;46:2917–2922.

172. Baylin SB, Fearon ER, Vogelstein B, et al. Hypermethylation of the 5′ region of the calcitonin gene is a property of human lymphoid and acute myeloid malignancies. Blood 1987;70:412–417.

173. Silverman AL, Park JG, Hamilton SR, Gazdar AF, Luk JD, Baylin SB. Abnormal methylation of the calcitonin gene in human colonic neoplasms. Cancer Res 1989;49:3468–3473.

174. Jones PA, Wolkowicz MJ, Rideout WM, et al. De novo methylation of

the MyoD1 CpG island during the establishment of immortal cell lines. Proc Natl Acad Sci USA 1990;87:6117–6121.

175. Antequera F, Boyes J, Bird A. High levels of de novo methylation and altered chromatin structure at CpG islands in cell lines. Cell 1990;62:503–514.

176. Knudson AG. Genetics of human cancer. Annu Rev Genet 1986;20:231–251.

177. Porfirio B, Tedeschi B, Vernole P, Caporossi D, Nicoletti B. The distribution of Msp1-induced breaks in human lymphocyte chromosomes and its relationship to common fragile sites. Mutat Res 1989;213:117–124.

178. Rideout WM, Coetzee GA, Olumi AF, Jones PA. 5-Methylcytosine as an endogenous mutagen in the human LDL receptor and p53 genes. Science 1990;249:1288–1290.

179. Nigro JM, Baker SJ, Preisinger AC, et al. Mutations in the p53 gene occur in diverse human tumor types. Nature 1989;342:705–708.

180. Kautainien TL, Jones PA. DNA methyltransferase levels in tumorigenic and nontumorigenic cells in culture. J Biol Chem 1986;261:1594–1598.

181. El Deiry WS, Nelkin BD, Celano P, et al. High expression of the DNA methyltransferase gene characterizes human neoplastic cells and progression stages of colon cancer. Proc Natl Acad Sci USA 1991;88:3470–3474.

182. Jones PA, Buckley JD. The role of DNA methylation in cancer. Adv Cancer Res 1990;54:1–23.

183. Monk M. Memories of mother and father. Nature 1987;328:203–204.

184. Monk M. Genomic imprinting. Genes Dev 1988;2:921–925.

185. Chandler LA, Ghazi H, Jones PA, Boukamp P, Fusenig NE. Cell 1987;50:711–717.

186. Knudson AG Jr. Mutation and cancer: statistical study of retinoblastoma. Proc Natl Acad Sci USA 1971;68:820–823.

187. Sapienza C. Genome imprinting and dominance modification. Ann NY Acad Sci 1989;564:24–38.

188. Solter D. Differential imprinting and expression of maternal and paternal genomes. Annu Rev Genet 1988;22:127–146.

189. Surani MA, Allen ND, Barton SC, et al. Developmental consequences of imprinting of parental chromosomes by DNA methylation. Philos Trans R Soc Lond Biol 1989;326:313–327.

190. Sapienza C. Genome imprinting and carcinogenesis. Biochim Biophys Acta 1991;1072:51–61.

191. Scrable H, Cavenee W, Ghavimi F, Lovell M, Morgan H, Sapienza C. A model for embryonal rhabdomyosarcoma tumorigenesis that involves genome imprinting. Proc Natl Acad Sci USA 1989;86:7480–7484.

192. Toguchida J, Ishizaki K, Sasaki MS, et al. Preferential mutation of paternally derived RB gene as the initial event in sporadic osteosarcoma. Nature 1989;338:156–158.

193. Mannens N, Slater RM, Heyting C, et al. Molecular nature of genetic changes resulting in loss of heterozygosity of chromosome 11 in Wilm's tumors. Hum Genet 1988;81:41–48.

194. Reeve AE, Housiaux PJ, Gardner RJM, Chewings WH, Grindley RM,

Millow LJ. Nature 1984;304:174–176.

195. Schroeder WT, Chao LY, Dao DD, et al. Nonrandom loss of maternal chromosome alleles in Wilm's tumors. Am J Hum Genet 1987;40:413–420.

196. Williams JC, Brown KW, Mott MG, Maitland NJ. Maternal allele loss in Wilms' tumor. Lancet 1989;1:283–284.

197. Dryja TP, Mukai S, Petersen R, Rapaport JM, Walton D, Yandell DW. Paternal origin of mutations of the retinoblastoma gene. Nature 1989;339:556–558.

198. Zhu X, Dunn JM, Phillips RA, et al. Preferential germline mutation of the paternal allele in retinoblastoma. Nature 1989;340:312–313.

199. Matsunaga E. Genetics of Wilm's tumor. Hum Genet 1981;57:231–246.

200. Knudson AG, Strong LC. Mutation and cancer: neuroblastoma and pheochromocytoma. Am J Hum Genet 1972;24:514–532.

201. Anderson DE. Genetic study of breast cancer: identification of a high risk group. Cancer 1974;34:1090–1097.

202. Moolgavkar SH, Day NE, Stevens RG. Two-stage model for carcinogenesis: epidemiology of breast cancer in females. J Natl Cancer Inst 1980;65:559–569.

203. Anderson DE. An inherited form of large bowel cancer. Cancer 1980;45:1103–1107.

204. Lovett E. Family studies in cancer of the colon and rectum. Br J Surg 1976;63:13–18.

205. Dumont-Herskowitz RA, Safaii HS, Senior B. Ovarian fibromatosis in four successive generations. J Pediatr 1978;93:621–624.

206. Fraumeni JF, Grundy GW, Creagan ET, Eversen RB. Six families prone to ovarian cancer. Cancer 1975;36:364–369.

207. Mahboubi AO, Ahlvin RC, Mahboubi EO. Familial aggregation of urothelial carcinoma. J Urol 1981;126:691–692.

208. Chaganti RSK, German J, eds. Genetics in clinical oncology. Oxford, UK: Oxford University Press, 1985.

209. Pathak S, Strong LC, Ferrell RE, Trindale A. Familial renal cell carcinoma with a 3:11 chromosome translocation limited to tumour cells. Science 1982;217:939–941.

210. Blattner WA, Dean JH, Fraumeni JF. Familial lymphoproliferative malignancy: clinical and laboratory follow-up. Ann Intern Med 1979;90:943–944.

211. Branda RF, Ackerman SK, Handwerger BS, Howe RB, Douglas SD. Lymphocyte studies in familial chronic lymphatic leukemia. Am J Med 1978;64:508–514.

212. Cohen HG, Shimm D, Paris SA, Buckley CE III, Kremer WB. Hairy cell leukemia–associated familial lymphoproliferative disorder. Ann Intern Med 1979;90:174–179.

213. Anderson DE. Clinical characteristics of the genetic variety of cutaneous melanoma in man. Cancer 1971;28:721–725.

214. Green GJ, Hong WK, Everett JR, Bhutani R, Amick RM. Familial intraocular malignant melanoma: a case report. Cancer 1978;41:2481–2483.

215. Den Otter W, Koten JW, van der Vegt BJH, et al. Hereditary cancer and its implications: a view. Anticancer Res 1990;10:489–496.

216. Knudson AG. Hereditary cancer, oncogenes, and antioncogenes. Cancer Res 1985;45:1437–1443.

217. Muller H. Dominant inheritance in human cancer. Anticancer Res 1990;10:505–512.

218. Weinberg RA. Finding the anti-oncogene. Sci Am 1988;259(3):44–51.

219. Den Otter W, Koten JW, van der Vegt BJH, et al. Oncogenesis by mutations in anti-oncogenes: a view. Anticancer Res 1990;10:475–488.

220. Li FP. Familial cancer syndromes and clusters. In: Haskell CM, ed. Current problems in cancer. Chicago: Year Book Medical Publishers, 1990.

221. Shelbourne E. Unstable DNA may be responsible for the incomplete penetrance of the myotonic dystrophy phenotype. Hum Mol Genet 1992;1: 467–473.

222. Brunner HG, Jansen G, Nillesen W, et al. Brief report: reverse mutation in myotonic dystrophy. N Engl J Med 1993;328:476–480.

223. O'Hoy KL, Tsilfidis C, Mahadevan MS, et al. Reduction in the size of the myotonic dystrophy trinucleotide repeat mutation during transmission. Science 1993;259:809–812.

224. Brook JD. Retreat of the triplet repeat? Nature Genet 1993;3:279–281.

225. Aaltonen LA, Peltomaki P, Leach FS, et al. Clues to the pathogenesis of familial colorectal cancer. Science 1993;260:812–816.

226. Fishel R, Lescoe MK, Rao MRS, et al. The human gene homolog *MSH2* and its association with hereditary nonpolyposis colon cancer. Cell 1993;75:1027–1038.

227. Leach FS, Nicolaides NC, Papadopoulos N, et al. Mutations of a *mutS* homolog in hereditary nonpolyposis colorectal cancer. Cell 1993;75:1215–1225.

228. Parsons R, Li G-M, Longley MJ, et al. Hypermutability and mismatch repair deficiency in RER+ tumor cells. Cell 1993;75:1227–1236.

229. Papadopoulos N, Nicolaides NC, Wei Y-F, et al. Mutation of a *mutL* homolog in hereditary colon cancer. Science 1994;263:1625–1629.

230. Richards RI, Sutherland GR. Heritable unstable DNA sequences. Nature Genet 1992;1:7–9.

CHAPTER 3

The *ras* Oncogene: From Basic Mechanisms to Therapeutic Intervention

Jeff Deyo
David Berger
Michael A. Tainsky

The *ras* oncogene is one of the most intensely studied molecules in the biomedical sciences. Since the discovery of an activated *ras* oncogene as the acutely transforming component of the Harvey sarcoma virus 15 years ago, and later the three cellular homologues H-*ras*, K-*ras*, and N-*ras*, numerous scientists from disparate fields have converged to study these molecules. Knowledge of the workings of *ras* continues to increase at a rapid rate, reflected in the thousands of articles published about *ras* in the last several years. This explosion of newly discovered information is beginning to clarify mechanistic interactions that were previously relegated to "black boxes" both upstream and downstream of *ras* and has placed *ras* in the middle of an intricate signal transduction pathway. The breadth and pace of research in this field has made the current study of *ras* tremendously exciting.

The model systems used to study *ras* span hundreds of millions of years of evolutionary time ranging from mating pheromone response in yeast to photoreceptor cell differentiation in *Drosophila* to growth factor response in human teratocarcinoma cells. The *ras* superfamily, itself part of the larger family of guanine-binding proteins, or G proteins, includes the major subfamilies *rho*, which is involved in cytoskeleton function and control of cell shape, and *rab*, which is involved in intracellular vesicle transport. The reader is referred to an excellent review article on the subject of the other subfamilies listed at the end of this chapter. In the last few years *ras* has emerged as a central player in intercellular signal transduction. The purpose of the first section of this chapter is to highlight *ras* as a component of the signal transduction pathways in model systems currently used to answer mechanistic questions about *ras* function. The next section will cover cell culture systems and discuss the invaluable contributions made with this technology in the study of *ras*. Finally, in the last section, a variety of novel and exciting approaches to designing cancer therapy guided by a logical consideration of *ras* biology will be reviewed.

In most systems, *ras* is a low copy number gene that encodes a small 21-kDa GTPase localized to the cytoplasmic side of the plasma membrane. It is capable of binding to and catalyzing the hydrolysis of guanine nucleotide triphosphate (GTP) to guanine nucleotide diphosphate (GDP) and inorganic phosphorus. In all systems the *ras* molecular switch in the GTP-bound state, and not the GDP-bound state, relays signal to downstream effectors. Thus, molecules that interact with *ras* and increase its GTPase activity decrease signal intensity, while molecules that increase the proportion of GTP-bound *ras* increase the resulting signal (see Fig. 3.1). The specific regulators of *ras* GTPase activity and the effectors downstream of *ras* differ among the systems in which they are studied and will be described below. The essentials of a signal transduction pathway starting at the plasma membrane and ending with changes in gene expression are emerging. The initiating event, ligand binding to a receptor tyrosine kinase or activation of a nonreceptor tyrosine kinase, results in the phosphorylation of a variety of substrates, including autophosphorylation in some cases. The subsequent assembly of a multimeric protein complex at the plasma membrane involves, at the least, the tyrosine kinase, an adaptor protein, and a nucleotide exchange factor. The exchange factor is an activator of *ras* that accelerates the rate of formation of GTP-bound *ras*. Termination of the signal by GTP hydrolysis is accelerated by the GTPase-activating protein (GAP). Assignment of the immediate downstream effector of *ras* GTP is tentative, but evidence points to an interaction with a serine/threonine kinase in the *raf* family of protooncogenes in several systems. The presumed *ras*-mediated mechanism of activating *raf* is unknown. Important substrates for *raf* phosphorylation include members of the mitogen-activated protein kinase (MAPK) cascade. Although a number of proven and proposed substrates exist for the MAPK network, in some systems one endpoint includes nuclear transcription factors responsible for the transcription of early response genes involved in proliferation. Thus, although the details vary from pathway to pathway, a core sequence of events that results in the transmission of information from the plasma membrane to the nucleus is emerging. Information about *ras*-mediated signal transduction derived from the study of these model systems is critical for developing new classes of specific therapies used in cancer patients.

MODEL SYSTEMS

Yeast

Saccharomyces cerevisiae Fundamental work in the molecular mechanisms in a variety of pathways has been accomplished using the budding yeast *Saccharomyces cerevisiae* because of its facility for both detailed genetic and biochemical analysis. Isolating yeast mutants for specific known genes is relatively simple, and once a mutant in a pathway is characterized, isolation of other elements in the same pathway can involve screening for a second mutant, which can rescue function of the first mutant. Further characterization must then be carried out to confirm their alignment and order in the pathway. The two *S. cerevisiae ras* homologues, *RAS1* and *RAS2*, are 90% homologous at the amino portion of the protein to the mammalian *ras*. Functionally, *ras1*$^-$ *ras2*$^-$ yeast mutants are nonviable but can be successfully rescued by overexpressing mammalian *ras*. This remarkable functional conservation between mammalian *ras* and yeast *RAS* was extended by others who demonstrated oncogenic transformation in mouse fibroblast NIH3T3 cells expressing a modified activated yeast *RAS1* gene. Thus, the intrinsic biochemical function of *ras* in the mammalian and yeast systems seems to be conserved, but as discussed below, the similarity in each signal transduction pathway involving *ras* is limited.

When grown on a medium lacking essential nutrients, normal budding yeast stop growing in stage G_1 of the cell cycle. Yeast with an activating mutation leaving the Ras protein impaired in GTPase function, however, do not arrest in G_1 when grown in nutrient-deficient media. This failure in growth arrest is analogous to that in human tumors and transformed cells in culture that fail to respond to environmental stimuli. These mutant yeast also fail to sporulate and do not survive well after storage. Investigators recognized that this phenotype was similar to that seen for mutants in the cyclic adenosine 3′,5′-monophosphate (cAMP) pathway. Later it was confirmed that yeast with an activated *RAS* gene do indeed have elevated cAMP levels. In this pathway, activated adenylate cyclase catalyzes the conversion of ATP to cAMP. Cyclic AMP binds to the regulatory subunit of cAMP-dependent protein kinase (protein kinase A, PKA), causing the derepression of PKA. Protein kinase A is a heterotetramer comprising two regulatory subunits and two catalytic subunits. The free catalytic subunits phosphorylate a variety of targets to effect cellular change. Real support for Ras involvement in the cAMP pathway in *S. cerevisiae* came from the suppression of lethality in *ras1*$^-$ *ras2*$^-$ cells carrying the *bcy1* mutation. *BCY1* codes for the cAMP-binding regulatory subunit of PKA, and *bcy1* mutants are defective in their normal negative regulation of PKA. PKA in *bcy1* mutant yeast functions independently of cAMP. Thus, the capacity of this mutant to rescue yeast lacking functional Ras clearly places Ras in a growth-control pathway with cAMP in *S. cerevisiae*. Unfortunately, the majority of research in higher eukaryotes argues against cAMP as the second messenger system downstream of *ras* in these organisms. In addition, although exquisite genetic and biochemical evidence exists to align Ras and the cAMP pathway into a single system in budding yeast, evidence for direct physical interaction is lacking.

Modulators of Ras function in *S. cerevisiae* include activating nucleotide exchange factors and inhibiting GAPs. The best-characterized nucleotide exchange factor in yeast is *CDC25*, which codes for the protein that catalyzes the exchange of Ras-bound GDP for GTP and increases the proportion of GTP Ras. Basal and glucose-induced cAMP synthesis are both deficient in yeast strains lacking a functional *CDC25* gene. Consistent with its role as an activator of Ras, Cdc25 has been reported to function as a transforming oncoprotein in focus-forming assays using murine NIH3T3 fibroblasts. Negative regulators of Ras in yeast include the *ras* GAP homologues *IRA1* and *IRA2*. Deletion mutants of *IRA1* and *IRA2* result in a constitutively active protein phenotypically similar to yeast expressing Ras activated by point mutation. The proteins in the pathway directly upstream of the exchange factors and GAPs are not well characterized in *S. cerevisiae*.

As indicated above, the cAMP pathway serves as an important effector system for Ras in budding yeast. The elements of the pathway immediately downstream of Ras have yet to be clearly defined; however, more distal elements have been identified. Surprisingly, elimination of the adenylate cyclase gene itself, *CYR1*, is not lethal, although the yeast cells grow slowly. That the *ras1⁻ ras2⁻* mutants are lethal suggests that Ras has effects mediated through other pathways. The *PDE1* and *PDE2* genes encode phosphodiesterases that serve in this pathway to quench signal from an activated adenylate cyclase by hydrolyzing cAMP to AMP. Essential to the activated pathway is the release of regulatory subunits from the catalytic subunits of PKA to allow phosphorylation of substrates. Important substrates of PKA include those for mobilizing energy reserves needed for cell division, factors needed for transcription of growth-promoting genes, and proteins involved in the elimination of the Ras/cAMP signal. The description of this signaling pathway, although still incomplete, provides an interesting context for the study of Ras and has initiated a great deal of very fruitful research. Thus, the budding yeast continues to serve as an excellent model system for molecular interactions detailed through powerful genetic and biochemical studies. *S. cerevisiae* is also the setting for an intricate and elegant new technique being used to detect protein-protein interactions, which will be discussed later in this section.

Schizosaccharomyces pombe Research using the fission yeast, *S. pombe*, often demonstrates, at least for some investigations, that this organism is more closely related mechanistically to metazoan animals than are the budding yeast described above. The *ras* pathway as studied in *S. pombe* is quite different from that in its distant cousin *S. cerevisiae*. Despite these differences, *S. pombe* still benefits from all the genetic power of *S. cerevisiae* with the same ease of generating and screening for mutants and subsequent assays for genes that suppress loss-of-function mutants.

In contrast to the budding yeast, there is only one *ras* locus in *S. pombe*, *ras1*, which encodes a 219 amino acid protein. The Ras1 protein of the fission yeast is not involved in growth regulation and does not affect cAMP levels. In *S. pombe*, mutations in the *ras1* locus lead to a triad of defects including sterility (defective conjugation) when haploid, abnormal cell shape, and aberrant meiosis and sporulation when diploid. The defect in mating leading to sterility has been studied extensively. Haploid fission yeast have two mating types, h^+ (P, plus) and h^- (M, minus), analogous to the "a" and "α" mating types of *S. cerevisiae*. Haploid yeast of one mating type produce a diffusible polypeptide factor that activates the cognate receptor found on the opposite mating type. After stimulation of the receptor, a cascade of events is initiated culminating in conjugation and the formation of a diploid yeast. The diploid h^+/h^- mating-type heterozygote is then capable of undergoing meiosis and sporulation to form the haploid phase once again. The response to receptor stimulation requires *ras1*. However, an activated Ras1 protein cannot induce cellular change in the absence of pheromone, indicating that Ras1 appears only to modify the incoming pheromone signal. After binding the appropriate ligand, the receptor is believed to interact with a heterotrimeric G protein. The α-subunit of the G protein, encoded by *gpa1*, transmits the signal to the next element of the pathway. The exact molecular mechanism connecting the occupied receptor and Ras1 is unknown, but evidence from other systems points to involvement of the putative nucleotide exchange factor Ste6. Ste6 shares amino acid homology over the carboxy-terminus with the *S. cerevisiae* protein Cdc25. Cdc25, as discussed above, is reported to function as a nucleotide exchange factor for GDP-bound Ras. This interaction results in an increased proportion of GTP-bound Ras (activation). Termination of the signal through Ras1 involves the GTPase-activating protein Gap1/Sar1.

Evidence from disruption experiments of *ras1* indicates that there are at least three pathways—conjugation, sporulation, and cell shape—in which downstream effectors may work. Genetic screens for suppressors of *ras1⁻* mutants have revealed a series of genes with incomplete abilities to rescue defects in sexual differentiation. The first such suppressor, *byr1*, is believed to code for a protein with serine/threonine kinase activity. A second suppressor, *byr2*, which is also suspected to encode a serine/threonine kinase, is believed to lie upstream of Byr1, based on the ability of *byr1* to rescue the sporulation defect of *byr2/byr2* null diploid mutants. A recently described third gene, *byr3*, can mediate the partial suppression of *ras1⁻* mutants and has homology to the mammalian cellular nucleic acid binding protein. All three of these gene products can rescue the sporulation defect of *ras1⁻* mutants, whereas *byr1* and *byr2* alone can rescue low levels of mating activity in these same mutants. These experiments show the considerable overlap of members in the sporulation and conjugation pathways of fission yeast. The importance of the above findings linking these serine/threonine kinases together with Ras1 is emphasized by the recent discovery of sequence homology between *byr1* and the mammalian MAPK/Erk kinase, MEK. MAPK has recently been placed in the *ras* signal transduction pathway in mammalian systems (see below). A potential protein kinase substrate for MEK or *byr1* is the *spk1* gene product shown to be 54% homologous to the mammalian MAPK/Erk and downstream of *byr1*. Additional support for the theory connecting *ras1* to the MAPK network comes from experiments demonstrating physical interaction between *byr2* and human *ras*. Using this technique, alluded to earlier, developed in *S. cerevisiae*, potential protein-protein interactions can be tested by taking advantage of the modular nature of transcription factors. Hybrid proteins are generated that, when the proteins of interest physically interact, reconstitute a working transcriptional activator capable of being assayed. The results of this technique have been outstanding in several instances. Thus, evidence is mounting that the *byr2-byr1* relation is analogous to the relation of *raf* to MEK. If these similarities hold, *ras1* of *S. pombe* will be linked to the extensive MAPK cascade described in detail in later sections of this chapter. Research using the fission yeast *S. pombe* as a model system to investigate signal transduction is progressing at a rapid rate and continues to identify and verify mechanisms of signal transduction studied in higher eukaryotic cells.

Caenorhabditis elegans

A valuable organism in the study of many molecular mechanisms related to development, the nematode *C. elegans* has the distinction of being the only organism with a complete map of its cellular fate. The migration and progeny of cells from ovum to mature organism has been charted. This remarkable scientific accomplishment, coupled with the ability to observe the effects of mutations in a small number of cells, has had an impact on the study of the role of *ras* in cellular differentiation using both molecular and genetic approaches. The *let-60 ras* gene encodes a 184 amino acid *ras* homologue in *C. elegans* that, for the first 164 residues, is 83% identical to the human N-*ras* protein.

Developmental studies of vulval organogenesis using the hermaphroditic variety of *C. elegans* have identified six vulval precursor cells (VPCs) needed to make up the mature structure. Each of the six bipotential cells is capable of adopting the relatively unspecialized hypodermal cell fate or the fate of the mature vulval cell. In normal development, three of the VPCs become hypodermal cells while the other three develop into mature vulval cells. This decision is controlled by the interplay of an inductive signal sent by the anchor cell of the gonad and an inhibitory signal sent by the surrounding hypodermis. The inductive signal acts to release the appropriate cells from inhibition. The *let-23* gene product codes for a putative receptor tyronsine kinase of the epidermal growth factor (EGF) subfamily of receptors, which when functionally eliminated results in loss of any inductive signal, and no vulval cells are made. An "activated," conditional mutant allele of *let-23*, *let-23 n1045*, when homozygous diploid, results in a hyperinducible phenotype; that is, in the presence of inductive signal and at the appropriate temperature, these worms have multiple vulva. However, a worm hemizygous for the *n1045* allele has a lower rate of VPC differentiation and, therefore, fewer vulval cells than wild type. Thus, the possibility exists that *let-23* controls both a positive, inductive response and a negative, inhibitory response. The biochemical nature of this seeming paradox of function is under investigation. One of many possible scenarios, however, is that the negative function of this receptor is more susceptible to the mutation. Thus, when the mutation is homozygous diploid, inhibitory function is lost, and the animal becomes multivulvate. When function is further diminished in the hemizygous state (only one copy of the mutant gene), there is not even enough positive function to carry out differentiation, and the animal is vulvaless. In addition, any amount of wild-type *let-23* function is sufficient to provide normal vulval induction. Certainly more research is needed to clarify this conundrum.

Upstream of the receptor is *lin-3*, the EGF-like ligand that is the inductive molecule secreted by the anchor cell. The inhibitory function of the hypodermis has only been characterized genetically as the *lin-15* gene product. The *let-23* receptor lies upstream of the *let-60 ras* molecule in *C. elegans*. *Let-60 ras* appears to mediate the positive inductive signal pathway from *let-23*. Activated *let-60 ras* results in a multivulva phenotype, even in the absence of inductive signal (in contrast to activated *ras1* function in *S.*

pombe, which does not provide all the function of the activated receptor), while dominant negative *let-60 ras* mutants result in a vulvaless phenotype. Although literature on *C. elegans* modulators is lacking, attention has been directed to the adaptor protein *sem-5*, which appears to function as a link between receptor and *ras* in this system. The *sem-5* protein contains one SH2 (src homology 2) domain (important in binding to specific phosphotyrosine residues of target proteins), flanked by two SH3 domains (implicated in protein-protein interactions probably by binding proline-rich residues of target proteins). On the basis of these conserved domains, and the observation that mutations in these domains impair function, *sem-5* has been postulated to be important in the assembly of protein components into functional complexes at the plasma membrane. In fact, a barrage of recent reports has shown that *sem-5* or its mammalian counterpart, Grb2 (for growth factor receptor–bound protein 2), work to couple activated (phosphorylated) receptor (and nonreceptor) tyrosine kineses to nucleotide exchange factors like hSos1 in mammalian cells. These and other related experiments serve to outline a simple, but highly conserved, pathway from receptor tyrosine kinase to *ras*. Once ligand is bound to receptor and the receptor is autophosphorylated at the appropriate residues, *sem-5*/Grb2 attaches to the receptor via the SH2 domain at the specific phosphotyrosine residue. Bound to *sem-5*/Grb2 via SH3 domains is the nucleotide exchange factor (in this case hSos1), now recruited from the cytosol to the plasma membrane and primed to interact with *ras*. This interaction results in the expected increase in *ras* GTP and ensuing signal transmission. In *C. elegans*, the signal transmission downstream of *let-60 ras* is becoming increasingly clear and, just as with events immediately upstream of *ras*, seems to be conserved across vast stretches of evolutionary time. The *lin-45* gene encodes the *C. elegans* homologue of the mammalian *raf* protooncogene. The *raf* family of serine/threonine kinases has been implicated in the *ras* signal transduction pathway in other systems, including *Drosophila* and humans. Genetic evidence places *Ce-raf-1/lin-45* after *let-60 ras* but before *lin-1*. The *lin-1* gene product has also been placed in this pathway on the basis of genetic studies showing that *lin-1* is required for expression of the vulvaless phenotype of *let-23* and *let-60* mutants. However, loss-of-function mutations in *lin-1* yield a multivulva phenotype, which implies that *let-60 ras* function normally inhibits *lin-1* action. Members of this pathway after *lin-1* are more speculative. Evidence from the *C. elegans* system indicates a strong conservation of function in the pathway immediately upstream and immediately downstream of *ras*. The consequence of these discoveries for investigators is to accelerate greatly the pace at which new information about signal transduction and cellular communication is uncovered.

Drosophila melanogaster

Studies of the fruit fly, one of the most complicated of the invertebrate organisms used to investigate molecular mechanisms of signal transduction and developmental biology, have benefited from this complexity and from years of intricate genetic studies, which make it a very powerful research instrument. The predominant organ system used in *Drosophila* to study *ras* is the retina, specifically photoreceptor cell differentiation. The *Drosophila ras* genes (*Dras1*, *Dras2*, and *Dras3*) were isolated nearly a decade ago. *Dras2* and *Dras3* have very limited sequence homology to mammalian *ras* or to *Dras1*. *Dras3* has since been renamed *rap1* to reflect its similarity to mammalian *rap* genes. *Dras1* and human H-*ras* are 75% identical. In the development of the fly compound eye, eight precursor cells, R1–R8, differentiate following a set sequence into an ommatidium, the functional unit of the fly retina. The last cell to mature is the R7 cell, and it is induced by local cell-cell interaction with the R8 cell. The details of this molecular interaction are coming increasingly into focus. The R7 cell expresses the *sevenless* (*sev*) gene, which codes for a receptor tyrosine kinase. This receptor is stimulated by the ligand expressed on the surface of the R8 cell coded for by the *bride of sevenless* (*boss*) gene. Loss of *sev* function results in the failure of the R7 precursor to adopt the R7 fate and instead default to a nonneuronal cone cell fate. Interestingly, ectopic expression of *boss*, which is normally restricted to the R8 cells of the eye, causes cone cells to differentiate into R7 cells. This result suggests that the signaling pathway is intact in these other cells but that their location away from direct contact with the R8 cell leaves them physically removed from the inductive signal. Consequently, these cells normally adopt the relatively undifferentiated cone cell phenotype.

As in several other systems, *ras* fits in downstream of the receptor. The *Dras1* gene was first put in this pathway on the basis of genetic and molecular evidence. An elegant genetic screen was set up to find mutants that reduced, but did not eliminate, function downstream of *sev*. The investigators postulated that receptor tyrosine kinases may share downstream signaling components; thus complete disruption of a component shared with a critical pathway would be lethal. *Dras1* was implicated in the R7 differentiation scheme using this screen and was concomitantly found to function downstream of another receptor tyrosine kinase, *DER*. Certain mutations in the *DER* locus cause embryonic lethality, adding credence to the investigators' hypothesis that complete disruption of signaling components would be a poor screen in this system.

Interposed between receptor and *Dras1* are at least two other molecules in a pattern described previously for *C. elegans*. The SH2/SH3-

containing adaptor protein in *Drosophila*, termed *drk* (for downstream of receptor kinases), binds to activated receptor via the SH2 domain. The adaptor *drk* also binds the nucleotide exchange factor *son of sevenless* (*sos*). The *sos* protein has a central domain of homology with the *S. cerevisiae* nucleotide exchange factor *CDC25* and has been shown to function as an exchange factor for mammalian *ras*. The function of *sos* is crucial to the developing R7 cell. Recent work indicates that these three molecules—receptor, adaptor, and exchange factor—may form a multimeric complex after ligand binding activates the receptor. Also reported was the observation that when antibodies were used to precipitate Grb2 (the mammalian homologue of *drk*), the murine mSos1 protein was present in the precipitated complex in both EGF-stimulated and control cells. This result indicates that *sos* and *drk* may form a complex that is constitutively present, even in unstimulated cells. Other modulators of *Dras1* function include *Gap1*, which appears to be the *Drosophila* homologue of the mammalian *ras* GAP acting as a negative regulator of *Dras1* function. Downstream of *Dras1* the pathway is less well resolved but seems to rely on several mechanisms and molecules previously discussed for other organisms. This includes *Draf*, the *Drosophila* homologue of the mammalian *raf*, which has been shown by genetic studies to function after *ras* but before the nuclear protein *sina*. Genetic screens have also identified a gene termed *Dsor1*, which appears to function downstream of *Draf* and has sequence homology to MAPKK of mammalian systems and *byr1* of yeast. Other components of the extensive MAPK cascade have yet to be implicated in R7 development. Thus, *Drosophila melanogaster* has proved to be a valuable source of insight into the workings of *ras*.

Xenopus laevis

The amphibian *X. laevis* is another excellent tool for research in developmental biology, cell-cycle control, and the molecular mechanisms of cellular signaling. Study of the frog *X. laevis* has contributed to our understanding of *ras* and its role in signal transduction. In this system the maturation of oocytes as estimated by germinal vesicle breakdown (GVBD) is assayed. Similar to the human oocyte before ovulation, the *Xenopus* oocyte is arrested in prophase of the first meiotic division. Stimulation of the oocytes in vivo by progesterone leads to progression past the first meiotic block and dissolution of the germinal vesicle.

A variety of experimental methods are used to produce GVBD artificially, but for the current discussion, induction mediated by insulin treatment of oocytes will be highlighted. It has been found that *ras* is involved in the insulin-stimulated pathway to GVBD and not the pathway of the physiologic inducer progesterone. Activated H-*ras* microinjected into late-stage oocytes led to rapid GVBD. These experiments, which demonstrated that the *Xenopus* oocytes mature without a change in cAMP levels, led to the demise of a theory based on evidence from *S. cerevisiae* that cAMP would be a downstream effector of *ras* in higher eukaryotes. Later, investigators showed that microinjection of antibody against *ras* prevents insulin-induced GVBD, but not GVBD induced by progesterone. Before GAP had been isolated, the first evidence for GTPase activity came from experiments in *Xenopus* showing that *ras*-dependent nucleotide turnover occurred at a much faster rate in vivo than expected from in vitro data. Subsequently, research from other model systems on modulators of *ras* function eclipsed this advance in *Xenopus*; thus research on modulators of *Xenopus ras* has lagged. Contrary to other systems, evidence accumulated in *Xenopus* suggests that *ras* GAP acts as an effector in this system, in addition to its GTPase-accelerating function. Antibodies directed against GAP inhibit GVBD and other biochemical parameters of maturation. Notably, GAP expressed by itself was ineffective at stimulating GVBD, which suggests that other signals from *ras* are necessary for maturation to proceed. GAP and *ras* are known to interact through the effector domain of *ras* (amino acids 32–40) and the carboxy-terminus of GAP. Another domain of GAP, the SH3 domain, was later shown to be important to signaling in this system when peptides to this region specifically blocked GVBD in response to insulin. Similar evidence for a role as an effector of *ras* has been presented for the phosphatidylcholine-hydrolyzing phospholipase C (PC-PLC). Microinjection of PC-PLC was able to induce oocyte maturation, while antibodies against PC-PLC specifically inhibited oocyte maturation induced by transforming *ras*. Production of diacylglycerol (DAG) from phosphatidylcholine breakdown is also observed in *Xenopus* oocytes after injection of activated *ras* protein. Injection of activated *ras* has been shown to effect other typical second messenger cascade systems such as MAPK kinase and effectors believed to be further downstream such as MAPK and *rsk*. A member of the pathway believed to be upstream of MAPK, *raf*, has also been implicated in *ras*-mediated oocyte maturation. Expression of a kinase-defective mutant of human *raf* blocked GVBD in oocytes containing an activated H-*ras* and, similarly, in oocytes expressing an activated EGF receptor. More research is needed to clarify the exact number and sequence of molecular events in the MAPK pathway after *ras*.

Among other potential downstream effectors, biochemical evidence has placed the ζ isoform of protein kinase C (ζPKC) downstream of PLC and *ras* in *Xenopus* oocyte maturation. The ζ isoform of PKC is somewhat unique among PKCs in that ζPKC is stimulated by acidic phospholipids but not by DAG, calcium, or phorbol ester. In addition, experiments have shown partial cellular transformation of NIH3T3 murine fibroblasts by ζPKC. The ease of physical manipulation and the capacity for biochemical investigation indicate that this complex model system has great potential for explicating pathway components downstream of *ras*. Portions of the *Xenopus* oocyte maturation pathway remain enigmatic, but ongoing research promises to resolve these questions. Meanwhile, the system has clearly contributed to the understanding of signal transduction and helps to highlight the diversity of the *ras* pathway.

Mammalian Systems

Study of a wide variety of mammalian organisms has contributed to our understanding of *ras*. These organisms include the standard rat and mouse, as well as the rabbit, dog, cat, and even cow and human. The combined information derived from research using these animals accentuates the conserved nature of *ras* function among different cells and organisms while reinforcing the concept of the diversity of cell signaling systems in which *ras* plays a role.

Reminiscent of the results of experiments conducted in lower eukaryotes, *ras* has been found to contribute to processes ranging from proliferation and transformation to differentiation and the disruption of differentiation. The underlying mechanism that unites these seemingly disparate functions is the processing of extracellular signals into a change in the pattern of gene expression necessary to respond to the environment. For an individual cell this may mean cell division in response to growth factor stimulation or the acquisition of a terminally differentiated phenotype after ligand binding. Either way, the "inside" is sensing the "outside" and making changes accordingly. The intent of this section is to examine some of the mammalian model systems that have been used in the search for answers to mechanistic questions about *ras* function. The contribution of *ras* to human tumors and current strategies involving *ras*-specific therapies will be reviewed in the last section of this chapter.

Reflective of its role in carcinogenesis, a great deal of research has gone into investigating *ras* as a promoter of growth. The immortalized murine NIH3T3 fibroblast system has been a workhorse for in vitro studies of *ras* and other oncogenes in higher eukaryotes and will be discussed below with other in vitro techniques. The human complement of *ras* genes includes H-*ras*, K-*ras*, and N-*ras*. Activating mutations at codons 12, 13, and 61 of each *ras* locus have been associated with the ability to malignantly transform cells in culture as well as with naturally occurring human tumors. Mutations at these residues reduce GTPase activity, leading to the activated *ras* phenotype. It has been recognized since its discovery that *ras* is in some way involved in cellular growth control. In mammalian cells, after *ras* was shown to be a transforming gene in its own right, it was shown that microinjection of antibody against *ras* could block cellular proliferation induced by various growth factors. Similarly, dominant negative mutant *ras* blocks growth-factor-mediated proliferation. This evidence places *ras* downstream of receptor tyrosine kinases (RTK). Recent reports have shed light on the intermediaries found connecting RTK and *ras*. In the presence of EGF, antiserum directed against one component brings down receptor, the adaptor protein Grb2, and the nucleotide exchange factor mSos (murine son of sevenless, homologous to the *Drosophila* protein described above). Support of these results comes from the specific inhibition of complex formation using a phosphopeptide corresponding to the Grb2 binding site on the activated receptor. As in other systems, the mammalian Sos has been shown to have *ras*-activating properties. Thus, a molecular mechanism explaining the activation of *ras* after receptor and nonreceptor tyrosine kinase activation is emerging.

Negative modulation of *ras* activity, as performed by GAP, is also a consideration in signal transduction. GAP fulfills a need in the *ras* system for the rapid elimination of signal transmission. One pertinent consideration is the abrogation of GAP function to allow signaling to proceed. Several possible mechanisms exist to explain how *ras* GAP is deactivated during signal transmission. GAP has been shown to be phosphorylated on tyrosine residues and to bind activated RTK through SH2 domains. This binding alone may be sufficient to interfere with GAP and *ras* interaction. Alternatively, GAP may be inactivated directly by interaction with a variety of phospholipid metabolites. Down-regulation of *ras* in humans does not rely on GAP alone. Human cells harbor a second *ras* GAP termed neurofibromin (NF-1). Mutations in the neurofibromin locus can lead to von Recklinghausen's disease, also known as neurofibromatosis (NF) type 1. NF type 1 is classically characterized by three major features: bodywide distribution of numerous neural tumors; multiple pigmented macules (classically occurring in the axillae)—some large enough to be designated café au lait spots—and Lisch nodules or hamartomas of the pigmented iris. Alternate splice variants of NF-1 have been shown to correlate with certain central nervous system malignancies. Work using T cell activation as a model system suggests that deactivation of GAP itself is sufficient for activation of *ras*. Similarly, a case of NF was recently traced to the inactivation of the NF-1 locus by insertion of an Alu sequence. The implications for inactivation of GAP and its functional homologue, NF-1, in human tumors is thus made compelling.

Recent reports have also helped to clarify the effector side of the *ras* signal transduction pathway. Evidence in mammalian systems has been accumulating that shows *raf* acts downstream of *ras*. Investigators have demonstrated that nonfunctional kinase-defective mutant *raf* blocks proliferation and transformation initiated by growth factors and activated *ras* in NIH3T3 cells. A complementary series of experiments showed that an inactive, dominant negative *ras* could block activation of *raf* by stimulated receptors. This same report showed that the dominant negative *ras* also inhibits activation of MAP kinase and the RSK protein. These results confirmed experiments showing an oncogenic variant of *ras* mediated activation of MAPK. These experiments were conducted using PC12 cells derived from a rat pheochromocytoma that differentiate in response to several stimuli including nerve growth factor treatment or expression of an activated *ras* oncogene. Experiments using PC12 cells demonstrate the diversity of *ras* function in higher eukaryotes. The discovery that *ras* could stimulate both MAPK and *raf* led to speculation that MAPK and *raf* could lie in the same pathway after *ras*. Subsequently, it was shown that activated *raf* could mediate the activation of MAPK through a protein known as MAPK kinase (MAPKK). Evidence for direct interaction between *raf* and MAPKK is still lacking. However, of great significance is the in vitro demonstration of a direct physical interaction between *ras* and *raf*. Should this interaction hold up under further scrutiny, it would be the first description of a direct physical interaction between *ras* and an effector known to be downstream of *ras*. Alternatively, other experiments have shown that there are *ras*-independent ways to activate MAPK. Investigators have also shown specific phosphatases for several components of the MAPK pathway that turn off their respective kinase by removing an activating phosphate. The negative function of the phosphatases is analogous to the action of GAP on the pathway, and similarly, these phosphatases represent potential tumor suppressors. Taken together, however, a core kinase-mediated transduction pathway is beginning to emerge (see Fig. 3.1 and Table 3.1). In other avenues of investigation, effectors have been proposed that interact with GAP. Treatment of cells with EGF leads to the phosphorylation and stable association of GAP and proteins currently termed p62 and p190: p62 has been cloned and encodes a nucleic acid binding protein that may be involved in mRNA processing; p190 has also been cloned and contains domains with homology to *rho* GAP (and was subsequently found to have *rho* GAP activity) and another domain with homology to a transcriptional repressor. If this molecule lives up to its potential, it may be another connection between *ras* and the transcriptional control of specific genes. The recent efforts probing into the mechanisms controlling *ras* and controlled by *ras* have been very rewarding. Although the ultimate number of variations on the *ras*-mediated signal transduction theme will undoubtedly be as great as the number of model systems used for these

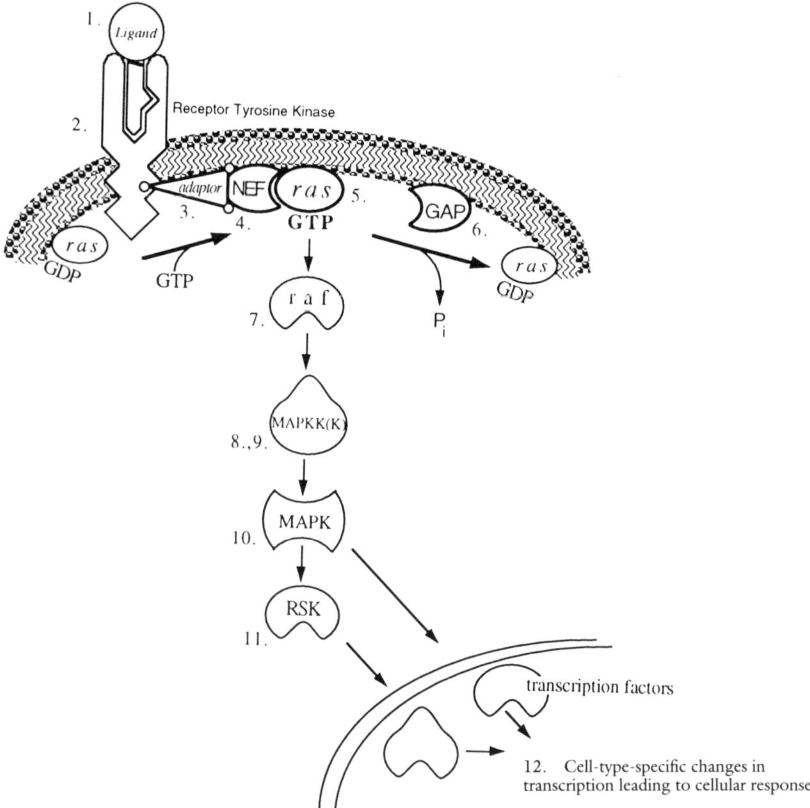

Fig. 3.1: Core elements of the signaling pathway involving *ras*. Components of this pathway are shared by several distantly related organisms. Some variations of this pathway are described in the text. NEF, nucleotide exchange factor; GAP, GTPase-activating protein; MAPK, mitogen-activated protein kinase; MAPKK(K), mitogen-activated protein kinase kinase (kinase). The ambiguity of the step between *raf* and MAPKK stems from the lack of proof for direct interaction. The numbers refer to those listed in Table 3.1.

studies, the conservation of certain components points to the necessity of biological systems to work within the confines of successful molecular mechanisms.

ANALYSIS OF *ras*-MEDIATED ONCOGENIC TRANSFORMATION IN VITRO

Mutations in genes either inherited or occurring in somatic cells are important effectors of the malignant phenotype. These mutations may be

Table 3.1: Components Identified in the Signal Transduction Pathways Involving *ras*

*No.**	*Component†*	*S. cerevisiae*	*S. pombe*	*C. elegans*	*D. melanogaster*	*X. laevis*	*Mammals*
1	Ligand		M-factor; mfm1, mfm2	*lin-3*	*Bride of sevenless*	Insulin, IGF-1	EGF and others
2	Receptor		map3	*let-23*	*Sevenless*	Insulin receptor	EGF receptor and others
3	Adaptor protein			*sem-5*	*drk*		Grb2, Shc
4	Nucleotide exchange factor, NEF	Cdc25, Scd25	ste6		*Son of sevenless*		hSos
5	GTPase-activating protein, GAP	Ira1 and Ira2	gap1/sar1		*gap1*		GAP, NF1
6	ras	Ras1, Ras2	ras1	*let-60*	*Dras1*	*ras*	*ras*
7	raf		byr2	*lin-45*	*Draf*	*raf*	*raf*
8	MAPKK(K)						MAPKK(K)
9	MAPKK/MEK		byr1		*Dsor1*	MAPKK	MAPKK/ MEK
10	MAPK/ERK		spk1			MAPK	MAPK/ERK
11	RSK						RSK
12	Cell response	Cell growth	Sexual differentiation	Vulval cell differentiation	Photoreceptor cell differentiation	Oocyte maturation	Growth or differentiation

* Numbers refer to numbered elements of Figure 3.1.

† A blank in the table indicates the component has not been identified in that organism, or does not function in the *ras*-mediated pathway of that system. See text for details.

as subtle as a single base change at a critical amino acid of a *ras* oncogene or as dramatic as a translocation of an otherwise unexpressed gene like c-*myc* to a transcriptionally active region of a chromosome. Also, epigenetic enhancement of gene expression can result in near permanent activation of genes (imprinting) controlling differentiation and proliferation. Although cancer is a disease of genetic damage, epigenetic events clearly play a role, in that nongenotoxic, hormonal effectors can alter the rate of carcinogenesis. In a subset of cancers this epigenetic component can be demonstrated by the hormonal sensitivity of tumors and a role for normal hormones in the etiology of the neoplasms. For example, osteosarcomas occur primarily during the accelerated growth period of the late teens. Breast cancer never occurs before pubarche and is affected by hormone-related phenomena such as age of menarche, parity, and menopause. These epigenetic, hormonal factors at the molecular level result in changes in gene expression. Estrogen and progesterone are known to bind and activate nuclear receptors that are members of the steroid hormone family of transcription factors and thus mediate changes in gene expression. With such a complex array of genetic and epigenetic phenomena that may contribute to tumorigenesis, it is difficult to assess whether a particular change is causal in tumor etiology.

One way to address these complex questions is to use in vitro cell models. Appropriate cell models allow the study of individual events of the multistep process of carcinogenesis. Using this approach the role of multiple signal transduction mechanisms in tumor development can be studied. For example, the result of gene transfer of oncogenic effectors such as *ras* oncogenes on growth and differentiation of normal or immortal human cells can be studied in combination with steroid or peptide hormones.

Several lines of evidence support the role of multiple genetic alterations in the neoplastic transformation of cells in vitro. It is obviously difficult to demonstrate directly the relevance of multiple phenomena to human tumorigenesis in vivo. Human tumor specimens are often available only at an advanced stage of the tumor development, and often multiple genetic and chromosomal changes have occurred, making it difficult to interpret the role of those changes relevant to tumorigenic progression. Tumor samples are often contaminated with normal cells, obscuring important genetic changes in the tumor cells. Therefore, it has been even more difficult to prove a causal role for most of the genetic elements in human neoplasia. To prove a statistically significant correlation for the role of an oncogene or a suppressor gene at some stage of human cancer, large numbers of samples must be analyzed. This type of large-scale retrospective study can involve more than 1000 tumor samples, and often the conclusions remain ambiguous. Therefore, the approach of establishing cell culture systems has been taken to study the mechanisms of oncogene-induced transformation of human cells. Human cells in culture are inherently more stable than rodent cells and represent a controllable means to study the interaction of oncogenes and tumor suppressor genes. They also provide the means to detect new genetic elements that are involved in human cancer. In vitro experimental systems that address multiple stages of the carcinogenic process have been difficult to develop because of the complexity of the genetic changes that occur during oncogenesis. The long-term objectives of establishing such systems have been to develop cell lines dependent on multiple genetic events for the induction of tumorigenesis, to characterize those genes and their effects on the biology of human cells, and to use the cell lines developed to isolate new oncogenes from human tumors or human suppressor genes that effect a transition in the multistep carcinogenic process.

Originally, studies of the biological effects of human oncogenes were limited to experiments with rodent cell lines such as NIH3T3. These cells are nontumorigenic in athymic nude mice but are far from normal. They can be considered on the verge of tumorigenicity, requiring only a single additional genetic event to gain the capacity to form tumors. In those systems the genes from human tumors that had biological activity very often were found to be members of the *ras* family of oncogenes. However, for technical and mechanistic reasons, the vast majority of DNA samples from human tumors do not have transforming activity in the NIH3T3 cell assay.

Human cell systems to study oncogenes in multistage carcinogenesis have been difficult to establish because human cells are inherently more resistant to tumorigenic transformation. Sager and coworkers were able to induce changes in cell morphology by transfecting simian virus 40 viral DNA into human foreskin fibroblasts; however, these transfected cells were nontumorigenic in nude mice. When the cloned T24 bladder carcinoma H-*ras* oncogene was transfected into the foreskin fibroblasts, no morphological transformation or tumorigenicity was detected. McCormick and coworkers were able to induce morphological transformation, focus formation, and anchorage independence by transfecting, into diploid human fibroblasts, a plasmid that produces very high levels of expression of an activated human H-*ras* oncogene. However, these cells did not form tumors in nude mice. In sharp contrast, Spandidos and Wilkie were able to induce tumorigenesis in rat embryo fibroblasts with the same plasmid. Clearly, rodent cells are more permissive for transformation and may not be a good model for human carcinogenesis. One of the first human cell transformation systems, a nontumorigenic human osteosarcoma cell line, TE-85, can be transformed to tumorigenicity by transfection with an activated *ras* oncogene. Attempts to transform TE-85 cells with other oncogenes that can transform immortal murine fibroblasts were unsuccessful, implying that *ras* transformation requires the abrogation of a very specific pathway that blocks the transforming activity in normal cells.

Another useful in vitro human cell transformation system uses human fibroblasts from patients with Li-Fraumeni syndrome that are spontaneously immortalized owing to the presence of germline p53 mutations found in those cells. Transfection of an activated H-*ras* oncogene can transform these spontaneously immortalized normal skin fibroblasts from patients with Li-Fraumeni syndrome to form tumors in nude mice, However, 20% of the colonies expressing *ras* are nontumorigenic, possibly implying the requirement for the loss of at least one additional tumor suppressor event. Human cells apparently possess some impedance factors that render them refractory to transformation, indicating that human cell systems to study carcinogenesis are more complicated to use than rodent cell systems. For example, the transfection of normal fetal foreskin fibroblasts with an activated *ras* oncogene failed to produce tumorigenic cells. When transfected with an activated N-*ras* oncogene driven from its endogenous promoter, human diploid fibroblasts were resistant to transformation, whereas transfection with the N-*ras* under the control of the viral long terminal repeat produced foci of morphologically transformed cells. H-*ras* or N-*ras* transfection induced tumorigenic transformation in v-*myc*-transfected human fibroblasts that have acquired an infinite life span in culture independent of the effect of v-*myc*.

Interesting systems have been developed using human keratinocytes to study the effect of *ras* on tumorigenesis and differentiation. The HaCAT cells of Boukamp and associates are immortal owing to a p53 mutation that probably was induced at the sun-exposed site from which the skin culture was established.

The PA-1 human teratocarcinoma cell line is a human cell culture system developed to study how an oncogene's action can fail to be regulated as cells progress through the transitions in multistage carcinogenesis. The cell line exhibits progression as it is passaged in culture. Initially the cells are nontumorigenic when injected into nude mice, from passage 30 to 90. As PA-1 cells are further passaged (after passage 90), they grow progressively more rapidly in culture and readily form neuroblastoma-like tumors in nude mice, with a latent growth period of 7 to 10 weeks. PA-1 cells are highly stable karyotypically, yet undergo progression in culture after establishment of the cell line at passage 30. This gives rise after further growth to spontaneously transformed tumorigenic cell lines. PA-1 cells at passage 100 and beyond contain an N-*ras* oncogene with an activating point mutation at amino acid position 12, from a glycine to an aspartate residue. The N-*ras* genes in early-passage preneoplastic PA-1 cells do not contain the activating mutation in the first exon of the gene. This activated N-*ras* oncogene had a causal role in the tumorigenesis of these cells, in that transfection of the activated *ras* oncogene into a nontumorigenic PA-1 cell line led to cells that could form neuroblastomas in athymic nude mice. Recipient cells for these transfection experiments were cloned and initially were not tumorigenic after expressing the transfected oncogene. Although initially resistant to *ras* (untransformable by *ras*), 20 to 30 passages (100 to 150 cell generations) later, they become responsive to the oncogene and can be transformed to form tumors after transfection with a *ras* oncogene.

This human cell culture system has allowed the study of how regulation of an oncogene's action by other genes changes as cells progress along the transitions in multistage carcinogenesis. Normal cells are resistant to the transforming activity of the dominant oncogenes. Because tumors arise whose genesis was dependent on the action of the oncogene and because continued expression of the *ras* oncogene is necessary for the maintenance of the tumorigenic phenotype, an obligatory step in carcinogenesis before tumor emergence is the loss of the ability of normal cells to resist the transforming activity of the oncogenes. The molecular genetic basis of the susceptibility to induction of transformation by a single oncogene is the loss of a tumor suppressor gene. The mechanism by which cells acquire the susceptibility to tumorigenic transformation by an activated *ras* oncogene is related to responsiveness to growth factors found in many normal tissues, EGF and basic fibroblast growth factor (bFGF). This loss of regulation of growth-factor-mediated stimulation of proliferation represents a new regulatory role for tumor suppressor genes during preneoplastic stages progressing toward tumorigenicity.

Immortalization as a Critical Step in Oncogenesis

Cancer is generally accepted to develop via a multistep mechanism involving both genetic and epigenetic changes in cellular regulatory pathways. Two hallmarks of cancer cells are genetic instability (chromosome damage) and an infinite life span in culture, which is an escape from cellular senescence, a process known as immortalization. At least four complementation groups of genes appear to be involved in escape from senescence. Both of these critical features of tumor cells, genetic instability and immortalization, can result from loss of cell cycle control. Immortalization is thought to be the rate-limiting step in the formation of a cancer cell.

Li-Fraumeni syndrome (LFS) is a hereditary cancer syndrome in which patients develop sarcomas, breast cancer, leukemia, and many other cancers. LFS fibroblasts contain germline p53 mutations. They become highly aneuploid in culture and then spontaneously immortalize; normal human cells from normal donors never spontaneously immortalize. These LFS fibroblasts have lost a G_1 checkpoint and cannot arrest in the G_1 phase cell cycle in the presence of PALA, a drug that starves the cells of nucleotides. Introduction of wild-type p53 into these cells restores this G_1 checkpoint. Other studies have shown that cells with no wild-type p53 have lost a checkpoint that allows the cells to stop and repair DNA damage from radiation.

The presence of a germline mutant p53 in LFS fibroblasts causes significant genomic instability leading to aneuploidy and other chromosomal abnormalities. Clearly, the dominant negative effects of the germline mutant p53, over the wild-type p53 present in heterozygous low-passage LFS fibroblasts, partially abrogates the G_1 checkpoint, causing the heterozygous cells to enter S phase inappropriately. When this happens continuously, the result is the accumulation of DNA damage and ultimately escape from senescence. Possibly other genes that are involved in cell cycle control are targets as inherited cancer genes. Immortal LFS fibroblasts are very close to a cancer cell in that they can be transformed by an activated *ras* oncogene to form tumors in athymic nude mice. Attempts to transform LFS fibroblasts with other oncogenes that can transform immortal murine fibroblasts have been unsuccessful, implying the abrogation of a very specific pathway that blocks the transforming activity in normal cells.

The mechanisms of immortalization vary depending on cell type. Normal fibroblasts can be induced to immortalize, albeit at a low frequency, by DNA viruses such as SV40, adenovirus, and human papillomavirus (HPV) due to the activity of the transforming genes of those viruses. Inactivation of the tumor suppressor genes p53 and RB by SV40 T antigen or HPV E6 and E7, respectively, also effect cell cycle control. Whereas fibroblasts require both HPV E6-mediated inactivation of p53 and E7-mediated inactivation of RB-1 to become immortal, mammary epithelial cells can be immortalized by E6 alone. SV40 T antigen immortalized human fibroblasts can be transformed by a *ras* oncogene. Immortalization in different cell types may require inactivation of different mechanisms of cell cycle control.

ras Oncogenes Can Alter Mechanisms that Regulate Differentiation

The initiation of differentiation requires changes in gene expression mediated by a cascade of transcriptionally regulated proteins, and we know that protooncogenes can participate in this process. Developmentally regulated transcription factors such as the AP-2, myogenin, and the homeobox family of transcription factors are proteins that control the synthesis of mRNA of other genes and play important roles in the differentiation of certain tumor cells. The differentiation of myoblasts into muscle cells can be inhibited by activated *ras* oncogenes. The differentiation of a hematopoietic precursor cell line by granulocyte-colony-stimulating factor into neutrophilic granulocytes can be blocked by a v-H-*ras* oncogene. However, the differentiation of PC-12 pheochromocytoma cells can be driven by a *ras* oncogene in a fashion similar to treatment with nerve growth factor.,

PA-1 human teratocarcinoma cells differentiate in response to retinoic acid in monolayer culture forming multiple cell types and thus are a good model for molecular mechanisms of embryonic development and differentiation. Variant cell lines have been engineered with a point-mutated N-*ras* oncogene that are unable to differentiate after treatment with retinoic acid owing to this constitutive proliferative signal. A mechanism that accounts for some or all of the resistance is overexpression of the transcription factor AP-2. The teratocarcinoma cells transformed by an activated N-*ras* oncogene are retinoid resistant. AP-2 is retinoid inducible in responsive teratocarcinoma cells. However, in retinoid-resistant *ras*-transformed cells, AP-2 mRNA is overexpressed. Although *ras*-transformed cells overexpress AP-2 mRNA, they have lost all endogenous transcription activity measured from a transcription reporter plasmid with three AP-2 DNA-binding sites adjacent to a basal-level promoter. Antisense inhibition of *ras* results in restoration of normal AP-2 activity, indicating that overexpression of AP-2 is an epigenetic effect dependent on the activated N-*ras* oncogene. These data also show that AP-2 is part of the signal transduction pathway that is mediated by *ras*. If AP-2 is constitutively overexpressed from an SV40 promoter, AP-2 activity is similarly inhibited, and the resulting cells have low AP-2 transcriptional activity. The mechanism of transcriptional inhibition is general for certain transcription factors. Others within this class of transcription factors can cross-inhibit transcription. This sort of pleiotropic effect on transcription has never been shown before and may account for how oncogenic transcription factors transform cells and block differentiation.

Several oncogene changes have been observed in breast cancer tumor specimens and cell lines. As discussed in the next section, the mutational activation of *ras* oncogenes is a frequent event in other cancers, such as pancreas, lung, and colon, but it is rare in breast cancer. However, this does not eliminate *ras* as a potential point of intervention in this disease. Overexpression of wild-type *ras* has been observed in breast cancer.

Overexpression of the EGF receptor has been observed as well as its ligand, transforming growth factor-α (TGF-α), which is overexpressed in 25% to 50% of breast carcinomas. Overexpression of the HER-2/neu receptor is frequently observed also. Activated *ras* oncogenes can stimulate TGF-α expression, and it can be mimicked by overexpression of wild-type *ras* or stimulation of growth factor pathways that signal through *ras* as a G protein. Both the Her-2/neu and EGF receptor are likely to signal through the *ras* pathway. TGF-α expression increases upon growth stimulation of breast cancer cells with estrogen. Growth inhibition by antiestrogens is accompanied by increased secretion of the growth inhibitory factor TGF-β. Cells that are hormone independent are insensitive to both factors.

To compare the effects of *ras* and other oncogenes in mammary cell differentiation, Heynes and associates transfected an activated human H-*ras*, TGF-α, activated rat neuT, and human HER-2/neu activated by a point mutation in the transmembrane domain, and induced tumorigenic transformation of HC11 murine mammary cells in nude mice. HC11 cells expressing the neuT and activated HER-2/neu genes synthesized β-casein in response to lactogenic hormones, but those expressing the H-*ras* or TGF-α lost responsiveness to lactogenic hormones. This inhibition of β-casein production occurs at the transcriptional level, and in the TGF-α-transformed cells is due to an autocrine activation of EGF receptor. Although HER-2/neu and EGF receptors are highly related, their activation has differential effects on mammary epithelial cell differentiation.

The same group found that hormone action on β-casein gene transcription requires the mammary-gland-specific transcription factor MGF. EGF-receptor activation antagonizes the effects of the lactogenic hormones on MGF activation and β-casein gene transcription during the induction of predifferentiated cells. HC11 cell lines transfected with an activated H-*ras* or v-*raf* oncogene developed transformed properties and were resistant to lactogenic hormone induction of MGF and β-casein gene transcription. Introduction and high expression of the int-2 gene, a member of the fibroblast growth factor gene family, does not interfere with MGF and β-casein induction. Overexpression of the c-*myc* gene, causing mammary epithelial cell transformation, also does not suppress MGF activity.

ras IN HUMAN TUMORS

Incidence of Mutations

Because of its central role in growth-promoting signal transduction in normal cells, it is not surprising that activating mutations in genes of the *ras* family are among the most common and well-studied genetic changes identified in human neoplasia. Codons 12, 13, and 61 of H-*ras*, K-*ras*, and N-*ras* are mutational hot spots in a wide variety of human preneoplastic and neoplastic lesions. Work in experimental tumor systems and the spectrum of mutations identified in human tumors indicate that the carcinogenic agent, the target organ, and the genetic background determine the specific genetic lesion.

The incidence of *ras* family mutation in human tumors varies widely (Table 3.2). The highest incidence of *ras* mutations occurs in pancreatic adenocarcinoma, in which 75% to 90% of tumors have mutations in K-*ras*. The majority of mutations in this tumor occur at codon 12. Also, *ras* is frequently mutated in adenocarcinoma of the lung in smokers, follicular thyroid carcinoma, seminoma, colon adenocarcinoma, and acute myelogenous leukemia. Its mutations occur infrequently in lung squamous carcinoma, melanoma, bladder carcinoma, oral squamous carcinoma induced by chewing tobacco, and cervical squamous carcinoma.

Mutations in the *ras* genes also have been identified in several preneoplastic conditions. This illustrates that although activation of *ras* is an important event, it is not the only event in the multistep process of malignant transformation. Up to 50% of histologically normal colon adenomas have been shown to harbor K-*ras* mutations. Microscopic follicular adenomas of the thyroid have frequent mutations of H-*ras*, K-*ras*, and N-*ras* similar to those in follicular carcinomas. In addition, 25% of patients with myelodysplastic syndromes, which may be precursors to myelogenous leukemias, have a *ras* mutation in their bone marrow cells.

Alternative Mechanisms of *ras* Activation

Other mechanisms of *ras* activation have been found in addition to genetic mutation. Patients with NF fail to produce neurofibromin, a functional GAP protein. This leads to an overaccumulation of *ras* in the GTP-bound activated state. Activation of *src* and *yes* kinases leads to an accumulation of *ras* GTP in human colon cancer. EGF receptor is up-regulated in several human malignancies, including breast cancer. Since EGF receptor also signals to the nucleus through *ras* accumulation of *ras* GTP, a critical role for *ras* is likely in tumors overexpressing EGF receptor. With the discovery of new important intermediate signaling proteins, guanine nucleotide exchange factors, and GAPs, it is conceivable that other oncogenes and tumor suppressor genes exist that also may lead to *ras* activation.

Another mechanism of *ras* activation may be gene amplification. The overexpression of normal $p21^{ras}$ can transform some cells, and *ras* is amplified in some human tumors. However, in tumors in which *ras* is overexpressed, no correlation could be made between the mutant phenotype and the degree of *ras* overexpression. While overexpression of

Table 3.2: Incidence of Activating *ras* Mutations in Human Tumors

Tumor	*Percentage Mutated*	*Gene Mutated*
Breast	0–8	K
Gynecologic		
Ovary	0–30	K
Cervix	0–20	K
Endometrium	12–46	K,N
Nervous System		
Neuroblastoma	0–16	N
Skin		
Keratocanthoma	10	
Basal cell	0–30	H,K
Squamous cell	0–40	H
Xeroderma pigmentosa	50	N,H,K
Melanoma	0–25	H,K,N
Lung		
Epidermoid	5–10	H,K
Adenocarcinoma	22–33	K
Thyroid		
Follicular	40–50	H,K,N
Undifferentiated	40–60	H,K,N
Papillary	0–20	H,K,N
Follicular adenoma	40–50	H,K,N
Gastrointestinal		
Esophagus	0–10	H,K
Stomach (USA/Asia)	0–10/1–41	N,K/H,K
Colon adenocarcinoma	40–50	K
Colon adenoma	40–50	K
Pancreas	75–90	K
Genitourinary		
Male germ cell tumors	11–43	K,N
Bladder	7–17	H
Kidney	7–13	H
Myeloid disorders		
Myelodysplastic syndrome	5–41	N,K
Acute myelogenous leukemia	30–65	N
Chronic myelogenous leukemia	17–50	N
Lymphoid disorders		
Acute lymphocytic leukemia	11–18	N
Non-Hodgkin's lymphoma	0	—
Hodgkin's lymphoma	0	—

$p21^{ras}$ may have a role in some human cancers, this role has yet to be clarified.

ras-Directed Cancer Therapy

Because *ras* activation occurs in such a large, diverse group of tumors, several approaches have been developed to inhibit *ras*-mediated signal transduction, with the hope of inhibiting tumorigenesis. These approaches have been divided into those that inhibit *ras* function and those that use the immune system to exploit the presence of a mutant *ras* protein. Several strategies are currently being examined to directly inhibit oncogenic *ras* function, including antisense *ras* therapy, *ras* GAP-binding and effector-domain inhibitors, and inhibitors of *ras* post-translational modification.

***ras* Antisense** Although antisense technology has been a widely used tool in molecular biology for several years, only recently has there been an explosion in research concerning the use of antisense therapy as treatment for disease. The principle of this technique relies on the capacity of complementary nucleic acids to bind to each other. The mRNA transcribed from the gene encoded in DNA is in the sense orientation. Antisense molecules complementary to a specific mRNA bind to it and prevent its processing and translation. The major advantage of antisense therapy is specificity for the target. Cancer is an ideal disease in which to examine the use of antisense therapy because for many tumor types a major genetic event resulting in the production of an abnormal protein is known. This is especially true for tumors with activation of the *ras* genes. Several approaches to antisense therapy of disease are currently under study. Thus far, the use of antisense oligonucleotides, antisense ribozymes, and large antisense constructs has been studied in *ras*-mutated cells.

Antisense Oligonucleotides Antisense oligonucleotides are small pieces of DNA from 8 to 50 nucleotides in length that inhibit gene expression by acting on pre-mRNA and mRNA targets. Although the mechanism of action of these oligonucleotides is thought to be mediated by RNase H, an endonuclease that cleaves the RNA strand of RNA-DNA duplexes, direct evidence for this mechanism is lacking. The most effective sites of antisense oligonucleotide targeting are the protein translation start site, the exon-intron borders, and the region adjacent to the mRNA cap site. This suggests that antisense oligonucleotides also function by inhibiting RNA from interacting with cellular components required for translation of the mRNA into protein. The advantages of antisense oligonucleotides over other forms of antisense therapy are ease of preparation, good heteroduplex formation, and elimination of the routine need for transformation.

The applicability of antisense oligonucleotides to the treatment of tumors with *ras* mutations has been studied in cell-free systems and in tissue culture. In a cell-free system a modified (methylphosphonate), anti-*ras* oligonucleotide complementary to the first 11 nucleotides of the initiation codon region of H-*ras* inhibited translation completely at an oligonucleotide concentration of 100 μM. Similar results were demonstrated in RS 485 cells, a transformed NIH3T3 cell line that overexpresses the normal human H-*ras*. Inhibition of normal human H-*ras* has been demonstrated in the same system using an 11-nucleotide oligonucleotide complementary to the splice junction of the first intron 1/exon 2 splice site of H-*ras*. The production of H-*ras* with a mutation at codon 12 has been selectively inhibited by an 8-nucleotide oligonucleotide in a cell-free system. These findings have been extended to tissue culture, where it has been demonstrated that an 11-nucleotide oligonucleotide complementary to the region around codon 61 of a mutant H-*ras* can inhibit mutant H-*ras* synthesis 97% in NIH3T3 cells transfected with the mutant H-*ras* gene. Normal H-*ras* synthesis was inhibited only 26%.

Unmodified oligonucleotides of 15 nucleotides inhibit *ras* expression in *ras*-transfected NIH3T3 cells. In these experiments antisense oligonucleotides complementary to the mRNA 5′ cap were most effective in inhibiting protein production (75% inhibition at 50 μM) followed by an initiation-codon-directed sequence (65% inhibition). The least effective was a target upstream of the initiation codon (40%). Despite demonstration of *ras* inhibition by several strategies, documentation of an effect on tumorigenesis has been limited. Treatment of these *ras*-transfected cells in vitro with a 15-nucleotide oligonucleotide complementary to the 5′ flanking sequence of H-*ras* inhibited subsequent growth of these cells in nude mice.

There are several hurdles to overcome in the use of antisense oligonucleotides in whole animal models and their subsequent use in human trials. Antisense oligonucleotides are degraded rapidly by endonucleases. Molecular stability can be improved by certain modifications; however, no currently available oligonucleotide has the required properties of stability, affinity, and permeability. Large doses of oligonucleotide are necessary to achieve the desired effect, and continuous, repeated dosing is required. The breakdown product of the oligonucleotides is thymidine, which in large doses is toxic to nucleotide production. Delivery of the necessary concentrations of oligonucleotides to target tissues is also currently difficult. Considerable work is needed to overcome these problems before antisense oligonucleotide therapy of cancer becomes a reality.

Antisense RNA The use of antisense RNA involves the introduction of sequences that can be transcribed and yield a desired antisense molecule inside target cells. The advantage of this system is that stable transfection of cultured cells with vectors expressing antisense RNA allows continuous generation of antisense sequences within the cell itself. The problems with this technique are the potential loss of the incorporated construct, inefficiency of incorporation, and the requirement of vectors to introduce the construct into cells. One popular method used to introduce antisense RNA is infection with retroviral vectors. The use of retroviral vectors is attractive because they preferentially infect dividing cells. The desired construct is introduced into a retrovirus rendered replication incompetent through deletion of envelope (env) sequences. The target cells are then infected with retrovirus in an attempt to introduce the antisense construct. High viral titers are necessary because the virus cannot replicate. In addition, the long-term consequences of this kind of viral infection are unknown.

The use of retrovirus-mediated introduction of antisense RNA has been investigated in a human lung cancer model. A wild-type 2-kb K-*ras* genomic DNA segment carrying second and third exons together with flanking intron sequences was cloned into an expression vector under the control of a β-actin promoter. This construct was introduced into H460 cells (human lung adenocarcinoma cells with a point mutation at codon 61 of K-*ras*) by electroporation. Cells that incorporated the construct had a 95% decrease in production of K-*ras*, with a minimal effect on the total level of $p21^{ras}$. H460 cells expressing the antisense construct in culture exhibited a threefold reduction in cell growth and failed to produce tumors in nude mice.

Subsequently, the antisense K-*ras* construct was introduced into a retroviral vector and evaluated in an orthotopic human lung cancer model. H460 cells were injected intratracheally into nude mice. Mice were treated for 3 days with virus-containing supernatant. Only 10% to 14% of mice treated with virus containing the antisense K-*ras* construct developed tumors compared with 73% to 100% of control mice. These studies constitute the first reported use of retroviral supernatants to mediate antitumor effects and the first successful use of an antisense construct to mediate therapeutic tumor regression in vivo. A phase I trial is under way using bronchoscopic introduction of retroviral antisense K-*ras* in the treatment of obstructing endobronchial lesions. Positive results from this trial would suggest that investigators study patients with unresectable adenocarcinoma of the pancreas, who may benefit from the instillation of this antisense construct into an obstructed pancreatic duct.

Antisense Ribozymes Consistent with their ability to fold into complex three-dimensional structures and bind small molecules in the presence of divalent metal ions, some RNA molecules can function as catalysts. Ribozymes are RNA molecules capable of catalyzing nucleic acid hydrolysis at specific residues. A group of self-cleaving ribozymes called "hammerhead' have a consensus secondary structure. The hammerhead catalytic motif can be incorporated into an antisense RNA, resulting in an enzymatic antisense RNA. The antisense component of this hybrid ribozyme lends specificity for its target; the ability to cleave multiple substrates derives from the catalytic component. In cell-free systems, a ribozyme with 10 nucleotides homologous to N-*ras* at the 3′ and at the 5′ ends was capable of cleaving an N-*ras* substrate. A ribozyme targeted to the activated codon 12 in exon 1 of the H-*ras* gene specifically cleaved the mutant H-*ras* RNA substrate but not a normal H-*ras* RNA substrate in vitro. Transfection of T24 bladder carcinoma cells with a construct for the mutant H-*ras* ribozyme decreased the level of H-*ras* mRNA in cells expressing the oncogene. T24 transformants grown as monolayers in tissue culture expressing the *ras* ribozyme displayed altered growth characteristics and morphology. Nude mice that were intravesicularly administered T24 cells transfected with the *ras* ribozyme lived significantly longer than mice that received control T24 cells. Histologic studies revealed that ribozyme clones produced tumor nodules with no evidence of vascular invasion, whereas control tumors consisted of cells with highly invasive properties.

The use of ribozymes in whole animals and humans has many of the same problems as the use of oligonucleotides. Delivery of the ribozyme to the target, the stability of the RNA within a cell, and the turnover of the ribozyme are major problems that must be solved before ribozymes can be successfully used as therapeutic agents.

Inhibitors of *ras* GAP-Binding/Effector Domain Amino acids 32 to 40 of $p21^{ras}$ are known as the binding/effector domain because this area is responsible for the binding of GAP and transmission of signal to downstream targets. This region is an attractive target for therapy because normal *ras* is relatively insensitive to mutations in this domain. While single amino acid changes in this region reduce the ability of *ras* to affect downstream targets, there is also a reduced ability of GAP to hydrolyze *ras* GTP. Therefore, the potential decrease in activity of *ras* is not seen because there is an accumulation of *ras* GTP. Oncogenic *ras* is already in its maximal GTP-bound state; therefore, it cannot compensate for a decrease in effector binding by further increasing the *ras* GTP level.

Microinjection of a biologically inactive form of *ras* with a mutation in the binding/effector domain competes with *ras* for GAP and downstream effector molecules. Thymidine labeling of cells containing an oncogenic *ras* was inhibited 88% at a concentration of 150 μg/mL with no inhibition of thymidine labeling of cells with a normal *ras*. In an NIH3T3 transformation assay, either transfection or microinjection of an effector domain double mutant *ras* (codon 33 and 34) decreased the transforming ability of a mutant *ras* but not of a mormal *ras*. This approach to *ras* inhibition holds significant promise because of its selectivity. However, clinical application of this therapy remains a prospect for the future.

Inhibitors of *ras* Post-translational Modification In order to function, both normal and oncogenic *ras* require attachment to the inner surface of the plasma membrane. This is accomplished by attachment of a farnesyl group to the cysteine of the carboxy-terminal sequence CAAX by farnesyl transferase. In this motif, A represents an aliphatic amino acid and X is any residue. Following attachment of the farnesyl group, the -AAX is cleaved. Many cellular proteins use this mechanism for membrane attachment. Thus, inhibiting farnesylation of *ras* will not preferentially inhibit oncogenic *ras* function. Despite the nonselectivity, inhibition of *ras* post-translational modification is currently a popular target for drug design. The sites of action within this pathway of the various agents undergoing study are illustrated in Figure 3.2.

Inhibitors of Farnesyl Diphosphate Production Farnesyl diphosphate, the natural substrate for farnesyl transferase, is synthesized in the cholesterol biosynthetic pathway from mevalonic acid (Fig. 3.2). It has been postulated that interfering with farnesyl production will result in a decrease in *ras* activity. Lovastatin, a drug used to lower systemic cholesterol levels, inhibits hydroxymethylglutaryl (HMG) coenzyme A reductase, which is the initial and rate-limiting step of cholesterol biosynthesis. In tissue culture lovastatin inhibits the growth of H-*ras*-transformed NIH3T3 cells, H-*ras*-transformed rat liver cells, and human pancreatic adenocarcinoma cell lines. Lovastatin also inhibits the growth of H-*ras*-transformed NIH3T3 cells and human pancreatic carcinoma cell lines grown in nude mice. In several reports lovastatin treatment inhibits the membrane association of *ras*. However, it has not been demonstrated that the inhibition of farnesylation of *ras* by lovastatin is the mechanism of action mediating the observed properties of tumor growth inhibition. Lovastatin has also been shown to inhibit the growth of NIH3T3 cells transformed by v-*raf*. Transformation by v-*raf* is independent of *ras*

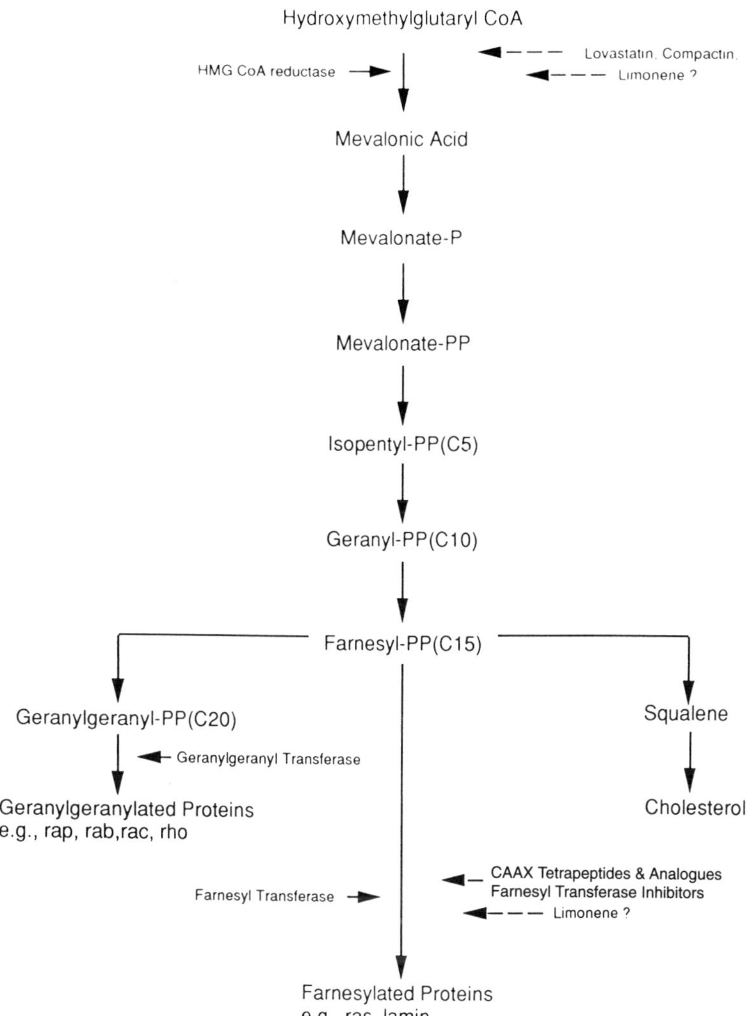

Fig. 3.2: The cholesterol biosynthetic pathway as related to *ras*. Enzymatic steps amenable to therapeutic intervention are indicated.

activity, consistent with its location downstream of *ras* in signal transduction. Because of the large number of cellular proteins modified by lipids synthesized along the cholesterol biosynthesis pathway, it is impossible to assign a direct *ras* effect to lovastatin. The use of lovastatin also is limited by the fact that the doses necessary to inhibit tumor growth are 10 times greater than those required to lower serum cholesterol levels.

Other agents that inhibit farnesyl diphosphate production are compactin, limonene, and its active metabolite, perillic acid (see Fig. 3.2 for site of action). Although these agents are effective in inhibiting tumor growth and *ras* membrane association in vitro, their growth-inhibiting properties cannot be attributed to a direct *ras* effect. Limonene is a natural product found in the peels of oranges. Its site of action in the cholesterol biosynthesis pathway is uncertain. Although limonene does inhibit HMG CoA reductase, this is not believed to be its main mechanism of action. Limonene is postulated to interfere with farnesyl transferase because of its selective effect on prenylation of 21- to 26-kDa proteins. However, the exact enzymology of limonene's effect is unknown.

Inhibitors of Farnesyl Transferase Since the purification of farnesyl transferase there has been an explosion of research investigating the results of inhibiting this enzyme. Although inhibitors of the CAAX farnesyl transferase also block the farnesylation of other essential proteins, dependence on oncogenic *ras* might render tumor cells more sensitive than normal cells to the action of a farnesyl transferase inhibitor. Several classes of agents are being evaluated as inhibitors of CAAX farnesyl transferase. A number of microbial inhibitors of farnesyl transferase have been identified. These include compounds from *Streptomyces* related structurally to manumycin, gliotoxins, pepticinnamins, and α-hydroxy-farnesyl phosphonic acid. These agents have all been shown in vitro to be effective, specific inhibitors of farnesyl transferase at low concentrations (0.3–5 μM). The effect of these compounds in vivo on *ras*-transformed cells has not been determined.

The most promising group of compounds being developed as inhibitors of farnesyl transferase are the tetrapeptides and their analogues. Peptides of the CAAX motif as small as four amino acids are capable of being substrates for farnesyl transferase. By modification of the amino acid sequence, or insertion of structurally related compounds for the two aliphatic amino acids, it is possible to develop tetrapeptides or peptidomimetics that are specific inhibitors of farnesyl transferase. The tetrapeptide CIIM was used as a template to develop the tetrapeptide analogue L-731,734 (Merck). This compound is a potent inhibitor of farnesyl transferase in vitro with a half-maximal inhibitory concentration of 282 nM. The anchorage-independent growth of Rat-1 cells transformed by v-*ras* was inhibited in a dose-dependent fashion. This compound had no effect on anchorage-independent growth of cells transformed by either v-*raf* or v-*mos*. Interestingly, anchorage-independent growth of Rat-1 cells transformed by v-*src* is also inhibited. Although *src* does not require farnesylation for activity, its transforming ability proceeds through a *ras*-dependent pathway. This finding suggests that tumor growth inhibition by inhibition of *ras* may be applicable to any tumor in which the *ras* pathway is activated, regardless of the method of *ras* activation.

Benzodiazepine peptidomimetics that are potent farnesyl transferase inhibitors have been designed. In these compounds the two aliphatic residues of the CAAX are replaced with a portion of a benzodiazepine to mimic the normal dipeptide turn. Unlike the tetrapeptides, these compounds readily enter cells to block *ras* farnesylation. These inhibitors are extremely potent, with an in vitro half-maximal inhibitory concentration of 0.26 nM. The benzodiazepine peptidomimetics reverse the transformed phenotype and inhibit the growth of Rat-1 cells with an activated H-*ras*. No effect on growth was seen in untransformed mouse myoblasts. These studies raised the possibility that partial inhibition of farnesyl transferase may allow cells to synthesize sufficient amounts of other farnesylated proteins, allowing growth while blocking the action of mutant oncogenic *ras*.

Immunization Against Mutant *ras* A novel approach to treatment of tumors with an oncogenic *ras* mutation is to determine if mutant *ras* oncoproteins result in the development of new immunologic epitopes that could be targets of a cellular immune response. T cells from healthy human volunteers can be induced to proliferate in an HLA-DR-restricted manner by repeated in vitro stimulation with a synthetic mutant *ras* protein. In vitro stimulation of human T cells with *ras* peptides with codon 12 or 13 mutations leads to HLA-DR or HLA-DQ proliferative responses. In Balb/c mice, cytotoxic T lymphocytes were generated by intraperitoneal administration of H-*ras* bearing an Arg-12 mutation. Nine of ten mice immunized with the mutant *ras* protein failed to develop tumors when challenged with activated H-*ras*-transfected NIH3T3 cells. Mice immunized and challenged with tumor cells bearing other *ras* mutations were not protected from developing tumors. Although these results are encouraging, the ability to raise mutation-specific responses needs to be tested in other genetic backgrounds.

Summary of *ras*-Directed Cancer Therapy As the complexity of the *ras* signaling pathway continues to be dissected, more potential targets of anti-*ras* therapy will become available. In addition, as we come to better understand intracellular signaling, the central role of *ras* in several pathways becomes evident. Therefore, the application of anti-*ras* therapy may extend to diverse tumors that do not have *ras* mutations. For example, it may be possible to apply anti-*ras* therapy in breast and lung tumors that overexpress EGF receptor. Tumors with mutations in the gene encoding neurofibromin (NF-1) may be amenable to anti-*ras* therapy. Any tumor in which the dominant pathway of intracellular signaling is through *ras* may be responsive to anti-*ras* therapy.

Diagnosis of *ras* Mutations

With the use of the polymerase chain reaction (PCR), the detection of *ras* mutations in tumors, stool, urine, bile, pancreatic secretions, and sputum has become possible. Most techniques involve initial purification of DNA from the material to be studied. As the mutational hot spots are localized to exons 1 and 2, PCR can be performed with primers designed to amplify these two exons separately. With the use of single-strand conformational polymorphism or denaturing gel electrophoresis, large numbers of samples can be analyzed for the presence of mutations by their mobility on a gel. Suspected mutations are confirmed with DNA sequencing. In another approach, PCR samples are blotted onto nitrocellulose membranes and exposed to labeled probes specific for the various mutations. Again, it is advisable to confirm the presence of a suspected mutation with sequencing. A relatively new technique called mutant-enriched PCR involves the use of specific oligonucleotide primers whose 3′ nucleotide matches the specific mutation. This technique requires the generation of multiple primers and multiple PCR reactions. All PCR products are then run on an ethidium-bromide-stained agarose gel in an attempt to identify a mutation.

Regardless of the technique used, the finding of a *ras* mutation may have diagnostic or prognostic significance. In the pancreas, it can be difficult to differentiate an area of pancreatitis from a pancreatic cancer. The ability to identify a mutant *ras* in the bile or pancreatic secretions of such a patient would be invaluable in making a diagnosis. With the progress being made in identifying oncogene and tumor suppressor gene changes in cancer, it may soon be possible to make a conclusive pretreatment diagnosis of cancer using molecular markers. In addition, the ability to detect minute levels of these oncogenes in patients will allow for the earlier diagnosis of recurrent disease.

Prognostic Significance of *ras* Mutations

The presence of mutations in codons 12, 13, and 61 of *ras* has been correlated with a poor prognosis in several human malignancies. In the myelodysplastic syndrome, patients with an N-*ras* mutation detected in their bone marrow had a significantly shorter survival and significantly higher chance of developing acute myelogenous leukemia than did patients with *ras*-negative bone marrow. Patients whose resected lung cancers harbor a K-*ras* mutation survive for shorter periods than patients with K-*ras*-negative tumors. This difference was magnified when patients were stratified by nodal status. Patients with Dukes' stage B and C colon cancer who have any activating K-*ras* mutation other than codon 12 GGT to GAT are at a very high risk of recurrence. In the same study, 71% of patients with recurrent disease had a mutant K-*ras*. In contrast, studies of pancreatic cancer have revealed no correlation between the presence of K-*ras* mutations and survival. However, the single study examining this question had only four patients in the control group. More work needs to be done on pancreatic cancer to clarify this question. Different *ras* mutations have differing abilities to transform NIH3T3 cells; thus, there may be prognostic significance to the presence of different mutations. Large numbers of tumors need to be analyzed with long-term follow-up of patients to answer these questions.

ras and Resistance to Chemotherapy and Radiotherapy

Despite the fact that mutations in *ras* may be associated with a poor prognosis in human tumors, there is no good evidence to show that *ras* contributes to resistance to standard chemotherapeutic agents. However, *ras* may contribute to radioresistance. It has been documented that *ras* oncogenes increase the resistance of NIH3T3 cells to ionizing radiation. Rat-1 cells transfected with mutant *ras* genes also are resistant to radiotherapy. Transfection of human osteosarcoma cells with the activated *ras* oncogene results in transformants that are more resistant to ionizing radiation than parental lines are. Interestingly, this increase in radioresistance is reversed by treatment with lovastatin.

Although *ras* does not appear to be responsible for resistance to standard cytotoxic chemotherapy, it may be important in retinoid response. PA-1 human teratocarcinoma cells with a mutant N-*ras* are resistant to treatment with all-*trans* retinoic acid. With the increasing use of retinoic acid in the treatment and chemoprevention of human cancers, this finding may have clinical relevance. Retinoic acid is used in the treatment and prevention of squamous malignancies of the oral mucosa and cervix. These tumors have a low but measurable incidence of *ras* mutations. It is possible that the resistance to retinoic acid observed in some squamous malignancies may be due to *ras* mutations.

If resistance to radiotherapy and retinoic acid prove to be related to the presence of *ras* mutations, anti-*ras* therapy may be useful as an adjunct to these treatment modalities. As in the case of human osteosarcoma cells transfected with *ras*, anti-*ras* agents may be effective radiosensitizers. Also, it may be possible to restore retinoid sensitivity through the use of anti-*ras* agents.

In summary, the participation of *ras* genes in cellular processes is rapidly becoming understood. Information from systems as diverse as yeast, nematodes, and humans has allowed sufficient understanding of these mechanisms to begin rational design of therapies for human malignancy.

BIBLIOGRAPHY

Aroian R, Sternberg P. Multiple functions of let-23, a *Caenorhabditis elegans* receptor tyrosine kinase gene required for vulval induction. Genetics 1991;128:251–267.

Barbacid M. ras genes. Annu Rev Biochem 1987;56:779–827.

Berra E, Diaz-Meco M, Dominguez I, et al. Protein kinase Cζ isoform critical for mitogenic signal transduction. Cell 1993;74:555–563.

Bischoff F, Yim SO, Pathak S, et al. Spontaneous immortalization of normal fibroblasts from patients with Li-Fraumeni cancer syndrome. Cancer Res 1990;50:7979–7984.

Blenis J. Signal transduction via MAP kinases: proceed at your own RSK. Proc Natl Acad Sci USA 1993;90:5889–5892.

Bollag G, McCormick F. Regulators and effectors of ras proteins. Annu Rev Cell Biol 1991;7:601–632.

Bos JL. ras oncogenes in human cancer: a review. Cancer Res 1989;49:4682–4689.

Boukamp P, Petrussevska RT, Breitkreutz D, Hornung J, Markham A, Fusenig NE. Normal keratinization in a spontaneously immortalized aneuploid human keratinocyte cell line. J Cell Biol 1988;106:761–771.

Boyd JA, Barrett JC. Genetic and cellular basis of multistep carcinogenesis. Pharmacol Ther 1990;46:469–486.

Broach J. Ras genes in *Saccharomyces cerevisiae*: signal transduction in search of a pathway. Trends Genet 1991;7:28–33.

Chardin P, Camonis J, Gale N, et al. Human Sosl: a guanine nucleotide exchange factor for Ras that binds to GRB2. Science 1993;260:1338–1243.

Chiao P, Bischoff F, Strong L, Tainsky MA. The current state of oncogenes and cancer: experimental approaches for analyzing oncogenetic events in human cancer. Cancer Metastasis Rev 1990;9:63–80.

Dickson B, Sprenger F, Morrison D, Hafen E. Raf functions downstream of Rasl in the sevenless pathway. Nature 1992;360:600–603.

Duchesne M, Schweighoffer F, Parker F, et al. Identification of the SH3 domain of GAP as an essential sequence for Ras-GAP-mediated signaling. Science 1993;259:525–528.

Egan S, Giddings B, Brooks M, Lászlό B, Sizeland A, Weinberg RA. Association of Sos Ras exchange protein with Grb2 is implicated in tyrosine kinase signal transduction and transformation. Nature 1993;363:45–51.

Farnsworth CL, Marshall MS, Gibbs JB, et al. Preferential inhibition of the oncogenic form of ras by mutations in the gap binding/effector domain. Cell

1991;64:625–633.

Feig LA. Strategies for suppressing the function of oncogenic ras protein tumors. J Natl Cancer Inst 1993;85:1266–1268.

Garcia de Herreros A, Dominquez I, Diaz-Meco M, et al. Requirement of phospholipase C–catalyzed hydrolysis of phosphatidylcholine for maturation of *Xenopus laevis* oocytes in response to insulin and ras p21. J Biol Chem 1991;266:6825–6829.

Gaul U, Mardon G, Rubin GM. A putative Ras GTPase activating protein acts as a negative regulator of signalling by the sevenless receptor tyrosine kinase. Cell 1992;68:1007–1019.

Georges BN, Mukhopadhyay T, Zhang Y, Yen N, Roth JA. Prevention of orthotopic human lung cancer growth by intratracheal instillation of a retroural antisense K-ras construct. Cancer Research 1993;53:1743–1746.

Hall A. Ras-related proteins. Curr Opin Cell Biol 1993;5:265–268.

Han M, Golden A, Yuming H, Sternberg P. *C. elegans* lin-45 raf gene participates in let-60 ras-stimulated vulval differentiation. Nature 1993;363:133–140.

Happ B, Hynes NE, Groner B. Ha-ras and v-raf oncogenes, but not int-2 and c-myc, interfere with the lactogenic hormone dependent activation of the mammary gland specific transcription factor. Cell Growth Differ 1993;4:9–15.

Horvitz H, Sternberg P. Multiple intercellular signalling systems control the development of the *C. elegans* vulva. Nature 1991;351:535–541.

Hughes D, Fukui Y, Yamamoto M. Homologous activators of ras in fission and budding yeast. Nature 1990;344:355–358.

Hywes NE, Taverna D, Harwerth FM, et al. Epidermal growth factor receptor, but Not c-erB2, activation prevents lactogenic hormone induction of the beta-casein gene in mouse mammary epithelial cells, Mol Cell Biol 1990;10: 4027–4034.

Imai Y, Miyake S, Hughes D, Yamamoto M. Identification of a GTPase-activating protein homolog in *Schizosaccharomyces pombe.* Mol Cell Biol 1991;11:3088–3094.

James GL, Goldstein JL, Brown MS, et al. Benzodiazepine peptidomimetics: potent inhibitors of ras farnesylation in animal cells. Science 1993;260:1937–1941.

Khosrhavi-Far R, Cox AD, Kato K, Der CJ. Protein prenylation: key to ras function and cancer intervention? Cell Growth Differ 1992;3:461–469.

Kohl NE, Mosser SD, deSolms J, et al. Selective inhibition of ras-dependent transformation by a farnesyltransferase inhibitor. Science 1993;260:1934–1937.

McCormick F. How receptors turn Ras on. Nature 1993;363:15–16.

Nadin-Davis SA, Nasim A, Beach D. Involvement of ras in sexual differentiation but not in growth control in fission yeast. EMBO J 1986;5:2963–2971.

Neckers L, Whitesell L. Antisense technology: biological utility and practical considerations. Am J Physiol 1993;265:11–12.

Olivier J, Raabe T, Henkemeyer M, et al. A Drosophila SH2-SH3 adaptor protein implicated in coupling the sevenless tyrosine kinase to an activator of ras guanine nucleotide exchange, Sos. Cell 1933;73:179–191.

Olson E, Spizz G, Tainsky MA. Expression of oncogenic forms of N-*ras* or H-*ras* prevents skeletal muscle differentiation. Mol Cell Biol 1987;7:2104–2111.

Pomerance M, Schweighoffer F, Tocque B, Pierre M. Stimulation of mitogen-activated protein kinase by oncogenic ras p21 in *Xenopus* oocytes. J Biol Chem 1992;267:16155–16160.

Powers S. Genetic analysis of ras homologs in yeast. Cancer Biol 1992;3: 209–218.

Roberts T. Cell biology. A signal chain of events. Nature 1992;360:534–535.

Rubin G. Signal transduction and the fate of the R7 photoreceptor in *Drosophila*. Trends Genet 1991;7:372–377.

Sager R, Tanaka K, Lau CC, Ebina Y, Anisowicz A. Resistance of human cells to tumorigenesis induced by cloned transforming genes. Proc Nath Acid Sc: USA 1983; 80:7601–7605.

Satoh T, Kaziro Y. Ras in signal transduction. Cancer Biol 1992;3:169–177.

Settleman J, Narasimhan V, Foster L, Weinberg RA. Molecular cloning of cDNAs encoding the GAP-associated protein p190: implications for a signaling pathway from ras to the nucleus. Cell 1992;69:539–549.

Shay JW, Wright WE, Brasiskyte D, Van Der Haegen BA. E6 of human papillomavirus type 16 can overcome the M1 stage of immortalization in human mammary epithelial cells but not in human fibroblasts. Oncogene 1993;8:1407–1413.

Sidransky D, Tokino T, Hamilton SR, et al. Identification of ras oncogene mutations in the stool of patients with curable colorectal tumors. Science 1992;256:102–105.

Slamon DJ, Clark GM, Wong SG, Levin WJ, Ullrich A, McGuire WL. Human breast cancer: correlation of relapse and survival with amplification of the HER-2/neu oncogene. Science 1987;235:177–182.

Smith D. The induction of oocyte maturation: transmembrane signaling and regulation of the cell cycle. Development 1989;107:685–699.

Smith JR, Pereira-Smith OM. Further studies on the genetic and biochemical basis of cellular senescence. Exp Gerontol 1989;24:377–381.

Spandidos DA, Wilkie NM. Malignant Transformation of early passage rodent cells by a single mutated oncogene. Nature 1984;310:469–475.

Strong L, Williams W, Tainsky MA. The Li-Fraumeni syndrome: from clinical epidemiology to molecular genetics. Am J Epidemiol 1992;135:190–199.

Takeda S, Ichii S, Nakamura Y. Detection of K-ras mutation in sputum by mutant-allele-specific amplification (MASA). Hum Mutat 1993;2:112–117.

Tanaka K, Nakafuku M, Satoh T, et al. *S. cerevisiae* genes IRA1 and IRA2 encode proteins that may be functionally equivalent to mammalian ras GTPase activating protein. Cell 1990;60:803–807.

Yin Y, Tainsky MA, Bischoff FZ, Strong LC, Wahl GM. Wild type p53 restores cell cycle control and inhibits gene amplification in cells with mutant p53 alleles. Cell 1992;70:937–948.

Zhang Y, Mukhopadhyay T, Donehower LA, Georges RN, Roth JA. Retroviral vector-mediated transduction of K-ras awulsense RNA into human lung cancer cells inhibits expression of the malignant phenotype. Human Gene Therapy 1993;4:451–460.

CHAPTER 4

The Retinoblastoma Susceptibility Gene, the p53 Gene, and Other Tumor Suppressors

E. Cristy Ruteshouser
Marc F. Hansen

The process of oncogenesis involves a series of genetic changes within a single cell. These changes generally occur within one or both of two groups of genes, the oncogenes and the tumor suppressor genes (sometimes referred to as "antioncogenes"). The oncogenes produce growth stimulatory proteins that are activated or overexpressed in tumor cells. The tumorigenic actions of these proteins are manifested despite the presence of the second, normal allele; that is, the oncogenes act in a dominant manner. One example is the product of the oncogene $p21^{ras}$, a GTPase, which has been implicated in the transformation of some mammalian cells (e.g., fibroblasts and epithelial cells). Mutated $p21^{ras}$ proteins are involved in the development of more than 30% of all human cancers (1).

The products of the tumor suppressor genes, on the other hand, suppress or regulate the growth of the cell. Therefore, the remaining allele can compensate for the loss of function of one allele of a tumor suppressor gene; that is, tumor suppressor mutations are recessive. In general, loss of growth control associated with tumorigenicity requires the loss of function of both alleles of a tumor suppressor gene. The recessive nature of tumor suppressor gene mutations is important because germline genetic alterations in a single allele of a tumor suppressor gene whose protein product has an important regulatory function are transmitted from one generation to another. Persons with such inherited genetic changes have a greater than normal risk of developing certain cancers because a single genetic lesion in a somatic cell can inactivate the remaining normal allele of the tumor suppressor gene and result in total loss of the growth control exerted by the protein product of this gene. These somatic events may therefore give rise to tumorigenic cells. In some human cancers, such as retinoblastoma, loss of function of a single tumor suppressor gene is apparently sufficient to initiate tumorigenicity in the cell; in many other cancers, loss of function of a single tumor suppressor gene is only one of several events required to initiate tumorigenesis.

Tumor suppressor genes have been identified, and several have been localized by fusion of tumorigenic and nontumorigenic cells, karyotype analysis to identify gross chromosomal deletions correlated with specific tumor types, and analysis of DNA from normal and tumor tissue to identify loss of maternal or paternal alleles in solid tumors. The observation that fusion of nontumorigenic human cells with tumorigenic human cells resulted in hybrids that were unable to form tumors (2) indicated that introducing genetic information from normal cells into cancer cells suppresses tumor formation and suggested that in many cases tumorigenicity arises through loss or mutation of genes that are involved in suppressing or regulating growth.

Karyotype analysis of metaphase chromosomes from tumor cells has provided further evidence for the existence of tumor suppressor genes. Although tumor cells frequently have multiple chromosomal abnormalities, suggesting that most such abnormalities are not specific to the development of a particular form of cancer, the analysis of large numbers of tumors revealed consistent areas of chromosomal deletions in specific tumor types. For example, approximately 5% of retinoblastoma tumors had deletions of the long arm of chromosome 13 that always included 13q14 (3), and in approximately 40% of sporadic colorectal carcinomas there was an interstitial deletion of chromosome 5q (4).

Finer analysis of loss of genetic material has been made possible by techniques analyzing restriction fragment length polymorphisms (RFLPs) and the variable number of tandem repeats (VNTRs). These techniques are used to identify common inherited differences such as point mutations, small deletions, or variation in the number of small tandem repeat elements in the maternal and paternal alleles. This information is then used to identify, in tumor tissues, the loss of the paternal or maternal allele (loss of heterozygosity), which reflects either total or partial chromosomal loss. This loss may be accompanied by duplication of the remaining chromosome. Loss of heterozygosity can result in the formation of a tumor in a person with one normal and one mutant allele of a tumor suppressor gene if it is the normal allele that is lost.

By these means many putative tumor suppressor genes have been identified and mapped on human chromosomes. Several of these suppressor genes have been isolated by determining the general location of the gene as described above and then using positional cloning methods (detailed below) to identify and clone the gene itself. The protein products of these genes have been identified and characterized. Some, such as the product of the retinoblastoma susceptibility gene (*RB1*), the product of the *p53* gene, and the product of the Wilms' tumor (*WT1*) gene, are nuclear proteins that bind DNA and evidently exert at least some of their effects on the transcription of other, largely unidentified, genes. However, not all tumor suppressor gene products are nuclear proteins. The product of the gene called *DCC* (for deleted in colon carcinoma) is located in the cell membrane and appears to be involved in cell-cell adhesion, and the product of the neurofibromatosis type 1 (*NF1*) gene functions in the cytoplasm and appears to activate the *ras* oncogene proteins. How these various tumor suppressor gene products control cellular proliferation is a topic of considerable interest at present, and new information is reported regularly.

In this chapter we will discuss in detail what is known about the structure and function of the products of the two most fully characterized tumor suppressor genes, *RB1* and *p53*. We will also briefly discuss the other known tumor suppressor genes, particularly the adenomatous polyposis coli (*APC*), *DCC*, *WT1*, *NF1*, *NF2*, and von Hippel–Lindau disease (*VHL*) genes, their respective protein products, and the roles of these proteins in the control of cell growth. Finally, we will conclude this chapter with a brief discussion of the clinical applications of our current knowledge of tumor suppressors.

RB1 AND THE Rb PROTEIN

Retinoblastoma

Retinoblastoma, a pediatric retinal tumor, occurs in two forms, familial and sporadic. The sporadic form is characterized by onset within the first 7 years of life and usually involves only a single tumor in one eye. The familial form generally has an onset within the first year of life and involves multiple tumors occurring in both eyes. Parents with the familial form of retinoblastoma pass the disease to 50% of their offspring, and the disease has a penetrance of about 90% to 95% in such families.

The two forms of retinoblastoma were used by Alfred Knudson to test his hypothesis that somatic and inherited mutations could have the same effect (5). The differences in age of onset, number of tumor foci, and heritability of the sporadic and familial forms of the disease demonstrated that familial retinoblastoma occurred with kinetics consistent with only one mutational event, while the sporadic form was consistent with two mutations. Knudson postulated that both forms of retinoblastoma occur by sequential mutation and that the difference between the two forms lies in the nature of the first mutation. In familial retinoblastoma, the first mutation occurs in the germline and so is inherited, while in sporadic retinoblastoma the first mutation occurs in a somatic cell. In both forms of the disease, a second somatic mutation is necessary for tumor formation to occur. That both forms are caused by mutations in the same gene is indicated by the observation that parents with the unilateral (sporadic) form of retinoblastoma can have offspring with the bilateral form.

The *RB1* Gene

The *RB1* gene was located by comparing the chromosomes of cells from retinoblastoma tumors and normal tissues of affected individuals. In the 1970s it was reported that approximately 5% of patients with bilateral retinoblastoma had a deletion in the long arm of chromosome 13 in normal cells (6). The extent of the deletion varied substantially, but the smallest region of overlap between the various deletions was 13q14. Subsequent analysis of tumors from patients with normal karyotypes demonstrated that in about 5% of cases, the tumor cells had deletions of the long arm of chromosome 13 that were not present in the normal tissues and that always included 13q14 (3). Further evidence that the retinoblastoma susceptibility gene was located on 13q14 was provided by the linkage of susceptibility to familial retinoblastoma and the gene for esterase D, which had been previously localized to 13q14 (7).

One fundamental question about cancer predisposition is how a constitutional mutation gives rise to focal tumors. The answer arose in large part from analysis of DNA sequence polymorphisms. It was found that germline mutations can change the DNA sites recognized and cut by bacterial restriction endonucleases and that these differences are inherited in a codominant mendelian manner (8). These differences were termed RFLPs. The number of possible polymorphisms is very large and was increased by the discovery of another large class of DNA sequence polymorphisms, the VNTRs. Using RFLP analysis, Cavenee and associates (9) demonstrated that in most cases of retinoblastoma there was a tumor-specific loss of constitutional heterozygosity consistent with the inactivation of a tumor suppressor gene on chromosome 13. The generation of tumors from a cell with a constitutional mutation therefore required a second mutation involving the loss of the remaining allele of a tumor suppressor gene. This loss in some cases occurred by mitotic error and resulted in loss of the entire chromosome carrying the remaining wild-type allele (Fig. 4.1). Further analysis of familial retinoblastoma cases revealed that the chromosome that was lost had been inherited from the unaffected parent (10), indicating that the chromosome that remained in the tumor carried the inactive allele of the tumor suppressor gene.

The *RB1* gene was identified and cloned using a series of chromosome 13q14-specific probes that were mapped by the use of human cell lines with heterozygous, cytogenetically visible deletions of chromosome 13 (11). These probes were used to screen retinoblastoma tumors from patients with no apparent constitutional deletions in chromosome 13 (12). This approach detected a small number of retinoblastoma tumors with a cytogenetically invisible homozygous deletion of the region homologous to one of the chromosome 13q14 probes. This probe was then used to clone a complementary DNA from a retinal library. In 10% of the retinoblastoma tumors tested, the cDNA detected homozygous deletions. Some of these deletions were intragenic, indicating that this gene, and not

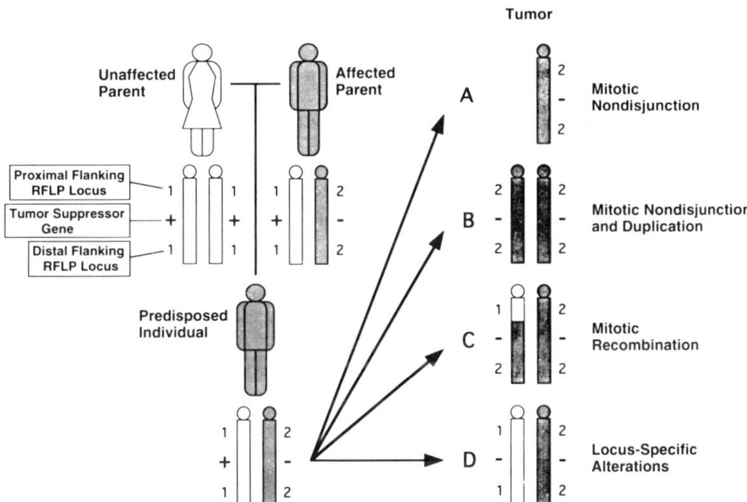

Fig. 4.1: Model for tumorigenesis by loss of expression of a tumor suppressor gene. A child inherits a chromosome carrying a mutation of a tumor suppressor gene. As a result of this mutation, the child is predisposed to tumorigenesis because all the cells in the body carry the mutation. A second somatic event occurs by one of four mechanisms: (A) mitotic nondisjunction, (B) mitotic nondisjunction and duplication of the chromosome carrying the mutant allele, (C) mitotic recombination, or (D) locus-specific alteration, which results in a homozygous mutant genotype in the cell, which then develops into a tumor.

a neighboring one, was involved in retinoblastoma (13). This gene was altered in 30% to 40% of retinoblastoma tumors from patients with both the familial and sporadic forms of the disease, demonstrating that both forms involve predisposing mutations in the same gene.

The *RB1* gene comprises 181 kilobase pairs (kb) of genomic DNA in 27 exons that encode a 4.7-kb messenger RNA. The mRNA codes for a nuclear phosphoprotein of 105 to 110 kDa (henceforth referred to as Rb) found in all cell types examined thus far. When retinoblastoma tumors were examined for expression of this mRNA, 60% to 70% either lacked expression or had altered expression (reviewed in 6). Immunohistochemical analysis revealed that all retinoblastoma tumors either lacked the Rb protein or expressed the protein inappropriately in the cytoplasm (14,15). Also, when *RB1* was introduced into tumor cells that lacked the gene, the cells' growth was inhibited and their tumorigenic potential was completely or partially suppressed (16,17).

Retinoblastoma is not the only tumor that occurs in persons with constitutive mutations in the *RB1* gene. More than 90% of patients with retinoblastoma currently survive the disease, and follow-up of these survivors has revealed that those with the inherited form of retinoblastoma are at risk of developing second cancers, most notably osteosarcomas. The incidence of second cancers may be as high as 15%, approximately 45% of which are osteosarcomas. Survivors of the sporadic (unilateral) form of retinoblastoma do not appear to be at increased risk of second cancers. A high percentage of the osteosarcoma tumors of the survivors of bilateral retinoblastoma were found to have homozygous *RB1* mutations (13,18,19). When patients with sporadic forms of osteosarcoma were examined for alterations in *RB1*, mutations were found in a high percentage of these tumors as well.

Studies of the families of patients with bilateral retinoblastoma have revealed an excess of nonocular cancers among first-degree relatives (20). The most frequent tumors are osteosarcomas and breast and lung cancer; melanoma is also observed at a frequency higher than expected. The sporadic forms of these cancers have been linked with loss of *RB1* expression; mutations in *RB1* have been reported in more than 95% of small cell lung cancers (SCLCs) and in about 20% of non-small cell lung cancers (NSCLCs). Mutations in *RB1* are also seen in about one-third of bladder cancers (21). The presence of *RB1* mutations in the sporadic forms of cancers appearing in families in which retinoblastoma cosegregates with other tumors strongly suggests that the predisposing mutation for these other tumors is likely to be in *RB1*. The familial nonocular tumors therefore appear to reflect the loss of *RB1* expression in other tissues.

The Rb Protein

Structure of Rb The product of the *RB1* gene is a 928 amino acid protein that is phosphorylated at multiple sites and apparently has three major functional domains: the oligomerization domain, located at the amino-terminus; the pocket domain, located between amino acids 379 and 792; and the DNA-binding and transcription factor interaction domain, located at the carboxy-terminus (Fig. 4.2). The pocket domain contains two recognizable protein-protein interaction motifs, a leucine zipper motif and a helix-loop-helix motif. The pocket domain has been shown to be involved in binding to viral oncoproteins and has two subdomains, A and B, separated by a spacer region. The Rb protein is phosphorylated at multiple sites, five of which are phosphorylated by the cdc2 kinase (22). The phosphorylation of Rb is critical to its activity. The viral oncoproteins, cyclins, and transcription factors that bind Rb all bind to the underphosphorylated form of the protein (23–25). As will be discussed below, the biological activity of the Rb protein is apparently regulated by its state of phosphorylation.

Rb and the Cell Cycle Figure 4.3 depicts the mammalian cell cycle. The decision to begin a new round of DNA replication and cell division involves a set of proteins known as the cyclins, which play a role in initiating both the transition between the G_2 and M phases and the transition between the G_1 and S phases in the cell cycle. The cyclins are the

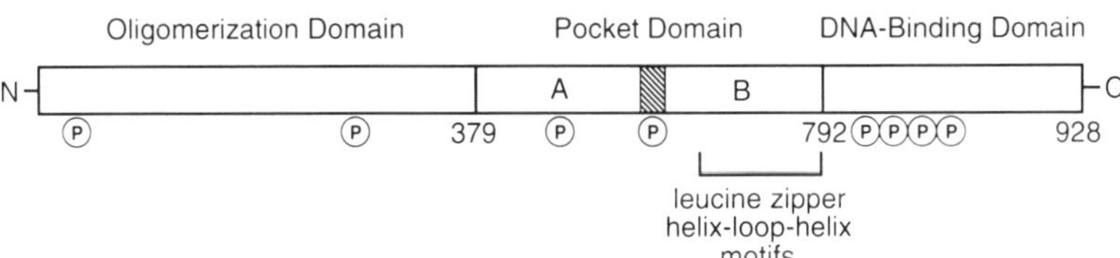

Fig. 4.2: The functional domains of the human Rb1 protein. The protein has 928 amino acids and three functional domains: the oligomerization domain, from the amino-terminus (N) to amino acid 379; the pocket domain, which is involved in interaction with viral oncoproteins and has two subdomains (A and B), from amino acid 379 to 792; and the DNA-binding domain, from amino acid 792 to the carboxy-terminus (C). The B subdomain of the pocket region contains a leucine zipper and a helix-loop-helix motif, both of which are protein-protein interaction motifs. The approximate locations of known phosphorylation sites are denoted by P. The spacer region between the A and B subdomains of the pocket domain is indicated by hatching.

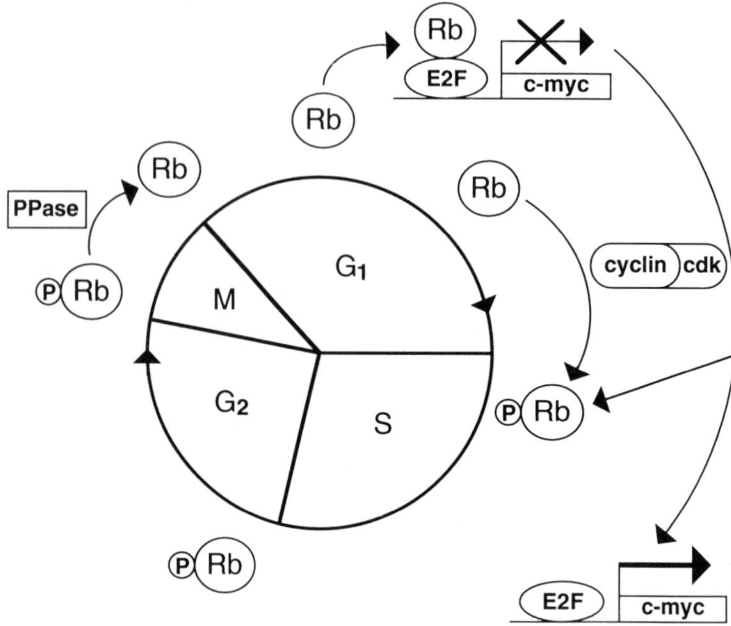

Fig. 4.3: The phosphorylation state of the Rb protein at each stage of the mammalian cell cycle. Unphosphorylated Rb, which can bind to the cellular transcription factor E2F, suppresses cell proliferation by maintaining the cell in G_1. Free E2F binds to and activates the promoter of the c-*myc* gene (which encodes a cellular proliferation factor); when E2F is in a complex with Rb, it is unable to activate transcription of c-*myc*. When the cell becomes committed to begin a new cell cycle and a new round of DNA replication, Rb is phosphorylated (P) by one of the cyclin-dependent kinases (cyclin-cdk), and free E2F is released. Rb becomes dephosphorylated again in M phase by the action of an M-phase phosphatase (PPase).

regulatory subunits of the cyclin-dependent kinases (cdks) and the protein kinase cdc2 (26,27). In mammals, there are three cyclins (cyclin A, cyclin E, and the D-type cyclins) that appear to be involved in the G_1-to-S-phase transition when the cell is committed to another cell cycle and hence to a new round of DNA replication.

Because inactivation of *RB1* leads to tumor formation in a number of cell types, it follows that the normal function of the Rb protein may be to constrain cell growth. In the absence of Rb, cells evidently lose their ability to respond to certain physiologic signals by ceasing to proliferate. Growth control of tumor cells lacking Rb can be restored by introducing normal *RB1* alleles. There are also considerable data to indicate that phosphorylation of Rb is critical to its suppression of cell growth. The phosphorylation of Rb varies as the cell progresses through the cell cycle; Rb is hypophosphorylated through early G_1 and then becomes hyperphosphorylated immediately before the G_1-to-S transition. It remains hyperphosphorylated through S, G_2, and M and loses most of its phosphate groups to become hypophosphorylated again as the cell reenters G_1. These findings, taken together with the observation that viral oncoproteins bind to hypophosphorylated Rb, suggest that the hypophosphorylated form of Rb has an activity that suppresses cell division that can be turned off in late G_1 by phosphorylation or by binding by viral oncoproteins. The discovery that microinjection of purified, unphosphorylated Rb into cells results in arrest of the cell cycle in G_1 provides further support for this hypothesis.

There is much evidence that the cyclin-cdk complexes are responsible for the hyperphosphorylation of Rb and thus for the release of the cell from the growth control imposed by the underphosphorylated form of Rb. The Rb protein is phosphorylated at serine and threonine residues present in sequence motifs similar to those that are phosphorylated by cdks. Overexpression of cyclins A, E, and D1 overcomes Rb suppression of proliferation, and in cells overexpressing cyclins A, E, and D2 the Rb protein is hyperphosphorylated (28,29). The Rb protein binds to one or more of the D-type cyclins; this physical interaction, interestingly, appears to involve the regions of Rb that are involved in binding by viral oncoproteins (25,29). The three D-type cyclins associate with cdk4, which has been shown to phosphorylate Rb in insect cells (30).

Rb and Viral Oncoproteins Four DNA tumor virus oncoproteins have been shown to bind to the retinoblastoma protein: the large-T antigen of simian virus 40 (SV40), the E1A protein of human adenovirus type 5 (Ad5), the E7 protein of human papillomavirus (HPV) types 16 and 18, and the EBNA-5 protein of Epstein-Barr virus (31–34). This interaction between viral growth-promoting oncoproteins and the growth-suppressing Rb suggests a simple mechanism for the oncogenic activity of the tumor virus proteins: that it is the inactivation of Rb function by binding of the viral oncoproteins that abrogates cell cycle control and results in cellular transformation.

As mentioned above, the viral oncoproteins bind to the pocket domain of Rb, and for SV40 T, Ad5 E1A, and HPV E7 this binding involves the hypophosphorylated form of Rb. SV40 T, Ad5 E1A, and HPV E7 have a common amino acid motif in the region essential for binding to Rb, but EBNA-5 appears to bind by a different mechanism. This common motif is also shared by the three D-type cyclins, which also bind to Rb.

Another protein that binds to Rb is the cellular transcription factor E2F, which was first identified as a DNA-binding activity that recognizes two specific sequence elements in the adenovirus E2 promoter and activates E2 expression. E2F binding sites are also present in the promoters of several genes that are activated in mid- to late-G_1 phase, such as *c-myc*, cdc2, and dihydrofolate reductase and other genes essential for DNA synthesis. The ability of E2F to activate transcription of these genes is blocked by the interaction of Rb with E2F (35). E2F appears to play an important role in signaling the cell to proceed into S phase, as overexpression of E2F activates DNA synthesis and microinjection of E2F into quiescent cells induces entry into S phase (36).

The ability of Rb to suppress the activation of E2F-dependent transcription appears to be essential to the growth-suppressing activity of Rb, because mutations (within the pocket and carboxy-terminal domains of Rb) that eliminate Rb binding to E2F also abrogate suppression of growth by Rb during G_1 (37). Significantly, the interaction between Rb and E2F is disrupted by the viral oncoproteins Ad5 E1A, SV40 T, and HPV E7 in vitro, and the E2F-Rb complex is absent in human cervical carcinoma cell lines that express HPV E7, such as HeLa and CaSki, despite the presence of wild-type Rb and E2F (38). It appears that disrupting the Rb-E2F complex is one of the major functions of the viral oncoproteins and part of the means by which they induce tumorigenesis in infected cells. Another target of these viral oncoproteins is the tumor suppressor protein p53; this interaction will be discussed in detail later in this chapter.

Rb as a Transcription Factor Rb activates transcription of the transforming growth factor-β1 (TGF-β1) and TGF-β2 genes and the c-*fos* gene and represses expression of the *c-myc* gene (39,40). TGF-β proteins induce G_1 arrest of many types of cells; therefore, activating these genes would suppress cell proliferation. The repression of c-*myc* appears to involve TGF-β1 and so may be a function of TGF-β induction by Rb. Binding of Rb to the c-*fos* and other promoters appears to require the presence of a promoter sequence termed the Rb control element, which is also bound by the well-characterized transcription factor Sp1, and Rb may regulate these promoters by binding Sp1 as it does E2F (41). Thus, Rb appears to have a role in modulating the expression of several genes involved in cell cycle control. The role is complex, however, because Rb has both positive and negative effects on the expression of a variety of genes and because the effect of Rb on expression of a particular gene tends to vary with cell type.

Rb-Related Proteins Rb is a member of a family of related proteins that cross-react immunologically and that interact with adenovirus E1A. This family includes the p107, p130, and p300 proteins. The best-characterized member of the Rb family (aside from Rb itself) is p107. This protein is also a nuclear phosphoprotein and is very similar to Rb in its structure and protein-binding properties. The gene for p107 has been cloned and mapped to chromosome 20q11.2. Like Rb, p107 binds not only E1A but also SV40 T, HPV E7, and E2F (42). Its role in tumorigenesis, if any, has not yet been established, although p107 appears to inhibit both E2F-mediated transactivation and cell proliferation (43).

The retinoblastoma susceptibility gene was the archetypal tumor suppressor gene and established the existence of this category of genes and their protein products involved in control of cellular proliferation. Since the discovery of *RB1*, intensive investigation of chromosomal deletions characteristic of various cancers has established the existence of a growing group of tumor suppressor genes. Much remains to be learned about the role of Rb in the cell and the disruption of Rb function in human cancers. The existence of a family of Rb-related proteins suggests that control of cell proliferation by these proteins may be much more complex than the current understanding of Rb function would indicate.

THE *p53* TUMOR SUPPRESSOR GENE AND PROTEIN

Unlike the *RB1* gene, the *p53* gene was originally thought to be an oncogene, and only years after its discovery was it found that the oncogenic form of p53 is actually a mutant; the wild-type protein is a tumor suppressor. The notion that p53 was an oncoprotein arose from the observation in the 1970s that antibodies against SV40 large-T antigen from animals bearing tumors produced by SV40-transformed cells coimmunoprecipitated a second protein of apparent molecular mass 53 kDa, called p53. From these observations it was concluded that p53 was encoded by the cell and not by SV40, that it interacted with SV40 large-T antigen, and that it was present in relatively high levels in transformed cells. Because antibodies against p53 were frequently present in animals bearing tumors, p53 was classified as a tumor antigen. It was also observed that the levels of p53 were frequently high in a variety of transformed cells, regardless of how the cells were transformed (44). Normal cells were observed to have low levels of p53 with a half-life of minutes, whereas transformed cells often had much higher levels of p53 with a half-life of several hours, indicating that the main mechanism of *p53* regulation was post-transcriptional.

The conclusion that p53 was an oncoprotein came from studies of the *p53* gene, which was isolated and characterized in the 1980s. Some *p53* cDNAs were able to immortalize cultured cells and contribute to their transformation. The classification of *p53* as an oncogene, however, was called into question by the discovery that the forms of p53 that acted as oncoproteins were mutant forms and that wild-type p53 did not transform cells. Even more significantly, wild-type p53 actually suppressed transformation, blocking oncogenesis. It was then found that in several types of tumors both alleles of the *p53* gene were mutant: one allele usually contained a missense mutation, and the other allele was absent owing to mitotic error or a mutation. In these tumors, therefore, there was a loss of heterozygosity at the *p53* locus, and this pointed to a role for *p53* as a tumor suppressor gene.

Cancers Involving *p53* Mutations

Mutations in the *p53* gene have been implicated in a variety of cancers by several lines of evidence, the most telling being the suppression of tumor cell growth through the introduction of a normal *p53* allele by chromosome 17 transfer or transfection of tumor cells with a *p53* cDNA. In these cases, growth was uniformly arrested in G_1 (45). Of human cancers, 50% to 80% express mutant *p53*, including bone, bladder, brain, breast, esophagus, stomach, liver, lung (SCLC and NSCLC), lymphoid system, ovary, and prostate cancers (44), making *p53* the most frequently mutated gene in human cancer and suggesting that there is some form of selection for the expression of mutant p53 proteins in cancer cells. Cancer cells that have mutant *p53* alleles and whose growth is suppressed by the introduction of wild-type *p53* alleles include colorectal carcinoma, glioblastoma, osteosarcoma, and breast and lung carcinoma.

Mutations in *p53* have also been linked to Li-Fraumeni syndrome, an inherited susceptibility disorder in which affected individuals are at increased risk of developing a variety of cancers including soft tissue sarcoma and cancers of the bone, breast, brain, and genitourinary tract. Affected individuals also tend to develop specific tumors at an earlier age than usual. Attempts to identify the predisposing mutation for Li-Fraumeni syndrome were complicated by the fact that affected individuals frequently die young and by the lack of common cytogenetic alterations in affected individuals (6). However, it was observed that fibroblasts from affected individuals tend to undergo spontaneous immortalization (46), with a concomitant loss of heterozygosity for chromosome 17. Chromosome 17 was also frequently lost in the sporadic forms of the tumors characteristic of Li-Fraumeni syndrome, and, as mentioned above, homozygous *p53* mutations had been implicated in the development of these types of tumors (47). It was then found that the spontaneously immortalized cultured fibroblasts from Li-Fraumeni family members showed a loss of heterozygosity for the region containing the *p53* locus. Sequencing the remaining *p53* allele from these fibroblasts revealed that it was mutated. From these same individuals, fibroblasts that had not yet become immortalized were heterozygous for *p53* mutations, as were other normal cells from the same patients. That heterozygous *p53* mutations are the predisposing event for Li-Fraumeni syndrome has subsequently been confirmed by analysis of many affected individuals from many families with Li-Fraumeni syndrome.

The *p53* Gene and Protein

The 20-kb *p53* gene has been localized to 17p13, consists of 11 exons, and encodes a 2.9-kb mRNA. The p53 protein is 393 amino acids in length and is, like Rb, localized to the cell nucleus and expressed in all tissues tested. The p53 protein is phosphorylated at several serine residues, perhaps by casein kinases I and II, and also appears to associate with a cell-cycle-regulated kinase like cdc2 (44). The p53 protein has three nuclear localization signals, one adjacent to the serine residue that is phosphorylated by cdc2 kinase; phosphorylation of this serine may affect the transportation of p53 from the cytoplasm into the nucleus of the cell (48). About 75% of *p53* mutations are missense mutations (altering only one amino acid), occurring mainly in the most highly conserved regions (44).

Structure of p53 The p53 protein has three domains (see Fig. 4.4): a region rich in acidic amino acids, containing the transcriptional activation region, at the amino-terminus; a highly conserved hydrophobic conformation

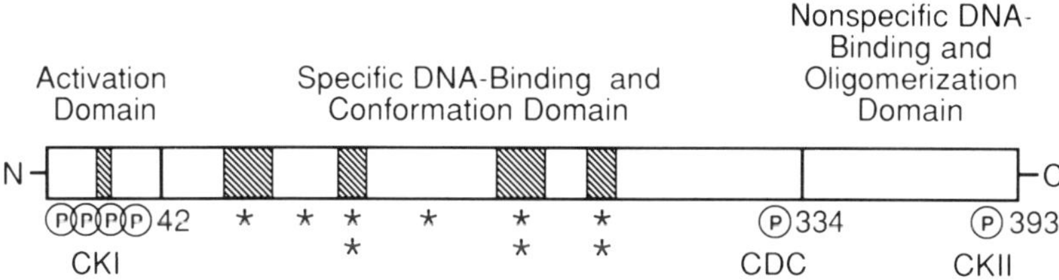

Fig. 4.4: The functional domains of the human p53 protein. The protein has 393 amino acids and three functional domains: the transcriptional activation domain, rich in acidic amino acids, from the amino-terminus (N) to amino acid 42; the sequence-specific DNA-binding and conformation domain, rich in proline residues and highly conserved between species, from amino acid 42 to 334; and the nonspecific DNA-binding and oligomerization domain, from amino acid 334 to the carboxy-terminus (C). The approximate location of amino acids frequently mutated in human cancers is indicated by asterisks; double asterisks indicate amino acids more frequently mutated (53). The hatched areas are highly conserved regions. The approximate position of known phosphorylation sites are indicated (P). CKI, CKII, and CDC indicate sites phosphorylated by casein kinase I, casein kinase II, and cdc2, respectively.

region rich in proline residues, with sequence-specific DNA-binding activity (49–51); and a region rich in basic amino acids at the carboxy-terminus (52,53). The conserved central DNA-binding domain contains most of the oncogenic *p53* mutations, and the carboxy-terminal domain contains nonspecific DNA-binding, protein-oligomerization, and nuclear-localization sequences. Wild-type p53 has a very short half-life in the cell, about 20 minutes. The mutant forms of the protein, in contrast, frequently have a much longer half-life, generally measured in hours, which results in an apparent elevation in *p53* expression.

The active form of the p53 protein appears to be a tetramer (54). This self-association may be responsible for the ability of some mutant forms of p53 to act, at least in cultured cells, in a dominant negative fashion. The presence of a single mutant p53 molecule in a heteromeric p53 tetramer may be sufficient to inactivate the growth regulatory effects of the entire complex. Because some mutations in the *p53* gene produce proteins that are more stable (i.e., have longer half-lives), the mutant forms would tend to be present in higher concentrations than wild-type p53 in the cells of heterozygotes. The mutant p53 protein, but not the wild-type protein, also interacts stably with the protein product of another cellular gene, *hsc70* (6). The formation of these stable complexes may also affect the formation of p53 oligomers so that the mutant and wild-type p53 proteins may become trapped in an inactive complex. The result of expressing both wild-type and mutant p53 in a single cultured cell may thus to a large extent be inactivation of the normal functions of wild-type p53.

Function of p53 When the wild-type *p53* gene is introduced into and expressed in malignant cells whose endogenous *p53* genes are inactivated, the cells often stop growing, indicating that p53 (like Rb) plays a role in the negative regulation of cell proliferation. In some cases, the activation of wild-type p53 has been shown to promote differentiation and to trigger apoptosis, or programmed cell death (55). The current hypotheses for the function of p53 have been reviewed by Prives and Manfredi (53). One hypothesis is that p53 responds to DNA damage by arresting the cell cycle in G_1 until the damage can be repaired or by triggering apoptosis if the damage is severe. That cells with damaged DNA must delay DNA synthesis until the damage is repaired is obvious; replication of damaged DNA could lead to mutations that cause tumorigenicity or, in cancer cells, to further mutations that might increase the malignant potential of the cells or make them resistant to chemotherapeutic agents.

This hypothesis is supported by evidence that overexpression of wild-type p53 blocks cell division in G_1 (44). Furthermore, while cells without p53 or with mutant p53 do not arrest in G_1 after DNA damage, introducing wild-type (but not mutant) *p53* into them reestablishes G_1 arrest in response to DNA damage (56). Cells from patients with the inherited disease ataxia telangiectasia (which causes hypersensitivity to DNA-damaging agents and cancer predisposition) do not increase their p53 levels, as do normal cells, in response to DNA-damaging agents such as ultraviolet light, gamma irradiation, and some cancer chemotherapeutic drugs. Furthermore, one of the targets of p53 is a gene (*GADD45*) whose expression is stimulated in response to DNA damage and which contains a p53-responsive element (48,53).

This latter finding supports the notion that p53 is a transcription factor or a protein that regulates transcription factors in the cell, as suggested by earlier work (reviewed by Levine, 44) including the demonstration by Weintraub and coworkers (57) that p53 activates transcription of the muscle creatine kinase gene. The p53 protein was shown some years ago to bind to double-stranded DNA; more recently this binding was shown to be sequence-specific, and a consensus binding site was identified (48). Analysis of a number of mutant p53 proteins with single point mutations indicated that all had lost the sequence-specific DNA-binding function. In addition, many p53 mutant proteins can act as dominant negatives to inhibit the DNA-binding activity of wild-type p53 (54).

In addition to its role in transcriptional activation, the p53 protein has been shown to repress transcription. The p53 protein suppresses the promoters of c-*fos* and c-*jun*, which are involved in promoting cellular growth, and the promoters of the human β-actin and interleukin-6 genes (48). Because p53 suppresses transcription from a number of unrelated promoters, it is possible that it acts on some general transcription factor. One candidate is the TATA-box binding protein, TBP, which binds to p53. In fact, the binding of TBP by p53 has been shown to repress transcription (58).

Another hypothesis for the function of p53 is that p53 induces cellular differentiation. This hypothesis is supported by evidence that wild-type p53 induces B cells to differentiate (53), and can activate the expression of the muscle creatine kinase gene, which is turned on during muscle cell differentiation (57). Transgenic mice completely lacking p53 have been made, however, and appear to develop normally (59), so the role of p53 in early development may not be essential. However, these p53-null mice are prone to a variety of tumors at a relatively young age (about 6 months); mice that are heterozygous for *p53* deletions also develop these tumors, but less frequently (48).

These hypotheses for the function of p53 in normal cells are not mutually exclusive, and the experimental evidence thus far suggests that in normal cells p53 suppresses the expression of genes like *jun* and *fos* that are involved in growth promotion while activating the expression of genes involved in repair, negative growth control, and differentiation.

Protein p53 and Oncoproteins The p53 protein, like Rb, interacts with the oncoprotein products of the SV40, adenovirus, HPV, and Epstein-Barr DNA tumor viruses. The p53 protein binds to SV40 large-T antigen, adenovirus E1B, HPV protein E6, and the EBNA-5 oncoprotein. This binding (except to EBNA-5) has been shown to inactivate wild-type p53 function (48). It is interesting to note that the region of SV40 large-T antigen that is required for complex formation with p53 is different from the region required for binding to Rb. It is also significant that the E6 proteins of HPV types 16 and 18, which are the very oncogenic forms of HPV implicated in the development of most human squamous cell carcinomas and cervical cancers, have a greater affinity for p53 than do the E6 proteins of other HPV types. These other HPV types, such as types 6 and 11, have low oncogenic potential and are commonly associated with benign lesions such as warts (60).

The interaction of p53 with either SV40 large-T antigen or the adenovirus E1B oncoproteins not only inactivates wild-type p53 but also extends the half-life of p53 and increases the steady-state levels of p53 in cells transformed with these viruses (33). The binding of p53 by SV40 large-T antigen prevents p53 binding to DNA (53). The E6 protein, however, has a different effect on p53. The levels of p53 in HPV-positive cervical carcinoma cell lines are quite low, and in fact HeLa cells (which contain HPV-18) have no detectable p53 (33). In vitro studies have shown that binding of p53 by E6 promotes the degradation of p53 by the ubiquitin-dependent proteolysis system (61).

Another oncoprotein that interacts with p53 is a cellular protein rather than the product of a viral gene. This protein is the product of the murine double minute 2 (*mdm2*) gene. Both rat and human homologues of the *mdm2* gene have been isolated, and both bind to p53 protein. Binding of p53 by mdm2, like p53 binding by the viral oncoproteins discussed above, strongly inhibits the ability of p53 to activate transcription (62). Interestingly, the p53 protein has been shown to activate expression of the *mdm2* gene, possibly indicating that *p53* and *mdm2* are regulated by a feedback loop mechanism (53). The *mdm2* gene and its protein product are especially interesting because *mdm2* has been shown to be amplified and its protein product overexpressed in many human sarcomas, including common bone and soft tissue forms (63), and also in metastatic osteosarcoma (64). In those tumors with amplification of *mdm2*, the *p53* gene has been shown to be wild-type.

Mutant Forms of p53 As mentioned above, mutations in *p53* have been found in 50% to 80% of types of human cancers, suggesting that there is powerful selection for the expression of these mutant proteins. However, mutant p53 does not appear to bind DNA, activate promoters with a binding site for wild-type p53, bind TBP, or suppress the activity of any known promoters (48). Mutant p53 does, however, function as an oncogene and can cooperate with the activated ras oncoprotein to transform cultured cells (44). Recently, it was observed that mutant p53 introduced into p53-null cell lines transformed the cells into a tumorigenic cell line. This result clearly indicates that the introduced p53 had a gain-of-function mutation (65). Therefore, mutant p53 may transform cells by inactivation of wild-type p53 in complexes with mutant p53 (as discussed above), by the oncogenic activity of the mutant protein itself, or by a combination of both activities.

Protein p53 and the WT1 Tumor Suppressor The p53 protein was recently shown to associate physically with the protein product of the Wilms' tumor 1 (*WT1*) gene, a tumor suppressor gene expressed in the developing kidney (66). Inactivation of the *WT1* gene leads to the development of the pediatric kidney cancer known as Wilms' tumor (67). The WT1 protein is a transcription factor that, in the presence of p53, binds to the early growth response gene 1 (*EGR1*) consensus sequence, mediating transcriptional repression. However, in the absence of p53, WT1 *activates*, rather than represses, the *EGR1* gene. In contrast to the effect of p53 on WT1, the WT1 protein *enhances* the ability of p53 to activate the muscle creatine kinase promoter (66). The interaction of the WT1 and p53 proteins, therefore, modulates the transactivating potential of both proteins.

Protein p53 and Multistage Tumorigenesis Although retinoblastoma is an obvious exception, in most cases carcinogenesis appears to require not one but a series of genetic changes. An example is colorectal carcinoma, which appears to develop in about four steps: from normal colonic mucosa to adenoma, to more advanced adenoma, to carcinoma in situ, to frank carcinoma. In small adenomas from patients with no predisposition to colon cancer, loss of the chromosomal region containing the familial adenomatous polyposis (*FAP*) gene is the most common genetic alteration. Such adenomas rarely (in less than 10% of cases) have *ras* mutations, but about 50% of advanced adenomas do. Deletions of the gene called *DCC*, for deleted in colon carcinoma, occur in about 50% of late adenomas and 70% of carcinomas. In contrast, loss of chromosome

17p, which contains the *p53* gene, almost never occurs in adenomas of any stage, including the very large, high-grade adenomas of carcinoma in situ. Loss of 17p is, however, the most frequent event in colorectal carcinomas, occurring in more than 75% of cases (68). It would appear, then, that loss or mutation of *p53* is a late event in the development of colorectal carcinoma.

This is not true of the tumors characteristic of Li-Fraumeni syndrome. That patients with Li-Fraumeni syndrome have germline *p53* mutations strongly suggests that alterations in *p53* are an early event in the generation of their tumors. It may be that *p53* mutations occur at different stages in development of different tumors. It is also possible that it is the accumulation of several genetic alterations, and not their order of occurrence, that is important for tumorigenesis. The expression of mutant p53 proteins in at least a subset of human cancers suggests that mutations in *p53* control the maintenance as well as the initiation of the oncogenic state.

OTHER TUMOR SUPPRESSOR GENES AND THEIR PROTEIN PRODUCTS

Although the *RB1* and *p53* genes and their protein products are the best-characterized tumor suppressors, much has been learned in recent years about an ever-increasing number of other tumor suppressor genes. Several tumor suppressor genes have been cloned, including *WT1*; *APC*; *DCC*; *NF1* and *NF2*; *VHL*; and the multiple endocrine neoplasia type 2A gene, *MEN2A*. Other tumor suppressor genes have been mapped but not yet cloned. The number of heritable conditions that predispose affected individuals to cancer has been estimated at about 50 (69), so this may reflect the minimum number of tumor suppressor genes. Obviously, more tumor suppressor genes remain to be mapped. We will here discuss some of the best-characterized cloned tumor suppressor genes and their protein products.

The *WT1* Gene

Wilms' tumor or nephroblastoma is a relatively common childhood solid tumor that arises in the kidney. Like patients with retinoblastoma, those with Wilms' tumor have a high frequency (10%) of bilateral tumors, which occur earlier than the unilateral tumors and often arise in families. Also like retinoblastoma, in individuals with bilateral Wilms' tumor the kinetics of tumor formation can be explained by a single genetic event, while in sporadic (unilateral) cases the kinetics suggest that there must be two independent genetic events.

In rare cases Wilms' tumor is part of a syndrome of congenital abnormalities called WAGR (for Wilms' tumor, aniridia, genitourinary abnormalities, and mental retardation) that is associated with a deletion of 11p13 (67). Sporadic Wilms' tumors are also associated with loss of heterozygosity for markers on 11p13, indicating that the Wilms' tumor susceptibility gene is within this region. Using a strategy similar to that used to clone *RB1*, a genomic DNA clone was isolated and found to be mutated in Wilms' tumors.

The Wilms' tumor gene, *WT1*, consists of 10 exons spanning 50 kb of DNA and encoding a 3-kb mRNA. The WT1 protein is 46 to 49 kDa in predicted molecular mass; the range may be due to the use of alternate splices. The amino-terminal portion of the protein is very rich in prolines and glutamines, as is the case with a number of transcription factors (70), and the carboxy-terminus contains four cysteine-histidine-type zinc-finger motifs. Such motifs are also a common characteristic of DNA-binding proteins (71). As mentioned above, WT1-p53 complexes bind to the EGR1 consensus sequence and mediate transcriptional repression; in the absence of p53, WT1 functions as a transcriptional activator of the EGR1 site (66).

WT1, unlike Rb and p53, is expressed in few tissues and is developmentally regulated (67). WT1 is found in the kidney, spleen, gonads, and uterus in mouse, and in the mouse kidney *WT1* expression is high during mid to late gestation but drops to low adult levels shortly after birth (72).

Like patients with bilateral retinoblastoma, patients with bilateral Wilms' tumors have constitutional mutations in one allele of *WT1* and have subsequently lost the normal allele in the tumors (73). The situation for Wilms' tumor is complicated, however, by the observation that some Wilms' tumors have also lost one copy of another portion of chromosome 11 (11p15) and by the observation that, for the familial forms of Wilms' tumor, both 11p13 and 11p15 have been excluded as the site of the *WT1* gene. This may indicate the existence of a second and possibly a third Wilms' tumor gene (67).

The *APC* Gene

Familial adenomatous polyposis (FAP) is a rare autosomal dominant syndrome characterized by the presence of hundreds to thousands of benign polyps in the large bowel. The polyps appear during late adolescence and are very likely to develop into adenocarcinoma of the colon (6). Patients with FAP have tumor-specific loss of heterozygosity on several chromosomes, including chromosomes 5, 17, and 18. Linkage analysis in FAP families has revealed that the predisposing locus is on chromosome 5 (74,75). Loss of heterozygosity at 5q21 has been observed in patients with no predisposition to colorectal cancer, and allelic losses of 5q are the most common genetic alteration in small, early adenomas in colon cancer patients without FAP (68).

Two cDNAs deleted in FAP patients were isolated from chromosome 5q21; the first was called *MCC*, for mutated in colon cancer, and the second *APC*, for adenomatous polyposis coli. No germline mutations in *MCC* have yet been detected in patients with FAP, but several germline *APC* mutations were detected (76,77). Mutations in *APC* have also been observed in tumors from patients with sporadic colorectal carcinoma (77). The *APC* gene has thus been implicated both in FAP and in sporadic colorectal carcinoma.

The protein product of the *APC* gene appears to be located in the cytoplasm, unlike the tumor suppressor proteins so far described in this chapter. The function of this protein is not yet understood, but effects of mutations in the *APC* gene have been reported. Loss of chromosome 5q occurs only rarely in FAP patients, indicating that the mutant allele may exert its effects through a reduction in the amount of the product of the *APC* gene. Patients with FAP have hyperproliferation of the colon lining (78), suggesting that the amount of protein produced by only one wild-type allele is insufficient to regulate colonic cell growth (6). This stimulation of growth of the colonic epithelium would not by itself produce polyps; additional somatic mutations in the cells of the epithelium are necessary for polyp formation. Mutations in *APC* may therefore be linked to tumorigenesis by causing hyperproliferation of colonic cells, which are then susceptible to further genetic alterations. This notion is supported by the observation mentioned above that mutations in *APC* are the most common genetic alteration observed in small adenomas of the colonic epithelium.

The *DCC* Gene

Allelic losses involving chromosome 18q have been observed in more than 70% of colorectal carcinomas, in almost 50% of late adenomas, and in very few early adenomas (4). A tumor suppressor gene was cloned from 18q21-qter and termed *DCC* (deleted in colorectal carcinoma). This gene spans at least 400 kb and encodes an mRNA of about 10 to 12 kb. The gene encodes a putative transmembrane protein with four immunoglobulin-like domains, multiple repeats of a fibronectin type III–related domain, and a cytoplasmic domain, and has significant homology to the neural cell adhesion molecule family (68). The product of the *DCC* gene may thus be a cell-cell adhesion molecule located in the outer membrane of the cell (79).

Loss of the *DCC* gene also appears to be involved in other cancers, such as gastric, pancreatic, and esophageal carcinomas. It was shown to be lost in many prostate tumors and to be underexpressed in most such tumors (80). *DCC* is also expressed in the nervous system (81). Thus far, the *DCC* gene is the only tumor suppressor gene for which no germline mutation has been found (82).

The *NF1* and *NF2* Genes

Neurofibromatosis is a common autosomal dominant disorder and can be subdivided into two distinct syndromes that are caused by mutations in two separate genes on different chromosomes. Von Recklinghausen neurofibromatosis (NF1) is the more common form (occurring in one in 4000 people) and is characterized by multiple café au lait spots, neurofibromas, Lisch nodules of the iris, and predisposition to certain malignant tumors. Neurofibromatosis type 2 (NF2) is much less common (occurring in one in 40,000 people) and is associated with bilateral schwannomas and increased risk of meningiomas. The *NF1* gene has been mapped to chromosome 17q11.2 and *NF2* to 22q12 (69,83).

NF1 and *NF2* both encode proteins that are not nuclear; the NF1 protein, called neurofibromin, is cytoplasmic and has been shown to associate with cytoplasmic microtubules, suggesting that neurofibromin may be involved in microtubule-mediated intracellular signal-transduction pathways (84). The NF1 protein has a region homologous to the GTPase-activating protein and appears to stimulate $p21^{ras}$ to hydrolyze GTP to GDP and hence inactivate the ras proteins (85). *NF1* appears to produce two differentially spliced mRNAs, encoding proteins of 2818 and 2839 amino acids with molecular masses of about 317 and 319 kDa, respectively (86). The NF2 protein has a predicted size of 587 amino acids (69 kDa) and may be located in the outer membrane of the cell, as is the product of the *DCC* gene. The NF2 protein is a member of the family of proteins that connect the cell membrane to the cytoskeleton, a structure often disorganized in tumor cells (87).

The *VHL* Gene

The von Hippel–Lindau (VHL) syndrome is a dominantly inherited susceptibility to hemangioblastomas of the central nervous system, renal cell carcinomas, and pheochromocytomas. The incidence of VHL is about one in 36,000. The *VHL* gene is located on chromosome 3p25, which is the minimal region of overlap of chromosome 3p deletions that have been detected in SCLC (83). The gene codes for two mRNAs, of 6 and 6.5 kb, which are both expressed in all adult tissues tested, including brain and kidney. The function of the protein is not yet understood, but homology to membrane proteins suggests that the VHL protein may be localized on the cell membrane and may be involved in signal transduction or cell adhesion (88).

OPPOSING ROLES OF TUMOR SUPPRESSOR PROTEINS AND ONCOPROTEINS

It has become clear that there are complex interactions between the protein products of tumor suppressor genes and the various oncogenes. As has been mentioned, the retinoblastoma and p53 proteins interact with the viral oncoproteins SV40 large-T antigen, adenovirus E1A and E1B, HPV types 16 and 18 proteins E6 and E7, and the EBNA-5 protein of Epstein-Barr virus. These interactions modulate the activity of the tumor suppressor proteins. The NF1 protein is also involved with an oncoprotein; it regulates the activity of the $p21^{ras}$ proteins, which are capable of transforming some types of mammalian cells. Many human cancers express mutant forms of $p21^{ras}$ (85). There is some evidence that the Rb protein interacts with cyclin D1 and regulates its activity (25). Cyclin D1, interestingly, appears to be the product of the *bcl-1* protooncogene (89,90) and is overexpressed in a variety of tumors (91).

The different classes of oncoproteins therefore appear to have counterparts in the various classes of tumor suppressor genes, and the protein products of these two groups of genes, for the most part, seem to act in opposite directions: the oncoproteins stimulate cell proliferation, and the tumor suppressor proteins suppress proliferation and promote differentiation. When the tumor suppressor proteins fail, through mutation or the action of tumor virus proteins, to oppose the actions of the oncoproteins, tumorigenesis is often the result.

CLINICAL APPLICATIONS OF RESEARCH ON TUMOR SUPPRESSOR GENES

The current knowledge of tumor suppressor genes and the link between mutations in them and predisposition to certain cancers has considerable significance for genetic counseling. Knowledge of genetic polymorphisms makes it possible in many cases to determine whether an individual carries a germline mutation in a tumor suppressor gene by examining the DNA of the individual and his or her relatives, including those who are known to possess the mutation. This approach is not informative in all cases, but in many instances it can reveal an individual's predisposition to certain cancers prevalent in the family.

Numerous studies have indicated a correlation between mutations in tumor suppressor genes and poor prognosis for survival of cancer patients. Certain mutations in *p53*, for example, correlate with quicker relapse and decreased long-term survival for patients with breast cancer (92). Mutations in *RB1* correlate with poor prognosis for patients with bladder cancers (21). As mentioned above, survivors of bilateral retinoblastoma have an increased risk of second cancers, most notably osteosarcoma. Knowledge of the mutation status of the tumor suppressor genes of a cancer patient may, therefore, be used to determine whether to employ an aggressive anticancer therapy. Determining whether patients with sporadic tumors have mutations in particular tumor suppressor genes may be of great use in determining the optimal treatment for them.

With the current knowledge of tumor suppressor genes and the existing arsenal of genetic engineering techniques, it is possible to envision the development of therapeutic approaches to restore tumor suppressor function in patients with mutations in these vital genes. Methods are already being tested that may eventually be used to deliver a wild-type tumor suppressor gene via a viral vector to cells lacking the gene. Adenovirus is a promising vector as it can efficiently infect nondividing cells and express large amounts of gene products; however, infection with adenovirus leads to cell lysis. Other viral vectors, including retroviruses, herpesvirus, and vaccinia virus, may be applicable to gene therapy. Nonviral gene-delivery methods are also being developed, including ligand-DNA conjugates, lipofection, and adenovirus-ligand-DNA conjugates (93). Such methods have already been tested for treatment of cystic fibrosis (94). New approaches to gene therapy are being continually developed and tested. As more is learned about tumor suppressor genes, their protein products, and the functions of these proteins in normal and tumor cells, more candidates for gene therapy will become available.

Research in the exciting field of the tumor suppressor genes and their protein products promises to increase understanding of the regulation of cell growth and differentiation pathways, the essential function of the cell. Tumor suppressor genes and their counterparts, the oncogenes, evidently function in an interlocking relationship, the tumor suppressors acting to suppress cell proliferation and promote cell differentiation, and the oncogenes acting to suppress differentiation and promote proliferation. Investigation of the complex interplay between tumor suppressor genes and oncogenes has provided much information about the genetic alterations that cause a cell to become tumorigenic and has, in addition, suggested promising approaches to the treatment of cancer.

REFERENCES

1. Bos JL. ras oncogenes in human cancer: a review. Cancer Res 1989;49:4682–4689.

2. Stanbridge EJ. Suppression of malignancy in human cells. Nature 1976;260:17–20.

3. Balaban G, Gilbert F, Nichols W, Meadows AT, Shield J. Abnormalities of chromosome 13 in retinoblastoma from individuals with normal constitutional karyotypes. Cancer Genet Cytogenet 1982;6:213–221.

4. Vogelstein B, Fearon ER, Hamilton SR, et al. Genetic alterations during colorectal tumor development. N Engl J Med 1988;319:525–532.

5. Knudson AG. Mutation and cancer: statistical study of retinoblastoma. Proc Natl Acad Sci USA 1971;68:820–823.

6. Hansen MF. Genetic predisposition and cancer. In: Hodges GM, Rowlatt C, eds. Developmental biology and cancer. Boca Raton, Florida: CRC Press, 1994.

7. Sparkes RS, Murphree AL, Lingua RW, et al. Gene for hereditary retinoblastoma assigned to human chromosome 13 by linkage to esterase D. Science 1983;219:971–973.

8. Hansen MF, Cavenee WK. Genetics of cancer predisposition. Cancer Res 1987;47:5518–5527.

9. Cavenee WK, Dryja TP, Phillips RA, et al. Expression of recessive alleles by chromosomal mechanisms in retinoblastoma. Nature 1983;305:779–784.

10. Cavenee WK, Hansen MF, Nordenskjold M, et al. Genetic origin of mutations predisposing to retinoblastoma. Science 1985;228:501–503.

11. Lalande M, Dryja TP, Schreck RR, Shipley J, Flint A, Latt SA. Isolation of human chromosome 13–specific DNA sequences cloned from flow sorted chromosomes and potentially linked to the retinoblastoma locus. Cancer Genet Cytogenet 1984;13:283–295.

12. Dryja TP, Rapaport JM, Joyce JM, Petersen RA. Molecular detection of deletions involving band q14 of chromosome 13 in retinoblastomas. Proc Natl Acad Sci USA 1986;83:7391–7394.

13. Friend SH, Bernards R, Rogeli S, et al. A human DNA segment with properties of the gene that predisposes to retinoblastoma and osteosarcoma. Nature 1986;323:643–646.

14. Xu HJ, Hu SX, Hashimoto T, Takahashi R, Benedict WF. The retinoblastoma susceptibility gene product: a characteristic pattern in normal cells and abnormal expression in malignant cells. Oncogene 1989;4:807–812.

15. Xu HJ, Sumegi J, Hu SX, et al. Intraocular tumor formation in RB reconstituted retinoblastoma cells. Cancer Res 1991;51:4481–4485.

16. Huang HJ, Yee JK, Shew JY, et al. Suppression of the neoplastic phenotype by replacement of the RB gene in human cancer cells. Science 1988;242:1563–1566.

17. Bookstein R, Shew JY, Chen PL, Scully P, Lee WH. Suppression of tumorigenicity of human prostate carcinoma cells by replacing a mutated RB gene.

Science 1990;247:712–715.
18. Hansen MF, Koufos A, Gallie BL, et al. Osteosarcoma and retinoblastoma: a shared chromosomal mechanism revealing recessive predisposition. Proc Natl Acad Sci USA 1985;82:6216–6220.
19. Hansen MF. Molecular genetics considerations in osteosarcoma. Clin Orthop 1991;270:237–246.
20. Strong LC, Herson J, Haas C, et al. Cancer mortality in relatives of retinoblastoma patients. J Natl Cancer Inst 1984;73:303–311.
21. Benedict WF. Altered RB expression is a prognostic clinical marker involved in human bladder tumorigenesis. J Cell Biochem Suppl 1992;16I:69–71.
22. Lees JA, Buchkovich KJ, Marshak DR, Anderson CW, Harlow E. The retinoblastoma protein is phosphorylated on multiple sites by human cdc2. EMBO J 1991;10:4279–4290.
23. Bandara LR, Adamczewski JP, Hunt T, La Thangue NB. Cyclin A and the retinoblastoma gene product complex with a common transcription factor. Nature 1991;352:249–251.
24. Weinberg RA. The retinoblastoma gene and gene product. In: Levine AJ, Franks LM, eds. Tumour suppressor genes, the cell cycle and cancer, vol 12. Cold Spring Harbor, New York: Cold Spring Harbor Laboratory Press, 1992.
25. Dowdy SF, Hinds PW, Louie K, Reed SI, Arnold A, Weinberg RA. Physical interaction of the retinoblastoma protein with human D cyclins. Cell 1993;73:499–511.
26. Hunt T. Cyclins and their partners: from a simple idea to complicated reality. Semin Cell Biol 1991;2:213–222.
27. Reed SI. The role of p^{34} kinases in the G_1 to S-phase transition. Annu Rev Cell Biol 1992;8:529–561.
28. Hinds PW, Mittnacht S, Dulic V, Arnold A, Reed SI, Weinberg RA. Regulation of retinoblastoma protein functions by ectopic expression of human cyclins. Cell 1992;70:993–1006.
29. Ewen ME, Sluss HK, Sherr CJ, Matsushime H, Kato J-Y, Livingston DM. Functional interactions of the retinoblastoma protein with mammalian D-type cyclins. Cell 1993;73:487–497.
30. Kato J-Y, Matsushime H, Hiebert SW, Ewen ME, Sherr CJ. Direct binding of cyclin E to the retinoblastoma gene product and pRb phosphorylation by the cyclin D–dependent kinase, cdk4. Genes Dev 1993;7:331–342.
31. Dyson N, Harlow E. Adenovirus E1A targets key regulators of cell proliferation. In: Levine AJ, Franks LM, eds. Tumour suppressor genes, the cell cycle and cancer, vol 12. Cold Spring Harbor, New York: Cold Spring Harbor Laboratory Press, 1992.
32. Livingston DM. Functional analysis of the retinoblastoma gene product and of RB-SV40 T antigen complexes. In: Levine AJ, Franks LM, eds. Tumour suppressor genes, the cell cycle and cancer, vol 12. Cold Spring Harbor, New York: Cold Spring Harbor Laboratory Press, 1992.
33. Münger K, Scheffner M, Huibregtse JM, Howley PM. Interactions of HPV E6 and E7 oncoproteins with tumour suppressor gene products. In: Levine AJ, Franks LM, eds. Tumour suppressor genes, the cell cycle and cancer,

vol 12. Cold Spring Harbor, New York: Cold Spring Harbor Laboratory Press, 1992.

34. Szekeley L, Selivanova G, Magnusson KP, Klein G, Wiman KG. EBNA-5, an Epstein-Barr virus-encoded nuclear antigen, binds to the retinoblastoma and p53 proteins. Proc Natl Acad Sci USA 1993;90:5455–5459.

35. Bagchi S, Weinmann R, Raychaudhuri P. The retinoblastoma protein copurifies with E2F-I, an E1a-regulated inhibitor of the transcription factor E2F. Cell 1991;65:1063–1072.

36. Johnson DG, Schwarz JK, Cress WD, Nevins JR. Expression of transcription factor E2F1 induces quiescent cells to enter S phase. Nature 1993;365:349–352.

37. Hiebert SW. Regions of the retinoblastoma gene product required for its interaction with the E2F transcription factor are necessary for E2 promoter repression and pRb-mediated growth suppression. Mol Cell Biol 1993;13:3384–3391.

38. Chellappan S, Kraus VB, Kroger B, et al. Adenovirus E1A, simian virus 40 tumor antigen, and human papillomavirus E7 protein share the capacity to disrupt the interaction between transcription factor E2F and the retinoblastoma gene product. Proc Natl Acad Sci USA 1992;89:4549–4553.

39. Hollingsworth RE Jr, Chen P-L, Lee W-H. Integration of cell cycle control with transcriptional regulation by the retinoblastoma protein. Curr Opin Cell Biol 1993;5:194–200.

40. Horowitz JM. Regulation of transcription by the retinoblastoma protein. Genes Chromosome Cancer 1993;6:124–131.

41. Udvadia AJ, Rogers KT, Higgins PDR, et al. Sp-1 binds promoter elements regulated by the RB protein and Sp-1-mediated transcription is stimulated by RB coexpression. Proc Natl Acad Sci USA 1993;90:3265–3269.

42. Cao L, Faha B, Dembski M, Tsai L-H, Harlow E, Dyson N. Independent binding of the retinoblastoma protein and p107 to the transcription factor E2F. Nature 1992;355:176–179.

43. Zhu L, van den Heuvel S, Helin K, et al. Inhibition of cell proliferation by p107, a relative of the retinoblastoma protein. Genes Dev 1993;7:1111–1125.

44. Levine AJ. The *p53* tumour suppressor gene and product. In: Levine AJ, Franks LM, eds. Tumour suppressor genes, the cell cycle and cancer, vol 12. Cold Spring Harbor, New York: Cold Spring Harbor Laboratory Press, 1992.

45. Stanbridge EJ. Functional evidence for human tumor suppressor genes: chromosome and molecular genetic studies. In: Levine AJ, Franks LM, eds. Tumour suppressor genes, the cell cycle and cancer, vol 12. Cold Spring Harbor, New York: Cold Spring Harbor Laboratory Press, 1992.

46. Bischoff FZ, Yim SO, Pathak S, et al. Spontaneous abnormalities in normal fibroblasts from patients with Li-Fraumeni cancer syndrome: aneuploidy and immortalizaton. Cancer Res 1990;50:7979–7984.

47. Hollstein M, Sidransky D, Vogelstein B, Harris CC. p53 mutations in human cancers. Science 1991;253:49–53.

48. Zambetti GP, Levine AJ. A comparison of the biological activities of wild-type and mutant p53. FASEB J 1993;7:855–865.

49. Bargonetti J, Manfredi JJ, Chen X, Marshak DR, Prives C. A proteolytic

fragment from the central region of p53 has marked sequence-specific DNA-binding activity when generated from wild-type but not from oncogenic mutant p53 protein. Genes Dev 1993;7:2565–2574.

50. Pavletich NP, Chambers KA, Pabo CO. The DNA-binding domain of p53 contains the four conserved regions and the major mutation hot spots. Genes Dev 1993;7:2556–2564.

51. Wang Y, Reed M, Wang P, et al. p53 domains: identification and characterization of two autonomous DNA-binding regions. Genes Dev 1993;7:2575–2586.

52. Raycroft L, Wu HY, Lozano G. Transcriptional activation by wild-type but not transforming mutants of the p53 anti oncogene. Science 1990;249:1049–1051.

53. Prives C, Manfredi JJ. The p53 tumor suppressor protein: meeting review. Genes Dev 1993;7:529–534.

54. Lane DP. p53, guardian of the genome. Nature 1992;358:15–16.

55. Fesus L. Biochemical events in naturally occurring forms of cell death. FEBS Lett 1993;328:1–5.

56. Kuerbitz SJ, Plunkett BS, Walsh WV, Kastan MB. Wild-type p53 is a cell cycle checkpoint determinant following irradiation. Proc Natl Acad Sci USA 1992;89:7491–7495.

57. Weintraub H, Hauschka S, Tapscott SJ. The MCK enhancer contains a *p53* responsive element. Proc Natl Acad Sci USA 1991;88:4570–4571.

58. Seto E, Usheva A, Zambetti GP, et al. Wild-type p53 binds to the TATA-binding protein and represses transcription. Proc Natl Acad Sci USA 1992;89:12028–12032.

59. Donehower LA, Harvey M, Slagle BL, et al. Mice deficient for p53 are developmentally normal but susceptible to spontaneous tumors. Nature 1992;56:215–221.

60. Tyring SK. Human papillomaviruses in skin cancer. Cancer Bull 1993;45:212–219.

61. Scheffner M, Werness BA, Huibregtse JM, Levine AJ, Howley PM. The E6 oncoprotein encoded by the human papillomavirus types 16 and 18 promotes the degradation of p53. Cell 1990;63:1129–1136.

62. Momand J, Zambetti GP, Olson DC, George D, Levine AJ. The *mdm-2* oncogene product forms a complex with the p53 protein and inhibits p53-mediated transactivation. Cell 1992;69:1237–1245.

63. Oliner JD, Kinzler KW, Meltzer PS, George DL, Vogelstein B. Amplification of a gene encoding a p53-associated protein in human sarcomas. Nature 1992;358:80–83.

64. Ladanyi M, Cha C, Lewis R, Jhanwar SC, Huvos AG, Healey JH. *MDM2* gene amplification in metastatic osteosarcoma. Cancer Res 1993;53:16–18.

65. Dittmer D, Pati S, Zambetti G, et al. Gain of function mutations in p53. Nature Genetics 1993;4:42–46.

66. Maheswaran S, Park S, Bernard A, et al. Physical and functional interaction between WT1 and p53 proteins. Proc Natl Acad Sci USA 1993;90:5100–5104.

67. Haber DA, Housman DE. Role of the *WT1* gene in Wilms' tumor. In: Levine AJ, Franks LM, eds. Tumour suppressor genes, the cell cycle and cancer, vol 12. Cold Spring Harbor, New York: Cold Spring Harbor Laboratory Press, 1992.

68. Fearon ER. Genetic alterations underlying colorectal tumorigenesis. In: Levine AJ, Franks AM, eds. Tumour suppressor genes, the cell cycle and cancer, vol 12. Cold Spring Harbor, New York: Cold Spring Harbor Laboratory Press, 1992.

69. Knudson AG. All in the (cancer) family. Nature Genet 1993;5:103–104.

70. Mitchell PJ, Tjian T. Transcription regulation in mammalian cells by sequence-specific DNA binding proteins. Science 1989;245:371–378.

71. Evans R, Hollenberg S. Zinc fingers: gilt by association. Cell 1988; 52:1–3.

72. Buckler AJ, Pelletier J, Haber DA, Glaser T, Housman DE. Isolation, characterization and expression of the murine Wilms' tumor gene (WT1) during kidney development. Mol Cell Biol 1991;11:1707–1712.

73. Huff V, Miwa H, Haber DA, et al. Evidence for WT1 as a Wilms' tumor (WT) gene: intragenic germinal deletion in bilateral WT. Am J Hum Genet 1991;48:997–1003.

74. Bodmer WF, Bailey CJ, Bodmer J, et al. Localization of the gene for familial adenomatous polyposis on chromosome 5. Nature 1987;328:614–616.

75. Leppert M, Dobbs M, Scambler P, et al. The gene for familial polyposis coli maps to the long arm of chromosome 5. Science 1987;238:1411–1413.

76. Groden J, Thliveris A, Samowitz W, et al. Identification and characterization of the familial adenomatous polyposis coli gene. Cell 1991;66:589–600.

77. Nishisho I, Nakamura Y, Miyoshi Y, et al. Mutations of chromosome 5q21 genes in FAP and colorectal cancer patients. Science 1991;253:665–669.

78. Lipkin M. Colonic cell proliferation in familial polyposis. Semin Surg Oncol 1987;3:165–170.

79. Marx J. Learning how to suppress cancer. Science 1993;261:1385–1387.

80. Gao Z, Honn KV, Grignon D, Sakr W, Chen YQ. Frequent loss of expression and loss of heterozygosity of the putative tumor suppressor gene DCC in prostatic carcinomas. Cancer Res 1993;53:2723–2727.

81. Lawlor KG, Narayanan R. Persistent expression of the tumor suppressor gene DCC is essential for neuronal differentiation. Cell Growth Differ 1992;3:609–616.

82. Fearon ER, Cho KR, Nigro JM, et al. Identification of a chromosome 18q gene that is altered in colorectal cancers. Science 1990;247:49–56.

83. Lasko D, Cavenee W, Nordenskjöld M. Loss of constitutional heterozygosity in human cancer. Annu Rev Genet 1991;25:281–314.

84. Gregory PE, Gutmann DH, Mitchell A, et al. Neurofibromatosis type 1 gene product (neurofibromin) associates with microtubules. Somat Cell Mol Genet 1993;19:265–274.

85. Polakis P, McCormick F. Interactions between p21^{ras} proteins and their GTPase activating proteins. In: Levine AJ, Franks LM, eds. Tumour suppressor genes, the cell cycle and cancer, vol 12. Cold Spring Harbor, New York: Cold

Spring Harbor Laboratory Press, 1992.
86. Bernards A, Haase VH, Murthy AE, Menon A, Hannigan GE, Gusella JF. Complete human NF1 cDNA sequence: two alternatively spliced mRNAs and absence of expression in a neuroblastoma cell line. DNA Cell Biol 1992;11:727–734.
87. Trofatter JA, MacCollin MM, Rutter JL, et al. A novel moesin-, ezrin-, radixin-like gene is a candidate for the neurofibromatosis 2 tumor suppressor. Cell 1993;72:791–800.
88. Latif F, Tory K, Gnarra J, et al. Identification of the von Hippel–Lindau disease tumor suppressor gene. Science 1993;260:1317–1320.
89. Williams ME, Swerdlow SH, Rosenberg CL, Arnold A. Chromosome 11 translocation breakpoints at the PRAD1/cyclin D1 gene locus in centrocytic lymphoma. Leukemia 1993;7:241–245.
90. Rosenberg CL, Motokura T, Kronenberg HM, Arnold A. Coding sequence of the overexpressed transcript of the putative oncogene PRAD1/cyclin D1 in two primary human tumors. Oncogene 1993;8:519–521.
91. Motokura T, Bloom T, Kim HG, et al. A BCL-1 linked candidate oncogene which is rearranged in parathyroid tumors encodes a novel cyclin. Nature 1991;350:512–518.
92. Thorlacius S, Borresen AL, Eyfjord JE. Somatic p53 mutations in human breast carcinomas in an Icelandic population: a prognostic factor. Cancer Res 1993;53:1637–1641.
93. Mulligan RC. The basic science of gene therapy. Science 1993;260:926–932.
94. Gao L, Wagner E, Cotten M, et al. Direct in vivo gene transfer to airway epithelium employing adenovirus-polylysine-DNA complexes. Hum Gene Therapy 1993;4:17–24.

CHAPTER 5

The HER-2/*neu* Gene in Human Cancers

Dihua Yu
Mien-Chie Hung

THE HER-2/*neu* GENE AND ITS PRODUCT

Injection of female BDIX rats with the chemical carcinogen *N*-ethyl-*N*-nitrosourea (a direct-acting alkylating agent) on day 15 after conception led to 50% reduction in the litter size. Between 4 and 10 months after birth, approximately half the offspring had symptoms of extreme nervous system disorders, and 93% of these animals developed central nervous system tumors such as neuroblastomas or glioblastomas (1). Clonal cell lines derived from the tumors were named B103, B104, etc. DNAs extracted from some of these neuroblastoma or glioblastoma cell lines were able to induce transformation of mouse embryo fibroblast NIH3T3 cells in a calcium phosphate focus-forming assay; in addition, the primary transfectants were tumorigenic in newborn NFS mice (2). Secondary transfectants were also developed in a subsequent cycle of transfection with the DNAs from the primary foci, and they induced tumors (fibrosarcomas) when injected into nude mice, which indicated the successful transfer of a transforming gene (3). The term *neu* was applied to this potential transforming oncogene, which was responsible for the original induction of rat neuroblastomas.

The *neu* gene is homologous to the avian erythroblastosis virus (AEV) transforming gene v-*erb*B (4) and homologous to, but distinct from, the epidermal growth factor receptor (EGF-R) gene, which is also related to the *erb*B gene (5–7). The mutation-activated *neu* oncogene was cloned from the rat neuroblastoma lines B103 and B104 (8), while its normal cellular counterpart, the *neu* protooncogene, was cloned from normal BDIX rat liver DNA by using v-*erb*B to probe genomic libraries (9).

The human homologue of the rat *neu* gene was cloned independently by several laboratories and by virtue of its homology with the retrovirally transduced oncogene v-erbB. The cellular counterpart of v-*erb*B is the c-erbB-1 gene, which encodes the epidermal growth factor receptor (EGF-R). The human *neu* gene was therefore named c-*erb*B2 or HER-2 or MAC117 or NGL (6,10–12). The third and fourth members of the *erb*B-related gene were cloned and were named c-erbB3 or HER3 and c-*erb*B4 or HER4, which encode predicted proteins with striking structural similarities to other members of this family (13–15).

The human HER-2/*neu* gene was mapped to chromosome 17q21 (6,7). It contains an open reading frame of 3765 nucleotides and has a major transcript of 4.8 kb (6,7,12), which translates into a 1255 amino acid polypeptide (6,10,11). The functional gene product of the HER-2/*neu* gene was termed p185 in accordance with its molecular weight (185,000 daltons), while the primary translation product has relative molecular mass (M_r) of 137,895 daltons (3,16). This difference in apparent molecular masses is due in part to *N*-linked glycosylation and probably *O*-linked oligosaccharides as well (17). Phosphorylation has also been found to be involved in the modification of the HER-2/*neu*-encoded protein (Kiyokawa and M.-C. Hung, unpublished observation).

The HER-2/*neu* gene sequence and its protein product are closely related to EGF-R and show an overall homology of 50% (5,10,18). Similar to EGF-R, *neu*-encoded p185 is a transmembrane glycoprotein having an intrinsic tyrosine kinase activity (6,17,19), and it could be grouped as a member of the growth factor receptor tyrosine kinase (RTK) gene family (20). The p185 consists of the following domains: 1) 16 of the first 19 amino acids following the first ATG codon are hydrophobic residues, thus representing a cleaved *signal peptide sequence*; 2) 640 residues constitute the *extracellular ligand-binding domain*, containing two cysteine-rich regions that may be important for ligand binding; 3) a second hydrophobic stretch from residues 650 to 680 suggests a *transmembrane domain*, and 4) the remaining carboxy-terminal 580 amino acids constitute the *intracellular/cytoplasmic domain* (6,10,18).

The extracellular domain of $p185^{neu}$ shows 44% homology with the ligand-binding domain of the EGF-R. Within the ligand-binding domain are two cysteine-rich regions that are completely conserved between these two erbB-related proteins. The cytoplasmic portion of the protein encompasses residues 727–986 and contains the tyrosine kinase domain with the ATP binding site. Some important structures in this region that are identical between the two proteins include threonine (Thr) 954, which is the site of phosphorylation mediated by protein kinase C (PKC), and several tyrosine residues (Y1023, Y1248, Y1139, and Y1222) for autophosphorylation (21). All of the autophosphorylation sites of *neu* reside in the carboxy-terminal tail (22).

A single point mutation from T to A at nucleotide position 2012 of the rat c-*neu* protooncogene is responsible for converting the *neu* protooncogene to the activated *neu* oncogene (23). This point mutation results in a change at amino acid residue 664 from valine (Val) to glutamic acid (Glu) in the transmembrane domain of the protein product and was shown to activate the intrinsic tyrosine kinase activity (23–26). The same mutation in human *neu*, derived through in vitro mutagenesis, was also shown to have increased tyrosine kinase activity similar to that observed in mutated rat *neu* (24,27–29). Our laboratory has also demonstrated a spontaneous mutation at the same position in dihydrofolate reductase (DHFR)-G8 cells, which are NIH3T3 transfectants containing about 100 copies of genomic c-*neu*, suggesting that amino acid 664 is a "hot spot" for activation (30). DHFR-G8 cells, though containing high copies of genomic c-*neu*, do not have an increase of p185 expression proportional to the gene copy number, probably owing to negative autoregulation of the *neu* gene transcription (31), and they remain untransformed (9). The c-*neu* cDNA construct, driven by the strong promoter of Rous sarcoma virus long terminal repeats (LTRs), is able to express a greater amount of p185 and can lead to transformation of NIH3T3 cells (32).

A similar point mutation has not been observed in human *neu*-encoded p185 at amino acid residue 659 (equivalent to Val-664 in rat *neu*), however, possibly because the codon at this position in the human *neu* gene would have to confer two mutations to allow the same change in amino acid sequence (33–35). On the other hand, overexpression alone of the human *neu* gene is sufficient to transform NIH3T3 cells (36,37). In addition, deletions of the amino-terminus of the human *neu* protein, as well as amino acid changes produced by site-directed mutagenesis at position 659 from Val to Glu or Asp, all lead to increased tyrosine kinase activity, and all produced high levels of transforming potency in NIH3T3 cells (38).

Understanding of the functions of the HER-2/*neu* gene product depends on the elucidation of its signal transduction pathway. Progress in the study of the signal transduction pathway of the *neu* gene has been hindered by the lack of success in identifying its ligand. However, several candidate ligands for HER-2/*neu* have been described (39–45), one of which is heregulin, which was detected from conditioned medium of MDA-MB-231 human breast carcinoma cells as a 45-kDa protein (HRG-α). It is presumed to interact specifically with HER-2-encoded p185 in breast cancer cells, it stimulates breast cancer cell proliferation in culture, and it increases tyrosine phosphorylation of p185 (41). Four members of the human heregulin family have been identified and named HRG-α, HRG-β1, HRG-β2, and HRG-β3, which are proteins of 640, 645, 637, and 231 amino acids, respectively. The rat version of this ligand was purified and cloned from the medium of *ras*-transformed rat1-EJ cells. It is similar in size to heregulin (44 kDa) and has similar biochemical properties except that it inhibits, rather than stimulates, breast cancer cell proliferation in culture and induces differentiation of AU-565 breast cancer cells. Therefore, it was named *neu* differentiation factor (NDF) (39,40). Another member of the HER-2 ligand family was cloned from chicken and is homologous to the human heregulin and rat NDF. It has been found to stimulate acetylcholine receptor synthesis as well as inducing tyrosine phosphorylation of HER-2 p185, and was named ARIA (for acetylcholine receptor inducing activity) (46). Three glial growth factors (GGF-I, GGF-II, and GGF-III) cloned from bovine genomic DNA were also found to be the alternatively spliced HER-2 ligands produced in the embryonic nervous system that can specifically activate the HER-2-encoded p185 receptor tyrosine kinase and are believed to have important functions in the development and regeneration of the nervous system (47). Several isoforms of heregulin including p45, gp30, and p75 also have been identified (15,42,48,49). Interestingly, heregulin was shown to directly bind to HER-4 and induce tyrosine phosphorylation of HER-4 (50). Most recently, estrogen was found to mimic the ligand activity of the HER-2/*neu*-encoded p185 that binds to and induces higher phosphorylation activity in p185 (51). Therefore, more extensive investigation is imperative to clarify the HER-2/*neu*-specific ligands.

Regardless of the frustration in identifying the ligand for HER-2/*neu*, great efforts have been made to understand the functions of the HER-2/*neu* gene. One approach is to take advantage of the similarity between the HER-2/*neu* gene and the EGF-R. A chimeric receptor, in which the intracellular domain of the HER-2/*neu* gene was fused to the extracellular domain of the EGF-R, was molecularly engineered. This chimeric molecule rendered the intracellular tyrosine kinase domain of the *neu* gene responsive to EGF, thus mimicking the ligand-induced events of the parental *neu* gene. Many mitogenic signaling events were triggered by EGF through this chimeric receptor (52), including the induction of tyrosine phosphorylation, DNA synthesis, and cell transformation (53), an increase in both intracellular Ca^{2+} concentration and hydrolysis of membrane polyphosphoinositides, as well as cell growth (54), the activation of *fos/jun* transcription factor complex, an increase in the glucose transporter messenger RNA, and an associated increase in glucose transport, followed by an increase in ornithine decarboxylase mRNA and its activity (55). Similar effects of EGF on tyrosine phosphorylation and the induction of mitogenic signals were demonstrated on a similar chimera consisting of the extracellular domain of EGF-R and the intracellular domain of human HER-2/*neu*. In addition, comparison of phosphorylation patterns among the chimeric and the parental constructs indicated that a 37-kDa protein may be a specific target of the human *neu* tyrosine kinase (56). The cell-specific mitogenic signaling activity of EGF-R and human HER-2/*neu* may reside in the 270 amino acid tyrosine kinase domain, and autophosphorylation sites located at the carboxy-terminus of both HER-2/*neu* and EGF-R may affect the intrinsic tyrosine kinase activity of these receptor molecules. Therefore, the carboxy-terminal region of p185 plays an important role in regulating the transformation activity (22,57,58).

Agonistic antibodies of *neu* were also used in studying the function of the gene product. The agonistic antibodies, but not the monovalent F_{ab} fragment, were effective in stimulating the tyrosine phosphorylation of p185, suggesting that dimerization of the receptor plays a role in signal transduction (59). The point-mutated p185 protein, but not the normal protein, has been shown to arrange on the cell membrane primarily in an aggregated form (29). In addition, down-regulation of human HER-2/*neu* protein by monoclonal antibodies has been facilitated by introduction of the single point mutation, which also suggests changed aggregation properties of the point-mutated protein (60). Stereochemical modeling by other groups also predicted an enhanced dimerization in the mutated *neu* protein owing to specific interhelical interaction (61). The increase in turnover rate and tyrosine phosphorylation in the mutated p185 protein (62), and the similarity in patterns of phosphorylation between the mutated and normal p185 proteins (59), all support the notion that the mutated p185 protein is actually mimicking the ligand-induced event.

Our laboratory has studied $p185^{neu}$ signal transduction and transformation by establishing revertant cell lines from the *neu*-transformed NIH3T3 cells using the chemical mutagen ethyl methanesulfonate (63). The revertants were nontransformed, and their $p185^{neu}$ proteins were underphosphorylated. The revertants also resisted transformation by *neu* and several other oncogenes (H-ras, N-ras, v-mos, v-abl, and v-fos). The results indicated that defective tyrosine phosphorylation of the $p185^{neu}$ may contribute to the inability in transformation, and *neu* may share common elements with several other oncogenes in their pathways for inducing cellular transformation (63).

neu-INDUCED TRANSFORMATION

The point-mutated rat *neu* gene is able to transform NIH3T3 cells upon transfection (9). Transformation is thought to be dependent on the tyrosine kinase activity of the p185 protein because *neu*-encoded p185 possesses tyrosine kinase activity (17,19,64) and increased tyrosine kinase activity is associated with the mutated p185 protein encoded by the activated *neu* oncogene (24,25,38). The correlations between tyrosine kinase activity and transforming ability have led to speculations that increased tyrosine kinase activity is part of the tumorigenic process for the *neu* gene. A direct link between tyrosine kinase activity and transforming potential was shown by an elegant experiment in which the mutated p185 oncoprotein was altered by site-directed mutagenesis to generate a mutant lacking a critical lysine in the ATP binding site of the tyrosine kinase domain, and the resulting protein, without a functional ATP binding site in the tyrosine kinase domain, could not mediate the transforming phenotype (28).

Oncogenic transformation of NIH3T3 cells by overexpression of the human *neu* gene has been reported by two groups, and it seems to be correlated with the level of *neu* expression in certain mammary carcinoma cell lines (36,37). Transgenic mice experiments demonstrated that the mutated rat *neu* oncogene induces mammary carcinoma in the entire epithelium of the mammary gland and benign bilateral epithelial hypertrophy and hyperplasia in the parotid gland and epididymis (65). In contrast to the nonstochastic mode of mammary tumor development demonstrated in this study, another group reported a stochastic appearance of mammary tumors in transgenic mice carrying MMTV/*neu*, in addition to epithelial hyperplasia of the epididymis salivary glands and seminal vesicles and dysplasia of the harderian gland (66). Besides mammary adenocarcinoma, transgenic mice carrying either the normal or the in vitro mutated human *neu* gene develop B cell lymphomas. This is probably due to the use of the immunoglobulin enhancer simian virus 40 early gene promoter as the transcriptional control unit in the transgene constructs, since the immunoglobulin enhancer would direct a strong expression in the lymphoid tissues.

Only some members of one transgenic family carrying the normal human *neu* gene developed pre-B cell lymphomas within 6 to 10 months, while all the members of four individual transgenic families carrying the mutant human *neu* gene developed clonal lymphomas neonatally. This suggests a difference in the oncogenic potential between the normal and the mutant human *neu*, similar to the differences between the normal and mutated rat *neu* gene.

Immunologic approaches have also been used to establish the role of the *neu* gene in transformation. Several monoclonal antibodies raised against p185 were able to specifically down-regulate the level of p185 and decrease cellular tyrosine kinase activity (67). When *neu*-transformed

NIH3T3 cells were exposed to these monoclonal antibodies reactive with the *neu* gene product, there was a rapid and reversible loss of both cell surface and total cellular *neu* protein. The cells reverted to the nontransformed phenotype, suggesting that expression of p185 is required for maintenance of transformation induced by the *neu* oncogene. A synergistic antitumor effect was shown when a panel of monoclonal antibodies reactive with distinct domains of the *neu*-encoded p185 were administered to nude mice that had been injected with *neu*-transformed B104-1-1 cells (NIH3T3 transfectants carrying approximately 10 to 20 copies of mutation-activated genomic rat *neu* oncogene). However, none of these antibodies individually was able to eradicate tumors (68,69). Furthermore, NFS mice immunized with a recombinant vaccinia virus that expressed the extracellular domain of the rat *neu*-encoded p185 protein were fully protected from tumor formation when challenged with *neu*-transformed NIH3T3 cells (70).

When the intracellular domain of the *neu* gene was fused to the extracellular domain of the immunoglobulin heavy chain, this chimeric molecule was able to transform cells only when the immunoglobulin light chain was coexpressed. This result indicates that *neu* gene expression on the cell surface is required for the p185 protein to function properly, because only a complete immunoglobulin molecule containing both heavy and light chains will be anchored on the cell surface (71). One possibility for this requirement is that certain substrates for the signal transduction pathway of the *neu* gene are available only on the inner cell surface. Another interesting observation from this study is that the immunoglobulin-*neu* fusion construct was able to transform cells regardless of whether the transmembrane domain was derived from the normal or mutated gene. This observation seems to contradict the fact that the single point mutation in the transmembrane region is important in transformation. However, considering the ability of antigen to modulate the activity of this fusion protein, and the consequent change in transforming effects, the mechanism that activates transformation of the normal *neu* to the oncogenic form may be a general uncoupling of the extracellular domain from the intracellular domain. The uncoupling here is due to the replacement of the extracellular domain of the *neu* gene product by the immunoglobulin gene product, and this change may result in the activation of the tyrosine kinase activity of the protein.

THE *neu* GENE IN HUMAN CANCER

The human HER-2/*neu* gene was originally isolated as an amplified v-*erb*-related sequence in human mammary carcinoma, salivary gland adenocarcinoma, and gastric cancer cell lines (10,11). Hence, it is not surprising to find this gene amplified or overexpressed in many different human primary tumors, especially adenocarcinomas (72). Abnormality of the HER-2/*neu* protooncogene has been studied most frequently in human breast cancer. The HER-2/*neu* gene has been found to be amplified or rearranged in many primary breast cancers and breast cancer cell lines (35,73,74). One of the early reports described a 2- to 20-fold amplification of the HER-2/*neu* gene in about 30% of primary human breast cancers in a study of 189 cases (73). A significant correlation between the level of HER-2/*neu* gene amplification and the time to relapse and the overall survival of patients was also suggested in this study. Numerous similar studies have also demonstrated the amplification and overexpression of HER-2/*neu* in mammary carcinoma (74–80). Several of these studies showed the HER-2/*neu* gene to be coamplified with other protooncogenes (74,75,81). For example, coamplification of the c-*erb*A protooncogene has been found in a high percentage of tumors that carry amplified HER–2/*neu* (74,82). Some studies also showed that amplification of HER-2/*neu* is correlated with the following phenomena in breast cancer: absence of estrogen and progesterone receptors (80,81,83,84), an enhanced degree of malignancy of the tumor (85), a shorter time to treatment failure (79), earlier relapse, shorter survival after relapse, and poorer overall survival (83,84). On the basis of all these studies, many have suggested that amplification of the HER-2/*neu* gene could be used as a prognostic indicator of poor clinical outcome (75,79,86,87). Some studies suggested a correlation of *neu* amplification with other prognostic factors such as nodal status or nuclear grading (73,87,88). This kind of correlation, however, was not reported by others (83,84). Although some investigators have concluded that amplification of the HER-2/*neu* gene is of no prognostic value at all in breast cancer (89,90), the body of evidence to support the significance of the HER-2/*neu* gene in the course of breast cancer is overwhelming (86). The clinical significance of the HER-2/*neu* gene in breast cancer was further determined by an extensive study involving long-term follow-up of 526 patients and the application of multiple techniques for analysis (91). The results supported a good correlation between the level of expression of the HER-2/*neu* oncogene and a poor clinical outcome of patients with breast cancer. HER-2/*neu* gene amplification or overexpression in human primary breast cancers was shown to be a powerful predictor of risk of recurrence (92). A similar correlation of HER-2/*neu* overexpression in primary tumors of the ovary with a poor clinical outcome was seen in patients with ovarian cancer (91,93). The human cancers in which the HER-2/*neu* gene has been reported to be amplified or overexpressed include ovarian cancer (91,93,94), gastric tumors (82,95), colon cancer (96), lung cancer (97,98), oral cancer (99), and cervical cancer (100). In contrast to amplification, a very low frequency of rearrangement of the HER-2/*neu* gene has been observed in breast cancer and gastric carcinoma (82,95). It has been reported that HER-2/*neu* gene amplification in gastric cancers correlates with a lower survival rate and may be used as an important prognostic indicator (101). All these reports provide very good evidence that the HER-2/*neu* gene plays an important role in the process of tumorigenesis of human cancers; this evidence is also supported by other studies described in the previous section.

Expression levels of HER-2/*neu*-encoded p185 also have been positively correlated with lymph node metastasis (85,102). A significant increase in the incidence of HER-2/*neu* gene amplification and overexpression has been reported in breast cancer patients with metastasis to more than three axillary lymph nodes. Studies of node-positive breast cancer patients revealed a positive correlation between the number of lymph node metastases and HER-2/*neu* overexpression; hence, amplification and overexpression of the HER-2/*neu* gene are significant prognostic indicators of both overall survival and time to relapse in node-positive patients (73,91). These data provide a link between the kinase oncogene HER-2/*neu* and metastasis of human breast cancer.

HER-2/*neu* GENE AND TUMOR METASTASIS

Cancer metastasis presents the most serious challenge to therapeutic intervention. Metastasis is a complex, multifaceted pathophysiologic process involving numerous interactions between tumor cells and host characteristics (103–107). Several clinical studies indicate that the average level of amplified oncogenes measured in the primary tumor can be correlated with clinical parameters of metastasis and recurrence (91,108–112). The oncogenes that can be used as molecular markers for determining the metastatic potential of primary tumors may play an important role in the metastatic process. The HER-2/*neu* gene, as discussed above, was found to correlate with the number of lymph node metastases in breast cancers, and therefore is considered to play an important role in tumor metastasis. However, owing to the limitations of clinical studies, the correlation found in clinical data is insufficient to prove that *neu* gene expression is responsible for aggressive disease behavior, such as metastasis. Therefore, it is important to study systematically the role of the *neu* gene in metastasis in an animal model system.

One direct approach to elucidate the role of a known oncogene in the metastatic cascade is to introduce the cloned oncogene (in this case the *neu* oncogene) into recipient nonmetastatic or low-metastatic cell lines and monitor the metastatic behavior of these cell lines (113). To address the role of the *neu* oncogene in metastasis, we introduced the mutation-activated rat *neu* into NIH3T3 cells and Swiss 3T3 cells and examined the metastatic properties of the *neu*-transformed 3T3 cells in vitro and in vivo in comparison with their parental 3T3 cells. In experimental metastasis assays in vivo, 3T3 cells transformed by mutation-activated rat *neu* oncogene induced metastatic lung nodules, while the parental 3T3 cells did not (Fig. 5.1). Monoclonal antibodies specific for the *neu*-encoded p185 protein abrogated the *neu*-induced metastatic properties. These data provided strong experimental evidence that *neu* expression is sufficient to induce metastasis in 3T3 cells (114). Important steps in the metastatic event are tumor cell adhesion to blood vessels and invasion of the basement membrane. Therefore, we compared the *neu*-transformed cells with the parental 3T3 cells for the ability to adhere to microvessel endothelial

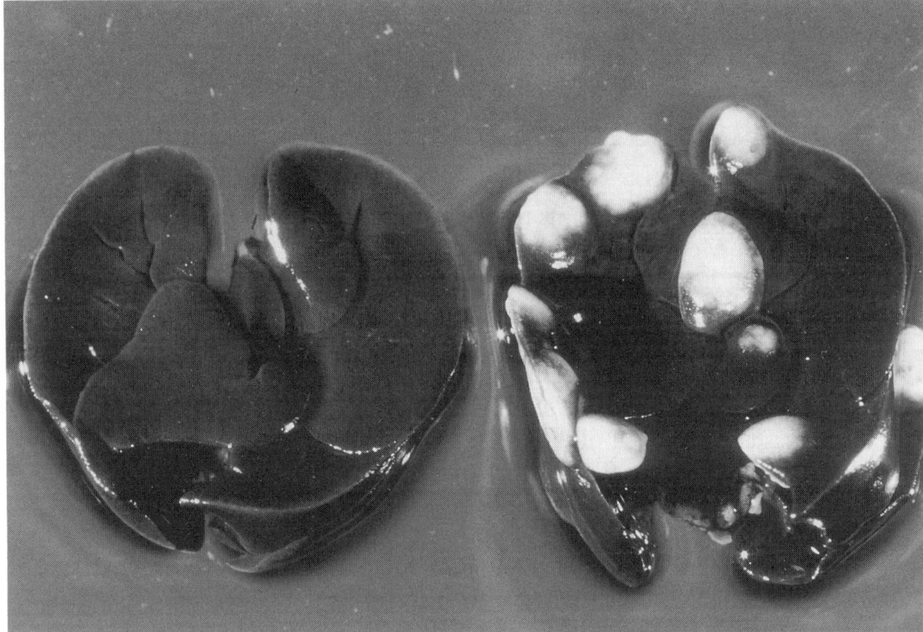

Fig. 5.1: Experimental metastasis assay. Gross appearance of lungs from mice injected with NIH3T3 cells (left) or B104-1-1 cells (right), showing formation of pulmonary metastasis in the lungs of nude mice injected with *neu*-transformed B104-1-1 cells, but not in mice injected with NIH3T3 cells.

cells, migrate through the layer of reconstituted basement membrane, and secrete enzymes that degrade the basement membrane. The *neu*-transformed 3T3 cells exhibited higher metastasis-associated properties in all three of the tested aspects. The results indicate that *neu* oncogene expression can induce metastasis in 3T3 cells by facilitating multiple steps in the metastatic cascade (114,115). A mutation-activated human HER-2/*neu* oncogene also was shown to increase the experimental metastatic potential of murine colon adenocarcinoma cells (116). Furthermore, experiments using transgenic mice indicate that introduction of a c-*erb*B-2/*neu* gene into mice can induce mammary tumors and metastasis (65,66,117,118). All these studies indicated that the HER-2/*neu* gene plays an important role in cancer metastasis.

Ovarian carcinoma is the most lethal tumor of the female genital tract and continues to be the major cause of mortality in female cancer deaths, largely as a function of early abdominal seeding of this neoplasm producing carcinomatosis (119). Therefore, studies investigating this process are necessary to determine the molecular mechanism that induces such aggressive phenotypes and to develop novel means of predicting and possibly treating these aggressive malignancies. We have found that amplification or overexpression of the c-HER-2/*neu* gene in primary ovarian cancer is a common phenomenon in such tumors across different populations (93,120). In addition, we have detected HER-2/*neu* gene overexpression in the ovarian carcinoma cell line SK-OV-3 (120). To further study the biological effect of HER-2/*neu* overexpression in SK-OV-3 cells, we injected such cells intraperitoneally into female nu/nu mice and found that this cell line forms extensive abdominal tumors and ascites. From the ascites in an injected mouse, we established the SKOV3.ip1 cell line and found that it expressed threefold more HER-2/*neu*-encoded p185 proteins than the parental SK-OV-3 cells. When transformation phenotypes of SK-OV-3 and SKOV3.ip1 cells were compared, SKOV3.ip1 cells showed higher rates of cell growth and DNA synthesis, formed more colonies in soft agar, produced larger subcutaneous tumors, and resulted in shorter survival times of nu/nu mice after intraperitoneal injection. These data indicate that the level of HER-2/*neu* overexpression may correlate with the degree of malignancy in these ovarian carcinoma cells (121).

Lung cancer is the leading cause of death from all malignancies in the United States (122). Non-small cell lung cancer (NSCLC) is among the most threatening of these types of malignancy because it occurs at high incidence, usually is resistant to chemotherapy, and is frequently disseminated upon diagnosis. To provide potential diagnostic, prognostic, and therapeutic reagents for lung cancers, intensive efforts have been made to identify the relation between specific gene alterations and the clinical behavior of lung cancers. We and others have reported on the abnormal expression of HER-2/*neu*-encoded $p185^{neu}$ in cell lines and primary tumors from human NSCLC (72,97,98,123–125). In addition, the extent of $p185^{neu}$ expression in primary NSCLC is correlated with poor clinical prognostic indicators including lymph node metastasis (123,125). These clinical correlations suggest that $p185^{neu}$ may play an important role in the malignancy of the human NSCLC. To systematically investigate the potential role of the HER-2/*neu* gene in lung cancer metastasis, we introduced the human HER-2/*neu* gene into the very low $p185^{neu}$-expressing NCI-H460 human NSCLC cells. We then examined the metastatic potentials among the parental NCI-H460 cells and the NCI-H460 stable transfectants that expressed increased levels of $p185^{neu}$. Compared with the parental NCI-H460 cells, the $p185^{neu}$-overexpressing NCI-H460 transfectants produced significantly more pulmonary and extrapulmonary metastatic tumors upon tail vein injection into nude mice. Enhanced tumor metastatic potential in vivo was accompanied by an increase of invasiveness in vitro. In addition, important steps in the invasion and metastasis process, such as the secretion of basement-membrane-degradative enzymes and migration through the layer of reconstituted basement membrane (Matrigel), were also enhanced in the $p185^{neu}$-overexpressing NCI-H460 transfectants. Moreover, scanning electron microscopy revealed that the $p185^{neu}$-overexpressing NCI-H460 transfectants had significantly more microvilli and membrane protrusions than the parental cells, which provided a clear structural feature for increased cell motility. The results demonstrate that $p185^{neu}$ can enhance the metastatic potential of NCI-H460 human lung cancer cells. The mechanism by which $p185^{neu}$ induces higher metastatic potential in human NSCLC is to promote the invasion steps of the metastatic cascade (D. Yu, S.-S. Wang, K. M. Dulski, G. L. Nicolson, C.-M. Tsai, and M.-C. Hung, manuscript submitted).

The above experimental results have provided solid evidence that the HER-2/*neu* oncogene plays a critical role in certain human malignancies and metastasis, including ovarian cancers and NSCLC. However, the role of HER-2/*neu* overexpression in human breast cancer metastasis has not been settled, and systematic studies measuring the metastatic potential of the human breast cancer cell lines that overexpress normal HER-2/*neu* gene after m.f.p. injection (orthotopic implantation of breast cancer cells can lead to better presentation of the metastatic phenotype) is critical to determine whether the HER-2/*neu* gene can induce higher metastatic potential in human breast cancers.

HER-2/*neu* GENE AND CANCER CELL CHEMOSENSITIVITY

The phenomenon of multidrug resistance (MDR) confers upon cells the ability to survive under lethal doses of many structurally unrelated antineoplastic agents. Several mechanisms have been suggested to be involved in MDR: overexpression of the transmembrane P-glycoprotein that promotes drug efflux, overexpression of the ATP-binding-cassette transmembrane transporter MDR-related protein (MRP), changes in regulation of phase I and phase II drug-metabolizing enzymes, and altered topoisomerase activity (126,127). Recent studies indicate that disorders in programmed cell death (apoptosis) may be one of the possible mechanisms whereby cancer cells can acquire or lose sensitivity to antineoplastic agents (128,129). In addition, several studies have identified other clinical and laboratory factors ("prognostic," "predictive," or "responsive" factors) that can be used to select patients who are likely to respond to chemotherapy (130). Among these studies, oncogene amplification or overexpression, especially HER-2/*neu* overexpression, has been shown to be a promising prognostic factor for breast cancer patients, as decribed above. One interesting report from a clinical study indicates that breast tumors with HER-2 overexpression are less responsive to adjuvant therapy regimens containing cyclophosphamide, methotrexate, and fluorouracil (CMF) than are those with a normal amount of the gene product (131). However, whether c-*erb*B-2 overexpression is a specific predictor of the response to chemotherapy is a new issue that needs to be intensively examined.

Most small cell lung cancers (SCLCs) are chemosensitive, whereas NSCLCs are usually chemoresistant. The HER-2/*neu* gene is differentially expressed in subpopulations of NSCLC, but not in SCLC, and has been linked to shortened survival (97,125). Overexpression of HER-2/*neu* gene therefore was speculated to be associated with intrinsic chemoresistance in NSCLC. Recently, the relation between chemoresistance and overexpression of the HER-2/*neu* gene on NSCLC cell lines was investigated. Twenty NSCLC cell lines established from untreated patients that expressed different levels of HER-2/*neu* mRNA were assayed for their in vitro sensitivity to chemotherapeutic drugs (doxorubicin, carmustine, cisplatin, melphalan, mitomycin, and etoposide), and chemosensitivity of these lines were demonstrated as their differences in IC_{50} values (i.e., the drug concentrations required to inhibit cell growth by 50%) (132). A statistically significant correlation between the IC_{50} values for all six drugs and the degree of HER-2/*neu* gene expression in all 20 cell lines ($r = 0.67$ to 0.86; $p < 0.005$) was reported, indicating that overexpression of HER-2/*neu* is a marker for intrinsic multidrug resistance in NSCLC cell lines and may be used as a predictor of therapeutic response in NSCLCs (132).

Chemotherapy, though demonstrated to benefit patients with advanced breast cancer, has only partial efficacy, inducing an objective response in 40% to 70% of patients (133,134). Failure of some breast cancer cells to respond to chemotherapy and the appearance of resistant cell populations upon relapse present a major obstacle to the ultimate success of cancer therapy. Therefore, it is extremely important to uncover the basis of resistance to anticancer drugs. In an attempt to investigate the relation between p185 expression and chemosensitivity in breast cancer cells, we tested a panel of established human breast cancer cell lines known to express the c-*erb*B-2-encoded p185 at low, intermediate, and high levels for their sensitivity to the chemotherapeutic drugs Taxol and Taxotere with the XTT assay (135). Our results showed that breast cancer cells that overexpressed c-*erb*B-2 mRNA at higher levels were generally less sensitive to Taxol and Taxotere than those that expressed lower levels of c-*erb*B-2 mRNA (unpublished data). In addition, the monoclonal antibody c-*neu*-Ab5, which recognizes the extracellular domain of p185 and downregulates p185, increased the chemosensitivity of p185-overexpressing breast cancer cells to Taxol (unpublished data), which suggests that the p185 protein in the p185-overexpressing breast cancer cells is responsible for the increased chemosensitivity. It will be critical to examine a large number of patients to determine whether overexpression of p185 correlates with increased chemoresistance in primary breast tumors and whether overexpression of HER-2/*neu* also may be used as a predictor of therapeutic response in breast cancers.

HER-2/*neu* AND ESTROGEN RECEPTOR

Several lines of evidence suggest that the estrogen receptor (ER) might be a negative regulatory factor of HER-2/*neu* expression. Loss of ER expression or function, like HER-2/*neu* overexpression, is strongly correlated with poor prognosis in breast cancer patients (136). Some studies of human breast cancer specimens have shown a correlation between overexpression of HER-2/*neu* and loss of functional ER (77,80,84). Finally, two studies using ER^+ human breast cancer cell lines, MCF-7 and T47D, found an inverse relation between estrogen stimulation and HER-2/*neu* expression (137,138). These facts suggest the possibility that estrogen stimulation of ER^+, but not ER^-, cells may lead to decreased

HER-2/*neu* expression through repression of the HER-2/*neu* gene by ER (139).

Our laboratory has demonstrated that ER negatively regulated the expression of the HER-2/*neu* gene protein product, p185^{neu}, in two ER$^+$, but not in an ER$^-$, breast cancer cell lines (139). We demonstrated that stable expression of ER in the ER$^+$ transfectants, which is established from a previously ER$^-$ human breast cancer cell line, is sufficient to confer the ability to respond to estradiol (E2) by down-regulating HER-2/*neu* expression at both the protein and RNA levels (139). This regulation occurs at the transcriptional level and requires the presence of both ER and E2, and a 140 base pair (bp) region of the HER-2/*neu* promoter represents an estrogen-responsive region within the HER-2/*neu* promoter and is required for this transcriptional regulation (139). Finally, gel mobility shift analysis demonstrated an alteration in the nuclear factor or factors binding to this promoter region in E2-stimulated versus E2-deprived breast cancer cells (139). This study provided the first evidence that the inverse clinical correlation between HER-2/*neu* and ER expression may be due to transcriptional repression of HER-2/*neu* by E2/ER.

In addition, when growth rates of ER$^+$, HER-2/*neu*-overexpressing breast cancer cells were assayed with or without addition of E2, cell growth was stimulated with, as compared to without, E2. However, when soft agar colony formation of these breast cancer cells was measured, there was less colony formation in the presence of E2 than in the absence of E2 (unpublished observations). As soft agar colony formation is an indicator of transforming potential, the results indicate that E2 can suppress transformation of ER$^+$, HER-2/*neu*-overexpressing breast cancer cells. A HER-2/*neu* cDNA was transfected into an ER$^+$, MCF-7 cell (line/Kern FG, personal communication). This ER$^+$, HER-2/*neu*-overexpressing cell line behaved differently from our ER$^+$, HER-2/*neu*-overexpressing breast cancer cells, in that E2 stimulated transformation phenotypes such as tumorigenicity. This is because their HER-2/*neu* was driven by a heterologous promoter that cannot be inhibited by E2/ER. Therefore, E2 will not suppress HER-2/*neu*-induced transformation in these cells. Our cell lines contain an endogenous HER-2/*neu* gene driven by its own promoter; therefore, p185 expression will be inhibited by E2/ER. Taken together, these results further strengthen the notion that HER-2/*neu* overexpression is critical for transformation of the phenotype in breast cancer cells.

Patients with ER$^+$ breast cancer are frequently treated with estrogen antagonists such as Tamoxifen (TAM). The rationale is that estrogen can stimulate growth of ER$^+$ tumor cells, and TAM, by blocking interaction of estrogen and ER, should inhibit growth of ER^+ tumor cells. However, 50% of these patients do not respond to hormonal treatment. On the basis of this clinical observation and our laboratory observation that estrogen has duel effects on the ER^+, HER-2/*neu*-overexpressing breast cancer cells (i.e., stimulation of cell growth but suppression of transformation), we hypothesize that TAM may also have dual effects on the ER^+, HER-2/*neu*-overexpressing cancer cells by blocking interaction of estrogen and ER. On the one hand, TAM may inhibit the growth of ER^+ tumor cells; on the other hand, TAM may enhance malignant transformation by increasing HER-2/*neu* overexpression. We predict that the subset of breast cancer patients who have ER^+ and HER-2/*neu*-overexpressing tumors may not respond to the TAM treatment. Therefore, TAM treatment will not be beneficial to this subset of breast cancer patients.

MOLECULAR MECHANISMS INVOLVED IN HER-2/*neu* OVEREXPRESSION

Among all the reports on the overexpression of the *neu* gene in human breast cancers, there are cases of overexpression without gene amplification (86). Certain established mammary carcinoma cell lines also express a high level of HER-2/*neu* RNA to an extent that could not be accounted for by the level of gene amplification. For example, breast cancer cell lines SK-BR-3 and BT474 overexpress HER-2/*neu* RNA 128-fold as compared with breast cancer cell line MCF-7; while the HER-2/*neu* gene amplification in these lines is only 4- to 8-fold that in MCF-7 cells (35). Many different mechanisms are possible for overexpression of a gene without gene amplification. One of the most studied mechanisms involves the regulation of gene expression at the transcriptional level. An understanding of the regulation of the HER-2/*neu* gene might allow us an insight into the defects that may lead to overexpression, which plays a very important role in tumorigenesis. A very general characterization of the human HER-2/*neu* promoter has been provided (140,141). Transcription of the *neu* gene is positively regulated by EGF, transforming growth factor-β1, and triiodothyronine (142); EGF, dibutyryl cyclic adenosine monophosphate, 1,2-*O*-tetradecanoylphorbol-1 3-acetate (TPA), and retinoic acid (143); in addition, if these factors are used in combination, they result in an additive or synergistic effect. On the other hand, the expression of human HER-2/*neu* has been shown to be inhibited by estrogens (137–139) and ligand-activated thyroid hormone and retinoic acid receptors (144).

Our laboratory has characterized the rat *neu* gene promoter and shown the presence of multiple *cis*- and *trans*-acting elements involved in regulation of the rat *neu* gene (145,146). We further studied the regulatory effects of several nuclear oncogenes, including the adenovirus type 5 E1A gene, the c-myc gene, and the simian virus large-T antigen, on *neu* gene expression (147–149). These gene products repress *neu* gene expression at the transcriptional level (147–149). C-myc was found to regulate HER-2/*neu* expression in a species-specific manner; that is, it inhibited HER-2/*neu* expression in mouse cells, but not in human breast cancer cells (150). In addition, we have found that the retinoblastoma tumor suppressor gene product RB also can regulate *neu* gene expression in a cell-type-specific manner by targeting two regions within the HER-2/*neu* regulatory sequence (151–153).

HER2/*neu* gene overexpression in the absence of gene amplification has been described both in primary tumors and in established cell lines (35,86,91), suggesting that mechanisms other than gene amplification can contribute to HER-2/*neu* overexpression in breast cancer. Several groups including ours have examined breast cancer cell lines that overexpress the HER2/*neu* gene but have little or no HER-2/*neu* gene amplification (154–156). Nuclear run-on experiments indicate that these breast cancer cell lines have an increased rate of HER-2/*neu* gene transcription. In one study, a novel transcription factor, OB2-1, was found to be required for HER-2/*neu* overexpression in several breast cancer cell lines (156). In our study, the enhanced HER-2/*neu* transcription rate in the MDA-MB453 cell line was shown to be due to activation of the gene in *trans* as determined by using HER-2/*neu* and promoter-CAT constructs (154). A 13-bp element on the HER-2/*neu* gene promoter that is required for the increased transcription rate was localized, and the sequence-specific interaction of this fragment with a nuclear protein complex was demonstrated. BT483 cells may possess an activation in *cis*, since activity from an exogenous HER-2/*neu*-CAT construct remains at a basal level when transfected into these cells, despite the increased transcriptional level of the endogenous HER-2/*neu* gene. In BT474 cells, transcriptional upregulation is due mainly to gene amplification and does not fully account for the levels of HER-2/*neu* mRNA, indicating that deregulation of HER-2/*neu* expression at the post-transcriptional level significantly contributes to HER-2/*neu* overexpression in this breast cancer cell line. Our results suggest that multiple activations of the HER-2/*neu* gene are involved in HER-2/*neu* overexpression in breast cancer lines and that multiple mechanisms may function simultaneously within a single cell line.

CANCER THERAPY BY TARGETING HER-2/*neu*

As discussed above, amplification and overexpression of the HER-2/*neu* gene in cancer patients has been correlated with poor clinical outcome.

And recent studies from our group and others have indicated that amplification and overexpression of the HER-2/*neu* gene may enhance metastatic potential in tumors that overexpress p185. Moreover, a relation between overexpression of $p185^{neu}$ and increased drug resistance in human breast cancers may exist that can cause p185-overexpressing tumors to respond poorly to chemotherapy. Therefore, more aggressive therapy targeting the HER-2/*neu* gene may be beneficial to those patients whose tumors express high levels of $p185^{neu}$ and who hence may be at higher risk of metastasis and chemotherapy failure.

Chemotherapy

Many cancer patients experience benefits from chemotherapy including prolonged disease-free intervals and improved overall survival rates. However, patients whose tumors express high levels of $p185^{neu}$ may have a poorer response to chemotherapeutic agents, and it will be beneficial to examine tumors (breast, ovarian, and lung tumors in particular) for their $p185^{neu}$ expression level before a particular chemotherapy regimen is chosen for a patient. For example, it has been reported that breast tumors with $p185^{neu}$ overexpression are less responsive to CMF-containing adjuvant therapy regimens than tumors with a normal level of $p185^{neu}$ (131). Hence, alternative regimens other than a CMF-containing regimen, or CMF-containing regimens at higher doses should be considered when treating breast cancer patients with $p185^{neu}$-overexpressing tumors. In addition, we have demonstrated that $p185^{neu}$-overexpressing breast cancer cell lines are more resistant to the new chemotherapeutic drugs Taxol and Taxotere; therefore, either higher doses of Taxol and Taxotere or other chemotherapeutic drugs should be given to $p185^{neu}$-overexpressing breast cancer patients.

Immunotherapy

The $p185^{neu}$ on the cell surface of HER-2-overexpressing tumor cells may be a good target for receptor-targeted immunotherapies. Down-regulation of the p185 protein by monoclonal antibodies that bind to the extracellular domain of the p185 protein has been shown to inhibit tumor formation and tumor metastasis in animal models (114,157). It will be important to examine whether immunotherapy with anti-p185 monoclonal antibodies that bind to the extracellular domain of the human p185 protein can suppress p185-overexpressing human cancer growth and metastasis in a clinical setting. In addition, anti-human p185 monoclonal antibody 4D5 was shown to enhance the sensitivity of $p185^{neu}$-overexpressing tumor cells to cisplatin, and we have shown that down-regulation of the p185 protein in p185-overexpressing BT474 and MDA-MB-361 breast cancer cells by treating these cells with anti-p185 antibodies could increase the chemosensitivity of the p185-overexpressing breast cancer cells to Taxol and Taxotere. These results suggest that combined immunotherapy and chemotherapy (e.g., anti-p185 antibodies and Taxol) may be a more effective treatment for p185-overexpressing breast cancers or other solid tumors that overexpress $p185^{neu}$.

Gene Therapy

Rapid advances in understanding the molecular basis of cancers and advances in the development of gene delivery systems suggest that gene therapy for human cancers may soon be a treatment choice.

The adenovirus E1A gene product has been identified as a viral suppressor gene for *ras*-induced metastasis, and we have found that E1A represses *neu* expression via a *cis* DNA element in the *neu* promoter (147,158). To investigate whether E1A can reduce the metastatic potential of *neu*-transformed cells, we introduced the E1A gene into the *neu*-transformed cells and developed cell lines that stably express E1A. We found that introduction of the E1A gene into *neu*-transformed 3T3 cells reduced the formation of experimental metastatic tumors (115). In addition, we examined the E1A transfectants for their adhesion, migration, and gelatinase-secreting ability. Transfer of the E1A gene into *neu*-transformed 3T3 cells inhibited all three of these metastasis-associated properties. The results indicate that the E1A gene is a metastatic suppressor gene for *neu*-induced metastasis that apparently functions by blocking multiple steps in the metastatic cascade. E1A also suppressed *neu*-induced transformation, including flattened cell morphology, reduced DNA synthesis rate, and reduced ability to induce tumors in nude mice (159). From these results we concluded that E1A suppresses transformation and metastasis of *neu*-transformed cells through repression of *neu* gene expression. However, E1A has been shown to repress also the transformation features of other human cancer cells that do not overexpress HER-2/*neu* (160). This observation raised the possibility that repression of *neu* gene expression in our *neu*+E1A cells might not be the only mechanism for transformation suppression. To study whether other molecular mechanisms might be involved in suppression of transformation by E1A in our *neu*+E1A cells, we reexpressed the p185 oncoprotein in the *neu*+E1A cells by transfecting them with a plasmid containing activated rat *neu* cDNA driven by a heterologous virus long terminal repeat promoter that cannot be repressed by E1A. All the features of transformed cells including cell morphology, DNA synthesis rate, colony formation in soft agar, and tumorigenicity in nu/nu mice were restored in the cell lines that reexpressed p185. However, the in vivo metastatic tumor formation by these p185-reexpressing cells was still inhibited by E1A, and those metastasis-associated properties were mostly inhibited by E1A (161). The data demonstrated that reexpression of p185 in *neu*+E1A cells can counteract the tumor-suppressing function of E1A but not the metastasis-suppressing function of E1A. We conclude from these results that repression of *neu* oncogene expression was the main molecular mechanism by which E1A suppressed tumor formation in *neu*-transformed 3T3 cells, and that E1A suppressed metastasis in *neu*-transformed 3T3 cells via multiple molecular mechanisms as well as repressing *neu*. In addition, our model system clearly demonstrated that tumorigenicity and metastasis are related but separable phenomena (Fig. 5.2).

We further examined whether E1A can abrogate malignancy in HER-2/*neu*-overexpressing human ovarian cancer cells. We introduced the E1A gene into HER-2/*neu*-overexpressing SKOV3.ip1 cells and found that the E1A-expressing ovarian cancer cell lines had decreased HER-2/*neu*-encoded p185 expression and reduced malignancy, including a decreased ability to induce tumors in nu/nu mice and reduced tumor dissemination (Fig. 5.3) (121). Therefore, we concluded that E1A is a tumor and metastasis suppressor gene for HER-2/*neu*-overexpressing human ovarian cancer cells.

We are currently exploring the potential of using liposome-mediated gene transfer technology for treating p185^{neu}-overexpressing human

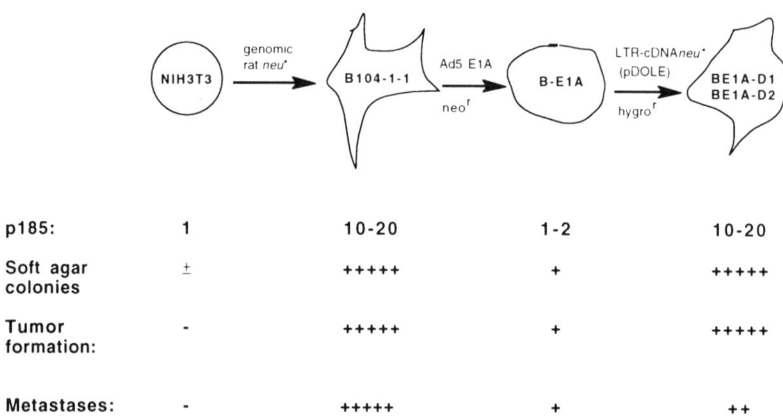

	NIH3T3	B104-1-1	B-E1A	BE1A-D1/BE1A-D2
p185:	1	10-20	1-2	10-20
Soft agar colonies	±	+++++	+	+++++
Tumor formation:	-	+++++	+	+++++
Metastases:	-	+++++	+	++

Fig. 5.2: Summary of 1) the effect of *neu* overexpression on enhancing tumorigenicity and metastasis; 2) the tumorigenicity and metastasis-suppressing function of E1A; and 3) the reexpression of p185, which counteracted the tumor-suppressing, but not the metastasis-suppressing, function of E1A.

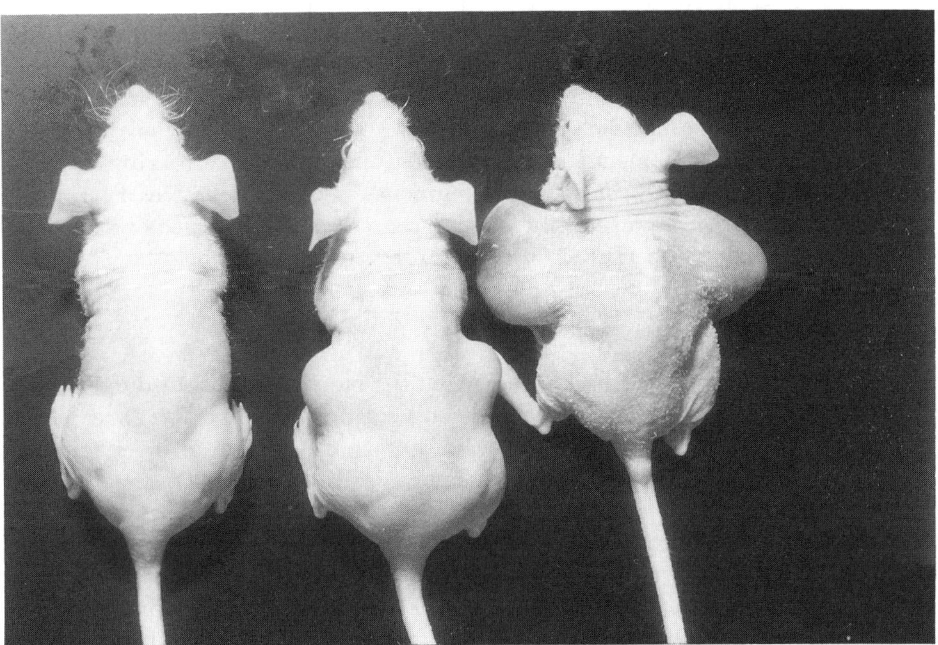

Fig. 5.3: Representative results of tumorigenicity assays demonstrating that E1A can function as a tumor suppressor for HER-2/*neu*-overexpressing human ovarian cancers. Mice were injected with 2×10^6 p185-overexpressing SKOV3.ip1 ovarian cancer cells (right), or with E1A-expressing ovarian cancer cells ip1.E1A1 (middle) or ip1.E1A2 (left).

ovarian cancers in a nude mice model (162). To examine whether E1A can function as a tumor suppressor gene for *neu*-overexpressing human cancer cells in living hosts, we used liposome-mediated direct gene transfer techniques to deliver the E1A gene into *neu*-overexpressing SKOV3 human ovarian cancer cells that have been growing in the peritoneal cavity of nude mice. We found that mice injected with cationic liposome plus E1A DNA complex survived much longer than those without appropriate treatment (unpublished observations). The results, although preliminary, indicate that liposome-mediated E1A gene transfer inhibited *neu*-overexpressing human ovarian cancer cell growth and dissemination, and led to prolonged survival of nude mice bearing *neu*-overexpressing SKOV3 cells. The results also suggest that liposome-mediated E1A gene therapy may serve as a powerful therapeutic reagent for HER-2/*neu*-overexpressing human ovarian cancers by directly targeting the HER-2/*neu*-oncogene.

Targeted delivery of DNA to specific cell types or organs for gene therapy requires different delivery systems. Recombinant adenoviral vectors and retroviral vectors have been used to transfer genes into a number of different cell types in vitro and in vivo and have been suggested to have clinical applications in gene therapy (163). Currently, we are investigating whether recombinant adenoviral vectors and retroviral vectors can be used to deliver HER-2/*neu*-suppressing genes, such as adenovirus E1A, into HER-2/*neu*-overexpressing human cancer cells, and hence exert therapeutic effects on HER-2/*neu*-overexpressing human cancers.

CONCLUSION

In this chapter, we have summarized the biochemical background of the HER-2/*neu* gene and clinical observations on HER-2/*neu* gene overexpression in different types of human cancers. From the studies of our group and others, we conclude that HER-2/*neu* gene overexpression may lead to a higher degree of malignancy, may enhance metastatic potential, and may increase the chemoresistance of human cancers. Hence, cancer therapy that targets HER-2/*neu* should be considered for cancer patients whose tumors overexpress HER-2/*neu*. The studies on the HER-2/*neu* gene are a nice example of how basic molecular biology and biochemical studies may provide critical information for clinical oncologists to benefit patients. The tight link between molecular biology and clinical oncology distinguishes traditional clinical oncology from current molecular oncology, which reflects the trends in the exciting new era of molecular medicine.

ACKNOWLEDGMENTS

This work is supported by National Cancer Institute research grants R29-CA60488 (to D.Y.) and RO1-CA58880, RO1-CA60856 (to M.C.H.).

REFERENCES

1. Schubert D, Heinemann S, Carlisle W, et al. Clonal cell lines from the rat central nervous system. Nature 1974;249:224–227.

2. Shih C, Padhy LC, Murray M, Weinberg RA. Transforming genes of carcinomas and neuroblastomas introduced into mouse fibroblasts. Nature 1981;290:261–264.

3. Padhy LC, Shih C, Cowing D, Finkelstein R, Weinberg RA. Identification of a phosphoprotein specifically induced by the transforming DNA of rat neuroblastomas. Cell 1982;28:865–871.

4. Schechter AL, Stern DF, Vaidyanathan L, et al. The *neu* oncogene: an *erb*-

B-related gene encoding a 185,000-Mr tumour antigen. Nature 1984;312:513–516.

5. Schechter AL, Hung M-C, Vaidyanathan L, et al. The *neu* gene: an *erb*B-homologous gene distinct from and unlinked to the gene encoding the EGF receptor. Science 1985;229:976–978.

6. Coussens L, Yang-Feng TL, Liao Y-C, et al. Tyrosine kinase receptor with extensive homology to EGF receptor shares chromosomal location with *neu* oncogene. Science 1985;230:1132–1139.

7. Fukushige S-I, Matsubara K-I, Yoshida M, et al. Localization of a novel v-*erb*B-related gene, c-*erb*B-2, on human chromosome 17 and its amplification in a gastric cancer cell line. Mol Cell Biol 1986;6:955–958.

8. Hung M-C, Schechter AL, Vaidyanathan L, Stern DF, Weinberg RA. Isolation of the molecular clone of the *neu* oncogene from the B103 rat neuro/glioblastoma cell line. In: Harris CC, ed. Biochemical and molecular epidemiology of cancer, vol. 40. New York: Alan L. Liss, 1986:391–395.

9. Hung M-C, Schechter AL, Chevray P-YM, Stern DF, Weinberg RA. Molecular cloning of the *neu* gene: absence of gross structural alteration in oncogenic alleles. Proc Natl Acad Sci USA 1986;83:261–264.

10. Yamamoto T, Ikawa S, Akiyama T, et al. Similarity of protein encoded by the human e-*erb*B-2 gene to epidermal growth factor. Nature 1986;319:230–234.

11. King CR, Kraus MH, Aaronson SA. Amplification of a novel v-*erb*B-related gene in a human mammary carcinoma. Science 1985;229:974–976.

12. Semba K, Kamata N, Toyoshima K, Yamamoto T. A v-*erb*B-related protooncogene, c-*erb*B-2, is distinct from the c-*erb*B-1/epidermal growth factor-receptor gene and is amplified in a human salivary gland adenocarcinoma. Proc Natl Acad Sci USA 1985;82:6497–6501.

13. Kraus MH, Issing W, Miki T, Popescu NC, Aaronson SA. Isolation and characterization of ERBB3, a third member of the ERBB/epidermal growth factor receptor family: evidence for overexpression in a subset of human mammary tumors. Proc Natl Acad Sci USA 1989;86:9193–9197.

14. Plowman GD, Whitney GS, Neubauer MG, et al. Molecular cloning and expression of an additional epidermal growth factor receptor-related gene. Proc Natl Acad Sci USA 1990;87:4905–4909.

15. Plowman GD, Culouscou J-M, Whitney GS, et al. Ligand-specific activation of HER4/p180^{erbB4}, a fourth member of the epidermal growth factor receptor family. Proc Natl Acad Sci USA 1993;90:1746–1750.

16. Drebin AJ, Stern DF, Link VC, Weinberg RA, Greene MI. Monoclonal antibodies identify a cell-surface antigen associated with an activated cellular oncogene. Nature 1984;321:545–548.

17. Stern DF, Heffernan PA, Weinberg RA. p185, a product of the *neu* proto-oncogene, is a receptor like protein associated with tyrosine kinase activity. Mol Cell Biol 1986;6:1729–1740.

18. Bargmann CI, Hung M-C, Weinberg RA. The *neu* oncogene encodes an epidermal growth factor receptor-related protein. Nature 1986;319:226–230.

19. Akiyama T, Sudo C, Ogawara H, Toyoshima K, Yamamoto T. The product of the human c-*erb*B-2 gene: a 185-kilodalton glycoprotein with tyrosine

kinase activity. Science 1986;232:1644–1646.

20. Yarden Y, Ullrich A. Growth factor receptor tyrosine kinases. Annu Rev Biochem 1988;57:443–478.

21. Hazan R, Margolis B, Dombalagian M, Ullrich A, Zilberstein A, Schlessinger J. Identification of autophosphorylation sites of HER2/*neu*. Cell Growth Differ 1990;1:3–7.

22. Margolis BL, Lax I, Kris R, et al. All autophosphorylation sites of epidermal growth factor (EGF) receptor and HER2/*neu* are located in their carboxyl-terminal tails. Identification of a novel site in EGF receptor. J Biol Chem 1989;264:10667–10671.

23. Bargmann CI, Hung M, Weinberg RA. Multiple independent activations of the *neu* oncogene by a point mutation altering the transmembrane domain of p185. Cell 1986;45:649–657.

24. Stern DF, Kamps MP, Cao H. Oncogenic activation of p185*neu* stimulates tyrosine phosphorylation in vivo. Mol Cell Biol 1988;8:3969–3973.

25. Bargmann CI, Weinberg RA. Increased tyrosine kinase activity associated with the protein encoded by the activated *neu* oncogene. Proc Natl Acad Sci USA 1988;85:5394–5398.

26. Bargmann CI, Weinberg RA. Oncogenic activation of the *neu*-encoded receptor protein by point mutation and deletion. EMBO J 1988;7:2043–2052.

27. Stern DF, Kamps MP. EGF-stimulated tyrosine phosphorylation of $p185^{neu}$: a potential model for receptor interactions. EMBO J 1988;7:995–1001.

28. Weiner DB, Kokai Y, Wada T, Cohen JA, Williams WV, Greene MI. Linkage of tyrosine kinase activity with transforming ability of the p185neu oncoprotein. Oncogene 1989;4:1175–1183.

29. Weiner DB, Liu J, Cohen JA, Williams WV, Greene MI. A point mutation in the *neu* oncogene mimics ligand induction of receptor aggregation. Nature 1989;339:230–231.

30. Hung M-C, Yan D-H, Zhao X. Amplification of the proto-*neu* oncogene facilitates oncogenic activation by a single point mutation. Proc Natl Acad Sci USA 1989;86:2545–2548.

31. Zhao X-Y, Hung M-C. Negative autoregulation of the *neu* gene is mediated by a novel enhancer. Mol Cell Biol 1992; in press.

32. Di Marco E, Pierce JH, Knicley CL, Di Fiore PP. Transformation of NIH 3T3 cells by overexpression of the normal coding sequence of the rat *neu* gene. Mol Cell Biol 1990;10:3247–3252.

33. Lemoine NR, Staddon S, Dickson C, Barnes DM, Gullick WJ. Absence of activating transmembrane mutations in the c-*erb*B-2 proto-oncogene in human breast cancer. Oncogene 1990;5:237–239.

34. Saya H, Ara S, Lee PS, Ro J, Hung M-C. Direct sequencing analysis of transmembrane region of human *neu* gene by polymerase chain reaction. Mol Carcinog 1990;3:198–201.

35. Kraus MH, Popescu NC, Amsbaugh SC, King CR. Overexpression of the EGF receptor-related proto-oncogene *erb*B-2 in human mammary tumor cell lines by different molecular mechanisms. EMBO J 1987;6:605–610.

36. Di Fiore PP, Pierce JH, Kraus MH, Segatto O, King CR, Aaronson SA.

*erb*B-2 is a potent oncogene when overexpressed in NIH/3T3 cells. Science 1987;237:178–182.

37. Hudziak RM, Schlessinger J, Ullrich A. Increased expression of the putative growth factor receptor p185^{HER2} causes transformation and tumorigenesis of NIH 3T3 cells. Proc Natl Acad Sci USA 1987;84:7159–7163.

38. Segatto O, King CR, Pierce JH, Di Fiore PP, Aaronson SA. Different structural alterations upregulate in vitro tyrosine kinase activity and transforming potency of the *erb*B-2 gene. Mol Cell Biol 1988;8:5570–5574.

39. Wen D, Peles E, Cupples R, et al. Neu differentiation factor: a transmembrane glycoprotein containing an EGF domain and an immunoglobulin homology unit. Cell 1992;69:559–572.

40. Peles E, Bacus SS, Koski RA, et al. Isolation of the neu/HER-2 stimulatory ligand: a 44 kd glycoprotein that induces differentiation of mammary tumor cells. Cell 1992;69:205–216.

41. Holmes WE, Sliwkowski MX, Akita RW, et al. Identification of heregulin, a specific activator of p185^{erbB2}. Science 1992;256:1205–1210.

42. Lupu R, Colomer R, Zugmaier G, et al. Direct interaction of a ligand for the erbB2 oncogene product with the EGF receptor and p185^{erbB2}. Science 1990;249:1552–1555.

43. Yarden Y, Peles E. Biochemical analysis of the ligand for the neu oncogene receptor. Biochemistry 1991;30:3543–3550.

44. Huang SS, Huang JS. Purification and characterization of the neu/erbB2 ligand-growth factor from bovine kidney. J Biol Chem 1992;267:11508–11512.

45. Dobashi K, Davis JG, Mikani Y, Freeman JK, Hamuro J, Green MI. Characterization of a *neu*/c-*erb*B-2 protein specific activating factor. Proc Natl Acad Sci USA 1991;88:8582–8586.

46. Falls DL, Rosen KM, Corfas G, Lane WS, Fischbach GD. ARIA, a protein that stimulates acetylcholine receptor synthesis, is a member of the neu ligand family. Cell 1993;72:801–815.

47. Marchionni MA, Goodearl ADJ, Chen MS, et al. Glial growth factors are alternatively spliced erbB2 ligands expressed in the nervous system. Nature 1993;362:312–317.

48. Culouscou J-M, Plowman GD, Carlton GW, Green JM, Shoyab M. Characterization of a breast cancer cell differentiation factor that specifically activates the HER4/p180^{erbB4} receptor. J Biol Chem 1993;268:18407–18410.

49. Lupu R, Colomer R, Kannan B, Lippman ME. Characterization of a growth factor that binds exclusively to the *erb*B-2 receptor and induces cellular responses. Proc Natl Acad Sci USA 1992;89:2287–2291.

50. Plowman GD, Green JM, Culouscou J-M, Carlton GW, Rothwell VM, Buckley S. Heregulin induces tyrosine phosphorylation of HER4/p180^{erbB4}. Nature 1993;366:473–475.

51. Matsuda S, Kadowaki Y, Ichino M, Akiyama T, Toyoshima K, Yamamoto T. 17β-Estradiol mimics ligand activity of the c-erbB2 protooncogene product. Proc Natl Acad Sci USA 1993;90:10803–10807.

52. Koskinen P, Lehvaslaiho H, MacDonald BH, Alitalo K, Bravo R. Similar early gene responses to ligand-activated EGFR and *neu* tyrosine kinases in

NIH3T3 cells. Oncogene 1990;5:615–618.

53. Lehvaslaiho H, Lehtola L, Sistonen L, Alitalo K. A chimeric EGF-R-*neu* proto-oncogene allows EGF to regulate *neu* tyrosine kinase and cell transformation. EMBO J 1989;8:159–166.

54. Pandiella A, Lehvaslaiho H, Magni M, Alitalo K, Meldolesi J. Activation of an EGFR/neu chimeric receptor: early intracellular signals and cell proliferation responses. Oncogene 1989;4:1298–1305.

55. Sistonen L, Holtta E, Lehvaslaiho H, Lehtola L, Alitalo K. Activation of the neu tyrosine kinase induces the fos/jun transcription factor complex, the glucose transporter and ornithine decarboxylase. J. Cell Biol 1989;109:1911–1919.

56. Lee J, Dull TJ, Lax I, Schlessinger J, Ullrich A. HER2 cytoplasmic domain generates normal mitogenic and transforming signals in a chimeric receptor. EMBO J 1989;8:167–173.

57. Di Fiore PP, Segatto O, Taylor WG, Aaronson SA, Pierce JH. EGF-receptor and erbB-2 tyrosine kinase domains confer cell specificity for mitogenic signaling. Science 1990;248:79–83.

58. Mikami Y, Davis JG, Dobashi K, et al. Carboxyl-terminal deletion and point mutations decrease the transforming potential of the activated rat *neu* oncogene product. Proc Natl Acad Sci USA 1992;89:7335–7339.

59. Yarden Y. Agonistic antibodies stimulate the kinase encoded by the *neu* protooncogene in living cells but the oncogenic mutant is constitutively active. Proc Natl Acad Sci USA 1990;87:2569–2573.

60. Van Leeuwen F, van de Vijver MJ, Lomans J, et al. Mutation of the human *neu* protein facilitates down-modulation by monoclonal antibodies. Oncogene 1990;5:497–503.

61. Sternberg MJ, Gullick WJ. *Neu* receptor dimerization. Nature 1989;339:587.

62. Huang SS, Koh HA, Konish Y, Bullock LD, Huang JS. Differential processing and turnover of the oncogenically activated *neu*/erb B2 gene product and its normal cellular counterpart. J Biol Chem 1990;265:3340–3346.

63. Reardon DB, Hung M-C. Downstream signal transduction defects that suppress transformation in two revertant cell lines expressing activated rat *neu* oncogene. J Biol Chem 1993;268:18136–18142.

64. Kadowaki T, Kasuaga M, Tobe K, et al. A M_r = 190,000 glycoprotein phosphorylated on tyrosine residues in epidermal growth factor stimulated KB cells is the product of the C-erbB-2 gene. Biochem Biophys Res Commun 1987;144:699–704.

65. Muller WJ, Sinn E, Pattengale PK, Wallace R, Leder P. Single-step induction of mammary adenocarcinoma in transgenic mice bearing the activated c-*neu* oncogene. Cell 1988;54:105–115.

66. Bouchard L, Lamarre L, Tremblay PJ, Jolicoeur P. Stochastic appearance of mammary tumors in transgenic mice carrying the MMTV/c-neu oncogene. Cell 1989;57:931–936.

67. Drebin JA, Link VC, Stern DF, Weinberg RA, Greene MI. Down-modulation of an oncogene protein product and reversion of the transformed

phenotype by monoclonal antibodies. Cell 1985;41:695–706.

68. Drebin JA, Link VC, Greene MI. Monoclonal antibodies reactive with distinct domains of the *neu* oncogene-encoded p185 molecule exert synergistic anti-tumor effects in vivo. Oncogene 1988;2:273–277.

69. Drebin JA, Link VC, Greene MI. Monoclonal antibodies specific for the *neu* oncogene product directly mediate anti-tumor effects in vivo. Oncogene 1988;2:387–394.

70. Bernards R, Destree A, McKenzie S, Gordon E, Weinberg RA, Panicali D. Effective tumor immunotherapy directed against an oncogene-encoded product using a vaccinia virus vector. Proc Natl Acad Sci USA 1987;84:6854–6858.

71. Flanagan JG, Leder P. Neu protooncogene fused to an immunoglobulin heavy chain gene requires immunoglobulin light chain for cell surface expression and oncogenic transformation. Proc Natl Acad Sci USA 1988;85:8057–8061.

72. Tal M, Wetzler M, Josefberg Z, et al. Sporadic amplification of the HER2/*neu* protooncogene in adenocarcinomas of various tissues. Cancer Res 1988;48:1517–1520.

73. Slamon DJ, Clark GM, Wong SG, Levin WJ, Ullrich A, McGuire WL. Human breast cancer: correlation of relapse and survival with amplification of the HER-2/*neu* oncogene. Science 1987;235:177–182.

74. Van de Vijver M, van de Bersselaar R, Devilee P, Cornelisse C, Peterse J, Nusse R. Amplification of the *neu* (c-*erb*B-2) oncogene in human mammmary tumors is relatively frequent and is often accompanied by amplification of the linked c-*erb*A oncogene. Mol Cell Biol 1987;7:2019–2023.

75. Varley JM, Swallow JE, Brammar WJ, Whittaker JL, Walker RA. Alterations to either c-erbB-2(*neu*) or c-*myc* proto-oncogenes in breast carcinomas correlate with poor short-term prognosis. Oncogene 1987;1:423–430.

76. Venter DJ, Tuzi NL, Kumar S, Gullick WJ. Overexpression of the cerbB-2 oncoprotein in human breast carcinomas: immunohistological assessment correlates with gene amplification. Lancet 1987; July:69–71.

77. Adnane J, Gaudray P, Simon M, Simony-Lafontaine J, Jeanteur P, Theillet C. Proto-oncogene amplification and human breast tumor phenotype. Oncogene 1989;4:1389–1395.

78. Garcia G, Dietrich M, Aapro M, Vauthier G, Vadas L, Engel E. Genetic alterations of c-myc, cerbB-2, and c-Ha-ras protooncogenes and clinical associations in human breast carcinomas. Cancer Res 1987;49:6675–6679.

79. King CR, Swain SM, Porter L, Steinberg SM, Lippman ME, Gelman EP. heterogeneous expression of erbB-2 messenger RNA in human breast cancer. Cancer Res 1989;49:4185–4191.

80. Zeillinger R, Kury F, Czerwenka K, et al. HER-2 amplification, steroid receptors and epidermal growth factor receptor in primary breast cancer. Oncogene 1989;4:109–114

81. Tsuchiya T, Ueyama Y, Tamaoki N, Yamaguchi S, Shibuya M. Co-amplification of c-myc and c-erbB-2 oncogenes in a poorly differentiated human gastric cancer. Jpn J Cancer Res 1989;80:920–923.

82. Yokota J, Yamamoto T, Miyajima N, et al. Genetic alterations of the c-*erb*B-2 oncogene occur frequently in tubular adenocarcinoma of the stomach and

are often accompanied by amplification of the v-*erb*A homologue. Oncogene 1988;2:283–287.

83. Thor AD, Schwartz LH, Koerner FC, et al. Analysis of c-erbB-2 expression in breast carcinomas with clinical follow-up. Cancer Res 1989;49:7147–7152.

84. Wright C, Angus B, Nicholson S, et al. Expression of c-*erb*B-2 oncoprotein: a prognostic indicator in human breast cancer. Cancer Res 1989;49:2087–2090.

85. Lacroix H, Iglehart JD, Skinner MA, Kraus MH. Overexpression of erbB-2 or EGF-receptor proteins present in early stage mammary carcinoma is detected simultaneously in matched primary tumors and regional metastasis. Oncogene 1989;4:145–151.

86. Guerin M, Barrois M, Terrier MJ, Spielmann M, Riou G. Overexpression of either c-myc or c-erbB-2/*neu* proto-oncogenes in human breast carcinomas: correlation with poor prognosis. Oncogene Res 1988;3:21–31.

87. Ro JS, el Naggar A, Ro JY, et al. c-erbB-2 amplification in node-negative human breast cancer. Cancer Res 1989;49:6941–6944.

88. Cline MJ, Battifora H, Yokota J. Protooncogene abnormalities in human breast cancer: correlations with anatomic features and clinical course of disease. J Clin Oncol 1987;7:999–1006.

89. Ali IU, Campbell G, Lidereau R, Callahan R. Lack of evidence for the prognostic significance of c-erbB-2 amplification in human breast carcinoma. Oncogene Res 1988;3:139–146.

90. Zhou D, Ahuja H, Cline MJ. Proto-oncogene abnormalities in human breast cancer: c-erbB-2 amplification does not correlate with recurrence of disease. Oncogene 1989;4:105–108.

91. Slamon DJ, Godolphin W, Jones LA, et al. Studies of the HER-2/*neu* proto-oncogene in human breast and ovarian cancer. Science 1989;244:707–712.

92. Nagai MA, Marques LA, Torloni H, Brentani MM. Genetic alterations in c-erbB-2 protooncogene as prognostic markers in human primary breast tumors. Oncology 1993;50:412–417.

93. Zhang X, Silva E, Gershenson D, Hung M-C. Amplification and rearrangement of c-*erb*B proto-oncogenes in cancer of human female genital tract. Oncogene 1989;4:985–989.

94. Berchuck A, Kamel A, Whitaker R, et al. Overexpression of HER-2/*neu* is associated with poor survival in advanced epithelial ovarian cancer. Cancer Res 1990;50:4087–4091.

95. Park J-B, Rhim JS, Park S-C, Kimm S-W, Kraus MH. Amplification, overexpression, and rearrangement of the *erb*B-2 protooncogene in primary human stomach carcinomas. Cancer Res 1989;49:6605–6609.

96. D'Emilia J, Bulovas K, D'Ercole K, Wolf B, Steele G Jr, Summerhayes IC. Expression of the c-*erb*B-2 gene product (p185) at different stages of neoplastic progression in the colon. Oncogene 1989;4:1233–1239.

97. Schneider PM, Hung MC, Chiocca SM, Manning J, Zhao XY, Fang K, Roth JA. Differential expression of the c-*erb*B-2 gene in human small cell and non-small cell lung cancer. Cancer Res 1989;49:4968–4971.

98. Weiner DB, Nordberg J, Robinson R, et al. Expression of the *neu* gene-encoded protein (p185*neu*) in human non-small cell carcinomas of the lung. Cancer Res 1990;50:421–425.

99. Hou L, Shi D, Tu S-M, Zhang H-Z, Hung M-C, Ling D. Oral cancer progression and c-erbB-2/neu proto-oncogene expression. Cancer Lett 1992;65:215–220.

100. Mitra AB, Murty VVVS, Pratap M, Sodhani P, Chaganti RSK. ERBB2 (HER-2/neu) oncogene is frequently amplified in squamous cell carcinoma of the uterine cervix. Cancer Res 1994;54:637–639.

101. Tsugawa K, Fushida S, Yonemura Y. Amplification of the c-erbB-2 gene in gastric carcinoma: correlation with survival. Oncology 1993;50:418–425.

102. Tauchi K, Hori S, Itoh H, Osamura RY, Tokuda Y, Tajima T. Immunohistochemical studies on oncogene products (c-erbB-2, EGFR, c-myc) and estrogen receptor in benign and malignant breast lesions—with special reference to their prognostic significance in carcinoma. Virchows Arch A 1989;416:65–73.

103. Liotta LA, ed. Mechanisms of cancer invasion and metastasis. Dordrecht, the Netherlands: Kluwer Academic Publishers, 1989.

104. Nicolson GL. Tumor cell instability, diversification and progression to the metastatic phenotype: from oncogene to oncofetal expression. Cancer Res 1987;47:1473–1487.

105. Fidler IJ, Hart IR. Biologic diversity in metastatic neoplasms—origins and implications. Science 1982;217:998–1001.

106. Schirrmacher V. Experimental approaches, theoretical concepts, and impacts for treatment strategies. Adv Cancer Res 1985;43:1–32.

107. Liotta LA, Rao CN, Barsky SH. Tumor invasion and the extracellular matrix. Lab Invest 1983;49:636–649.

108. McGuire WL, Tandon AK, Allred DC, Chamness GC, Clark GM. How to use prognostic factors in axillary node-negative breast cancer patients. J Natl Cancer Inst 1990;82:1006–1015.

109. Lidereau R, Callahan R, Dickson C, Peters G, Escot C, Ali IU. Amplification of the *int*-2 gene in primary human breast tumors. Oncogene Res 1988;2:285–291.

110. Tandon AK, Clark GM, Chamness GC, Chirgwin J, McGuire WL. Cathepsin D and prognosis in breast cancer. N Engl J Med 1990;322:297–302.

111. Tandon AK, Clark GM, Chamness GC, McGuire WL. Association of the 323/A3 surfact glycoprotein with tumor characteristics and behavior in human breast cancer. Cancer Res 1990;50:3317–3321.

112. Tandon AK, Clark GM, Chamness GC, Ullrich A, McGuire WL. HER-2/neu oncogene protein and prognosis in breast cancer. J Clin Oncol 1989;7:1120–1128.

113. Hart IR, Goode NT, Wilson RE. Molecular aspects of the metastatic cascade. Biochim Biophys Acta 1989;989:65–84.

114. Yu D, Hung MC. Expression of activated rat *neu* oncogene is sufficient to induce experimental metastasis in 3T3 cells. Oncogene 1991;6:1991–1996.

115. Yu D, Hamada J-I, Zhang H, Nicolson GL, Hung M-C. Mechanisms of *neu* oncogene induced metastasis and abrogation of metastatic properties by the

adenovirus 5 E1A gene products. Oncogene 1992;7:2263–2270.

116. Yusa K, Sugimoto Y, Yamori T, Yamamoto T, Toyoshima K, Tsuruo T. Low metastatic potential of clone from murine colon adenocarcinoma 26 increased by transfection of activated c-*erb*B-2 gene. J Natl Cancer Inst 1990;82:1633–1636.

117. Suda Y, Aizawa S, Furuta Y, et al. Induction of a variety of tumors by c-*erb*B2 and clonal nature of lymphomas even with the mutated gene (Val659–Glu659). EMBO J 1990;9:181–190.

118. Guy CT, Webster MA, Schaller M, Parsons TJ, Cardiff RD, Muller WJ. Expression of the neu protooncogene in the mammary epithelium of transgenic mice induces metastatic disease. Proc Natl Acad Sci USA 1992;89:10578–10582.

119. Piver MS, Baker TR, Piedmonte M, Sandecki AM. Epidemiology and etiology of ovarian cancer. Semin Oncol 1991;18:177–185.

120. Hung M-C, Zhang X, Yan D-H, et al. Aberrant expression of the c-*erb*B-2/*neu* protooncogene in ovarian cancer. Cancer Lett 1992;61:95–103.

121. Yu D, Wolf JK, Scanlon M, Price JE, Hung M-C. Enhanced c-*erb*B-2/*neu* expression in human ovarian cancer cells correlates with more severe malignancy that can be suppressed by E1A. Cancer Res 1993;53:891–898.

122. Silverberg E, Lubera JA. Cancer statistics. CA Cancer J Clin 1988; 38:5–22.

123. Kern JA, Schwartz DA, Nordberg JE, et al. $p185^{neu}$ expression in human lung adenocarcinomas predicts shortened survival. Cancer Res 1990;50: 5184–5191.

124. Sozzi G, Miozzo M, Tagliabue E, et al. Cytogenetic abnormalities and overexpression of receptors for growth factors in normal bronchial epithelium and tumor samples of lung cancer patients. Cancer Res 1991;51:400–404.

125. Shi D, He G, Cao S, et al. Overexpression of the c-*erb*B-2/*neu*-encoded p185 protein in primary lung cancer. Mol Carcinog 1992;5:213–218.

126. Moscow JA, Cowan KH. Multidrug resistance. J Natl Cancer Inst 1988;80:14.

127. Cole SPC, Bhardwaj G, Gerlach JH, et al. Overexpression of a transporter gene in a multidrug-resistant human lung cancer cell line. Science 1992;258:1650–1654.

128. Lowe SW, Ruley HE, Jacks T, Housman DE. p53-dependent apoptosis modulates the cytotoxicity of anticancer agents. Cell 1993;74:957–967.

129. Dive C, Hickman JA. Drug-target interactions: only the first step in the commitment to a programmed cell death. Br J Cancer 1991;64:192–196.

130. Harris JR, Lippman ME, Veronesi U, Willett W. Breast cancer. N Engl J Med 1992;327:319–328, 390–398, 473–480.

131. Gusterson BA, Gelber RD, Goldhirsch A, et al. Prognostic importance of c-erbB-2 expression in breast cancer. J Clin Oncol 1992;10:1049–1056.

132. Tsai C-M, Chang K-T, Perng R-P, et al. Correlation of intrinsic chemoresistance of non-small-cell lung cancer cell lines with HER-2/neu gene expression but not with ras gene mutation. J Natl Cancer Inst 1993;85:897–901.

133. Canellos GP, De Vita VT, Gold LG, Chabner BA, Schein PS, Young RC. Combination chemotherapy for advanced breast cancer: response and effect

on survival. Ann Intern Med 1976;84:389–392.

134. Kiang DT, Gay J, Goldman A, Kennedy BJ. A randomized trial of chemotherapy and hormonal therapy in advanced breast cancer. N Engl J Med 1985;313:1241–1246.

135. Alley MC, Scudiero DA, Monks A, et al. Feasibility of drug screening with panels of human tumor cell lines using microculture tetrazolium assay. Cancer Res 1988;48:589–601.

136. Fuqua SAW, Fitzgerald SD, Chamness GC, et al. Variant human breast tumor estrogen receptor with constitutive transcriptional activity. Cancer Res 1991;51:105–109.

137. Read LD, Keith D Jr, Slamon DJ, Katzenellenbogen BS. Hormonal modulation of HER-2/*neu* protooncogene messenger ribonucleic acid and p185 protein expression in human breast cancer cell lines. Cancer Res 1990;50:3947–3951.

138. Dati C, Antoniotti S, Taverna D, Perroteau I, De Bortoli M. Inhibition of c-*erb*B-2 oncogene expression by estrogens in human breast cancer cells. Oncogene 1990;5:1001–1006.

139. Russell KS, Hung MC. Transcriptional repression of the *neu* protooncogene by estrogen. Cancer Res 1992;52:6624–6629.

140. Ishii S, Imamoto F, Yamanashi Y, Toyoshima K, Yamamoto T. Characterization of the promoter region of the human c-erbB-2 protooncogene. Proc Natl Acad Sci USA 1987;84:4374–4378.

141. Tal M, King CR, Kraus MH, Ullrich A, Schlessinger J, Givol D. Human HER2 (neu) promoter: evidence for multiple mechanisms for transcriptional initiation. Mol Cell Biol 1987;7:2597–2601.

142. Fernandez Pol JA, Hamilton PD, Klos DJ. Transcriptional regulation of proto-oncogene expression by epidermal growth factor, transforming growth factor beta 1, and triiodothyronine in MDA-468 cells. J Biol Chem 1989;264:4151–4156.

143. Hudson LG, Ertl AP, Gill GN. Structure and inducible regulation of the human c-erb B-2/*neu* promoter. J Biol Chem 1990;265:4389–4393.

144. Hudson LG, Santon JB, Glass CK, Gill GN. Ligand-activated thyroid hormone and retinoic acid receptors inhibit growth factor receptor promoter expression. Cell 1990;62:1165–1175.

145. Suen TC, Hung MC. Multiple *cis*- and *trans*-acting elements involved in regulation of the *neu* gene. Mol Cell Biol 1990;10:6306–6315.

146. Yan D, Hung M. Identification and characterization of a novel enhancer for the rat neu promoter. Mol Cell Biol 1991;11:1875–1882.

147. Yu D, Suen T-C, Yan D-H, Chan LS, Hung M-C. Transcriptional repression of the *neu* protooncogene by the adenovirus 5 E1A gene products. Proc Natl Acad Sci USA 1990;87:4499–4503.

148. Suen T, Hung M. c-*myc* reverses *neu*-induced transformed morphology by transcriptional repression. Mol Cell Biol 1991;11:354–362.

149. Matin A, Hung M-C. Negative regulation of the neu promoter by the SV40 large T antigen. Cell Growth Differ 1993;4:1051–1056.

150. Miller SJ, Hung M-C. HER-2/*neu* overexpression counteracts the

growth effects of c-myc in breast cancer cells. Int J Oncol 1994;4;965–969.

151. Yu D, Matin A, Hung M-C. The retinoblastoma gene product suppresses *neu* oncogene-induced transformation via transcriptional repression of *neu*. J Biol Chem 1992;267:10203–10206.

152. Yu D, Matin A, Hinds PW, Hung M-C. Transcriptional regulation of *neu* oncogene by RB and E1A in rat-1 cells. Cell Growth Differ 1994;5;431–438.

153. Matin A, Hung M-C. The retinoblastoma gene product, RB, represses *neu* expression through two regions within the *neu* regulatory sequence. Oncogene 1994;9:1333–1339.

154. Miller SJ, Suen TC, Sexton TB, Hung MC. Both transcriptional and post-transcriptional deregulation contribute to overexpression of neu in breast cancer. Intl J Oncol 1994;4:599–608.

155. Pasleau F, Grooteclaes M, Gol-Winkler R. Expression of the c-erbB2 gene in the BT474 human mammary tumor cell line: measurement of c-erbB2 mNRA half-life. Oncogene 1993;8:849–854.

156. Hollywood DP, Hurst HC. A novel transcription factor, OB2-1, is required for overexpression of the proto-oncogene c-erbB-2 in mammary tumor lines. EMBO J 1993;12:2369–2375.

157. Shepard HM, Lewis GD, Sarup JC, et al. Monoclonal antibody therapy of human cancer: taking the HER2 protooncogene to the clinic. J Clin Immunol 1992;11:117–127.

158. Yan D, Chang L-S, Hung M. Repressed expression of the HER-2/c-*erb*B-2 proto-oncogene by the adenovirus E1a gene products. Oncogene 1991;6:343–345.

159. Yu D, Scorsone K, Hung M-C. Adenovirus type 5 E1A gene products act as transformation suppressors of the *neu* oncogene. Mol Cell Biol 1991;11:1745–1750.

160. Frisch SM. Antioncogenic effect of adenovirus E1A in human tumor cells. Proc Natl Acad Sci USA 1991;88:9077–9081.

161. Yu D, Shi D, Scanlon M, Hung M-C. Reexpression of *neu*-encoded oncoprotein counteracts the tumor-suppressing but not the metastasis-suppressing function of E1A. Cancer Res 1993;53:5784–5790.

162. Felgner PL, Ringold GM. Cationic liposome-mediated transfection. Nature 1989;337:387–388.

163. Li Q, Kay M, Finegold M, Stratford-Perricaudet LD, Woo SLC. Assessment of recombinant adenoviral vectors for hepatic gene therapy. Hum Gene Ther 1993;4:403–409.

CHAPTER 6

Tumor Angiogenesis and the Role of Paracrine Mediators

Richard S. Morrison
Alan M. Yahanda

Angiogenesis is the process by which new blood vessels are produced and recruited during normal tissue growth and in certain pathological conditions. Tumor development presents a striking example of a pathological process that is dependent on the development of new blood vessels. There is considerable evidence that tumor growth and metastasis are dependent on angiogenesis (1). The induction of new capillary blood vessels by solid tumors is essential for expansion of a developing tumor mass. In the absence of angiogenesis, tumor growth is restricted to a volume of only a few cubic millimeters. This is due to limitations in the diffusion of nutrients and metabolites both to and from the tumor mass.

The ability of tumors to accrue new vasculature is a complex process that is recognized as a distinct and essential stage of neoplastic progression. Many tumors exhibit a prevascular phase during early tumor development in which the growth of the tumor is restricted to a few cubic millimeters. After sufficient numbers of tumor cells have acquired the ability to induce the development of new blood vessels, the tumor can expand progressively and disseminate metastatic cells. The separation of tumor growth into these two distinct phases has been documented for several types of cancer, including melanoma (2), carcinoma of the cervix (3), and breast cancer (4). Although angiogenesis is necessary for tumor growth, it alone is not sufficient; tumor growth also requires the acquisition of enhanced proliferative and invasive properties by cells. Curtailing the angiogenic process in tumors, however, can have significant impact on tumor growth and metastasis (5–7).

The process of angiogenesis is complex, and it is tightly controlled in normal tissues. The steps that are essential to new blood vessel formation include 1) retraction of pericytes; 2) degradation of the extracellular matrix surrounding the capillaries; 3) migration and proliferation of endothelial cells; 4) formation of capillary tubes by endothelial cells; and 5) the formation of anastomoses and the establishment of blood flow between sprouting vessels and existing vessels.

The means by which tumors are able to break down or circumvent the mechanisms that normally control angiogenesis are equally complex. Tumor cells have evolved different means of recruiting new blood vessels (additional details are contained in several excellent reviews; 8–11). These include 1) production of diffusible factors that directly activate endothelial cell migration and proliferation (angiogenic factors); 2) production of cytokines that facilitate recruitment of macrophages or mast cells, which in turn produce angiogenic factors; 3) down-regulating expression of angiogenic inhibitors, such as thrombospondin (12); 4) production of proteolytic enzymes that facilitate degradation of the basement membrane and the release of matrix-associated growth factors such as basic fibroblast growth factor (bFGF); and 5) production of diffusible factors that elevate the expression of endothelial-cell-derived growth factors such as bFGF (13) and platelet-derived growth factor receptor-β (14).

It is clear that as our understanding of the cellular and molecular mechanisms underlying the angiogenic process becomes more complete, so will our understanding of the process of tumorigenesis. This review will focus on two mitogenic proteins, bFGF and vascular endothelial growth factor (VEGF), which are potent paracrine stimulators of endothelial cell growth and are important mediators of tumor progression. Due to the biological potency of both and the specificity of VEGF for endothelial cells, these two growth factors are important targets for therapies aimed at preventing tumor growth by inhibiting angiogenesis.

BASIC FIBROBLAST GROWTH FACTOR

Basic fibroblast growth factor belongs to a family of structurally related polypeptide mitogens (15,16). Nine members of this family have been identified on the basis of amino acid sequence homologies (17–28; Table 6.1). Basic fibroblast growth factor is a multifunctional protein recognized primarily for its mitogenic and angiogenic properties. On the basis of cell culture studies, bFGF has been shown to be mitogenic for a wide range of cell types derived from mesoderm and neuroectoderm. In addition to the many in vitro studies, bFGF has also been demonstrated to be active in numerous in vivo models of angiogenesis and wound healing (15,16). It also exhibits potent neurotrophic actions, promoting the survival and differentiation of neurons in the peripheral and central nervous systems (29–32). In general, proteins of the fibroblast growth factor (FGF) family

Table 6.1: Fibroblast Growth Factor Family of Heparin-Binding Growth Factors

Growth Factor	*Reference*
FGF-1 (acidic FGF)	Jaye et al. (17)
FGF-2 (basic FGF)	Abraham et al. (18)
FGF-3 (*int*-2)	Moore et al. (19)
FGF-4 (kFGF/hst)	Delli-Bovi et al. (20); Sakamoto et al. (21)
FGF-5	Zhan ct al. (22)
FGF-6	Marics et al. (23)
FGF-7 (keratinocyte growth factor [KGF])	Rubin et al. (24); Finch et al. (25)
FGF-8 (androgen-induced growth factor)	Tanaka et al. (26)
FGF-9	Naruo et al. (27); Miyamoto et al. (28)

express an extremely diverse range of actions. They have the capacity to induce, inhibit, and maintain the differentiation of many cell types in culture. Furthermore, there is substantial evidence that FGF family members display distinct temporal patterns of expression, with some members playing an important role during early development (33).

Basic FGF was one of the first FGF family members to be purified and characterized. This was accomplished by assaying its mitogenic activities on fibroblasts. It has subsequently been purified from a large variety of sources and has been identified in all organs, solid tissues, tumors, and cultured cells examined (15,16). However, the in vitro distribution of bFGF may not accurately reflect its distribution by cells in vivo. The expression of bFGF and certain of the FGF receptors (FGFRs) in some cultured cells may represent an adaptive response that allows survival outside the organism. Other FGF family members exhibit a more restricted tissue distribution (34–42).

The purification and sequencing of bFGF facilitated the cloning of its complementary DNA (18,43,44). The cloned cDNA for bFGF encoded a protein consisting of 155 amino acids. Western blotting and immunoprecipitation experiments, however, often demonstrated larger molecular weight forms of 22.5, 23, and 24 kDa. These isoforms, as well as the smaller 18-kDa form, can be transcribed from a single human bFGF messenger RNA transcript (45,46). The higher molecular weight proteins represent amino-terminal extensions of the 18-kDa bFGF, and appear to be initiated at leucine codons upstream of the first methionine. Potential upstream CTG start sites are also present in the bFGF message of other species, including rat (47). Moreover, the different molecular weight forms have distinct intracellular localizations, suggesting that they may have unique activities (48,49). Only the 18-kDa form appears to be released or exported from cells (50). The mechanism by which bFGF is released from cells is unclear, as the protein lacks a classic signal sequence necessary for its secretion.

Basic FGF is derived from a distinct, single copy gene on chromosome 4 (51). The human bFGF gene is composed of three exons interrupted by two large introns (18,51). Multiple bFGF mRNA transcripts are produced that differ in the length of their 3′ untranslated region. This is consistent with primer extension analysis conducted in the presence of multiple bFGF mRNA species. Such studies depict a single product, suggesting that the 5′ ends of all the bFGF mRNAs must be identical (52). Sequence analysis of the human bFGF gene promoter region demonstrates that it lacks typical TATA or CAAT boxes common to many gene regulatory regions. There is an element responsive to 1,2-O-tetradecanoylphorbol-1 3-acetate (TPA) that may serve as a potential binding site for transcription factor AP-1. This is supported by the reported induction of bFGF mRNA (53) and protein (54) in cells treated with TPA. A more comprehensive understanding of the mechanisms regulating bFGF expression could have a direct impact on modulating tumor cell growth and angiogenesis.

FIBROBLAST GROWTH FACTOR RECEPTORS

The biological responses of FGFs are mediated through specific, high-affinity, transmembrane receptors. Four structurally related genes encoding high-affinity FGFRs have been identified (55–59). In addition to high-affinity binding sites, cells exhibit low-affinity FGF binding sites (60), which have been characterized as extracellular heparan sulfate proteoglycans (61,62). Binding to the low-affinity, glycosaminoglycan sites appears to be obligatory for FGF binding to high-affinity receptors and for biological activity (63). Cells deficient in heparan sulfate biosynthesis are not able to bind or respond to bFGF (64,65). However, the addition of either free heparin or heparan sulfate restores high-affinity binding of bFGF (64). These results demonstrate that heparin-like, low-affinity sites play an important role in the regulation of bFGF activity and in the response of cells to bFGF.

Structural features common to members of the FGFR family include a signal peptide, two or three immunoglobulin-like loops in the extracellular domain, a hydrophobic transmembrane domain, and a highly conserved tyrosine kinase domain split by a short kinase insert sequence.

Overall, the proteins encoded by the four FGFR genes are strikingly similar. The most closely related proteins are FGFR1 and FGFR2 (72% amino acid identity), whereas FGFR3 and FGFR4 are the least closely related (55% identity). Given the structural similarities, it is not surprising that each of the FGFRs can bind several different types of FGFs. However, there are several reports of cell and tissue-specific expression of FGF receptors and responsiveness to different FGF family members (24,66–75). One mechanism for generating this selective responsiveness to different FGF family members would be to alter the ligand-binding specificity or affinity through alternative splicing of RNA, thereby producing several receptor isoforms from a single gene.

Structural variants of FGFR1 and FGFR2 are, in fact, generated by alternative splicing of their RNA transcripts (56,76–78). The divergent receptors generated by this process manifest different ligand-binding specificities and affinities (74,79–82; for recent review, see 83). One common structural variation that involves the second half of the third immunoglobulin-like disulfide loop of FGFR1 and FGFR2 dramatically alters these receptors' ligand-binding properties (74). Another splicing variant results in FGFRs containing either two or three immunoglobulin-like domains in the extracellular region (56,67,84,85). Alternative RNA splicing involving both the first and third immunoglobulin-like domains is subject to cell- and tissue-specific processing that reflects the changing FGF requirement which occurs during tissue growth and differentiation (24,56,67,74). Changes in ligand-binding affinity and specificity resulting from alternative splicing are also likely to be important in some types of human cancers that rely on FGF family members to sustain growth and invasiveness (86,87).

The FGFRs appear to be differentially expressed in diverse tissue types and during different periods of development. Studies that have examined the distribution of FGFRs have relied principally on Northern blotting, the RNase protection assay, and in situ hybridization to demonstrate the presence of messenger RNA transcripts. In general, FGFR1 and FGFR2 appear to be broadly distributed, while FGFR3 and FGFR4 exhibit more restricted patterns of distribution (Table 6.2). For example, in the developing embryo FGFR1 transcripts are predominant in the central nervous system (CNS) and in mesenchyme (68–71,73). FGFR2 transcripts are also observed in the CNS and in epithelium (68–71,73,74). FGFR3 transcripts are predominantly expressed in the CNS and cartilagenous rudiments of developing bone (75). In contrast to the other FGFRs, which are expressed to some degree in the CNS, FGFR4 transcripts are observed in developing endoderm, the myotomal compartment of somites, and in myotomally derived skeletal muscle (58,72). The unique temporal and

Table 6.2: Fibroblast Growth Factor Receptors

Receptor	*Chromosome Localization*	*Embryonic Distribution*	*Amino Acid Homology (R1)*
FGFR1 (flg)	8p12	Mesenchyme, CNS	—
FGFR2 (bek)	10q26	Epithelium-skin, Develop. organs, CNS	72%
FGFR3	4p16.3	CNS, cartilage	62%
FGFR4	5q33-qter	Epithelium (gut), myotomes, skeletal muscle	55%

spatial patterns of expression exhibited by different FGFR family members strongly suggest that they have distinct roles in tissue development, maintenance, and pathology.

RELATION OF BASIC FGF AND FGF RECEPTORS TO TUMOR GROWTH AND TUMOR-DEPENDENT ANGIOGENESIS

A role for bFGF in tumor development is supported by observations that transfection of cells with the bFGF gene increases autocrine growth in monolayer culture and in soft agar (88–92). In addition, several oncogenes discovered in human tumors encode proteins structurally related to the FGF family (20,22,23,25,93,94). Elevations in bFGF and acidic FGF have been detected in several human tumors, including glioblastomas (95–102), gastric carcinoma (103,104), renal cell carcinoma (105), and pancreatic cancer (106). The overexpression of one or more of these FGF family members is associated with increased malignancy in glioblastomas (96,97) and pancreatic cancer (106). Elevated levels of bFGF have also been demonstrated in the urine of patients with bladder cancer and, purportedly, in the urine of patients with cancer outside the urinary tract (107). Amplification of the bFGF-related *int*-2 and *hst* genes, as well as FGFR1 and FGFR2, has also been observed in a small percentage of patients with breast tumors (108,109). These results strongly suggest that the overexpression of FGF family members by tumor cells may contribute to their accelerated growth and progression.

Any contribution made by FGF proteins to disease progression could involve either a direct stimulation of tumor cell proliferation and invasiveness, or concomitantly, increased vascularization resulting from enhanced endothelial cell proliferation and migration. There is recent evidence suggesting that bFGF expression in tumor cells can directly influence cell growth. Antisense primers specific for bFGF have been shown to suppress the growth of both cultured melanoma (110) and glioblastoma cells (111,112). Furthermore, comparison of an invasive human bladder carcinoma cell line with a noninvasive bladder carcinoma cell line demonstrated that only the invasive one produced bFGF protein (113). Addition of exogenous bFGF to the invasive bladder carcinoma cell line induced protein tyrosine phosphorylation while addition of bFGF to the noninvasive cell line had no effect, suggesting that an autocrine loop composed of bFGF protein and functional FGF receptors might contribute to the tumorigenic properties of the invasive cell line. These results are consistent with reports that various neutralizing antibodies against human bFGF inhibit the subcutaneous growth of human glioblastoma cells (114), rat C6 glioma cells (115), and murine K1000 cells (Balb/c 3T3 cells transformed with a leader-sequence-fused bFGF gene) in nude mice (116).

In the study by Gross and associates (115), neutralizing antibodies for bFGF failed to suppress the growth of DLD-2 colon carcinoma cells in nude mice. However, addition of human recombinant bFGF given intraperitoneally daily for 18 days doubled the tumor weight formed by DLD-2 cells. Similarly, DLD-2 cells in vitro were not stimulated to divide by the addition of bFGF and did not express high-affinity bFGF binding sites. Taken together, these results suggest that bFGF could have augmented tumor growth via its effect on angiogenesis. Xenografts derived from both C6 and DLD-2 cells displayed [^{125}I]bFGF binding that was concentrated on vascular-like structures. However, a distinction between [^{125}I]bFGF binding to endothelial cells versus basement membranes was not definitive. Several studies also suggested that bFGF was not essential as an autocrine or paracrine growth factor (117,118). The results of in vivo studies are sometimes difficult to interpret owing to the complexity of tumor growth and angiogenesis. Moreover, because the angiogenic process is probably dependent on multiple factors and pathways, it is not surprising that modulating the action of only one factor produces somewhat ambiguous results.

The precise role played by bFGF in the process of tumor-dependent angiogenesis has not been conclusively defined. Many investigators have implied and assumed that tumor-derived bFGF promotes angiogenesis via a direct effect on endothelial cells. This is based, in part, on the large body of evidence demonstrating direct stimulation of endothelial cell growth by bFGF in vitro. However, if this actually occurs in vivo, then it is essential that endothelial cells express FGFRs in vivo. To the best of our knowledge, there is little or no direct evidence demonstrating in vivo expression of FGFRs by endothelial cells in normal tissues, in tissues following injury, or in tumors. A significant number of studies have examined the distribution and cellular localization of the four FGFR family members using in situ hybridization (13,58,68–75,119–123). None of these studies specifically described the expression of any FGFR family member in endothelial cells. Logan and coworkers (13) demonstrated focal elevation of bFGF and the FGFR1 in glial cells by in situ hybridization following localized cortical brain injury. FGFR1 expression was absent from endothelial cells, but was observed in astroglial cells that closely surrounded capillaries. In contrast to FGFR1, bFGF mRNA was clearly observed in capillary endothelial cells. A study analyzing injury to the skin demonstrated large increases in several FGF family members during wound healing (74). While three different FGFRs were observed in skin (FGFR1–3), no specific mention was made of endothelial cell expression. However, Peters and colleagues (73) demonstrated FGFR1 expression in aortic endothelial cells at embryonic day 12.5, but did not detect either FGFR1 or FGFR2 in small blood vessels. Similarly, there do not appear to be reports of FGFR3 or FGFR4 expression in endothelial cells. It is remarkable that none of these studies specifically described the expression of any FGFR in endothelial cells in normal tissues or in response to injury using in situ hybridization. This suggests that, in vivo, bFGF may not be a direct mitogen for endothelial cells. Alternatively, the levels of FGFR mRNA expression may be below the detection level associated with in situ hybridization histochemistry, or endothelial cells could express other FGFR family members not yet identified.

The absence of FGFR expression in normal or injured tissue does not rule out expression within tumor capillary endothelial cells. Tumor cells could specifically induce FGFR expression in endothelial cells by direct cellular interactions or by the release of diffusible factors. The platelet-derived growth factor receptor-β mRNA was not detectable in the vessels of normal human brain, but was expressed in the endothelial cell proliferation in glioblastomas (14). The induction of this mRNA in malignant glial tumors suggests that the phenotype of endothelial cells can change in response to a neoplastic state. Similar changes could occur for FGFRs. However, in our own analysis of malignant glial tumors, among the most vascularized human solid tumors, FGFR1 immunoreactivity was strongly expressed by glioblastoma cells but was absent in endothelial cells (124). It is conceivable that other FGFR family members are expressed by endothelial cells in these as well as other solid tumors. Further analysis of FGFR expression by a variety of tumors should determine whether FGFRs are expressed in tumor-derived endothelial cells. Until such data are forthcoming, it can only be concluded that the induction of angiogenesis in tissues in vivo following FGF administration (125,126), or following direct gene transfer (127), is mediated indirectly by other cell types.

The lack of detectable FGFR expression in vivo by endothelial cells in tumors is in direct contrast to the expression of receptors for another potent endothelial cell mitogen, vascular endothelial growth factor (VEGF). The expression and distribution of VEGF and its cognate receptors appear to fulfill many of the requirements expected of a paracrine-mediated pathway. In addition, this pattern of expression appears to be maintained by the appropriate cells in vivo, suggesting that VEGF may be an important mediator of endothelial cell growth in normal tissues and in a variety of human neoplasms.

VASCULAR ENDOTHELIAL GROWTH FACTOR

The absence of a classic signal sequence in bFGF and the inability to document the presence of FGF receptors on endothelial cells in vivo led many investigators to search for other paracrine mediators of endothelial cell growth. In 1983, Senger and coworkers described a factor secreted by line 10 guinea pig hepatocarcinoma cells that induced ascites formation when injected intraperitoneally in animals. The factor was named vascular permeability factor (VPF), as it was found to be a potent mediator of microvascular permeability (128). The VPF cDNA coded for a 189 amino acid polypeptide that was closely related to the B chain of platelet-derived growth factor (PDGF) (129). Leung and coworkers simultaneously reported the cloning of a specific endothelial cell mitogen from bovine folliculostellate cells and human HL60 leukemia cells that they named vascular endothelial growth factor (VEGF). In addition to its ability to induce endothelial cell proliferation in vitro, it was able to induce an angiogenic response in chick chorioallantoic membrane. Coincidentally, VEGF had virtually the same nucleotide and amino acid sequences as VPF. They also described two other cDNA clones that coded for distinct VEGF isoforms of 121 and 165 amino acids (130). This same group subsequently described the cloning of a fourth isoform from a fetal liver cDNA library that was composed of 206 amino acids (131). The 165 amino acid isoform ($VEGF_{165}$) appears to be the most abundantly expressed species in most cells.

It is now known that VEGF represents a family of polypeptides encoded by a single gene containing eight exons and seven introns. The four isoforms result from alternative splicing of exons 6 and 7. The $VEGF_{189}$ and $VEGF_{206}$ forms contain both of these exons. The 17 amino acid insertion in $VEGF_{206}$ results from a variant splice donor site in the 5′

end of exon 7. If exon 6 is spliced out, $VEGF_{165}$ results. If both exons 6 and 7 are removed, $VEGF_{121}$ is generated (132).

Further analysis of the human VEGF gene revealed that there was a single major transcriptional start site located 1038 bases upstream of the ATG initiation codon. The promoter region has a number of potential binding sites for transcription factors, including three Sp-1 sites, four Ap-1 sites, and two Ap-2 sites. The validity of these possible Ap-1 and Ap-2 binding sites was confirmed by demonstrating increased VEGF expression after treating cells with phorbol ester (132).

$VEGF_{165}$ is a dimeric, heparin-binding glycoprotein with a molecular weight of approximately 46kDa. Under reducing conditions, the dimeric complex dissociates to yield two 23-kDa monomers. The protein is basic (isoelectric point of 8.5), and is both acid- and heat-stable. In contrast, $VEGF_{121}$ is an acidic protein that binds poorly to heparin. This results from the absence of the basic, 44 amino acid region encoded by exon 7. The biochemical properties of $VEGF_{189}$ and $VEGF_{206}$ are similar to those of $VEGF_{165}$ (133–137).

The addition or deletion of amino acids in the four VEGF isoforms also results in significant changes in their biological function. Both $VEGF_{121}$ and $VEGF_{165}$ are secreted by cells and have similar mitogenic properties. The longer species, $VEGF_{189}$ and $VEGF_{206}$, are predominantly cell-associated despite the presence of the same signal sequence found in the two smaller forms. It appears that the 24 amino acid insertion from exon 6 is responsible for the longer forms of VEGF not being secreted (131). The mitogenic activity of the longer proteins was initially thought to be minimal, as conditioned media from transfected cells expressing these species failed to induce endothelial cell proliferation (138,139). It was reported that $VEGF_{189}$ was actually secreted from cells where it was bound to either the cell surface or the extracellular matrix. In this study, cells transfected with an expression vector for $VEGF_{189}$ were treated with suramin, a compound known to interfere with the binding of growth factors with their receptors. The conditioned medium from the treated cells, which was documented to contain the $VEGF_{189}$ transcript, had mitogenic activity for vascular endothelial cells similar to that of the shorter VEGF species (138). Thus, the physiologic importance of $VEGF_{189}$ may be in situations of cellular death, injury, or repair in which the sequestered factor is released from the cell surface or from degraded extracellular matrix.

All four VEGF species have the ability to increase vascular permeability. Conditioned medium from cells expressing the individual VEGF transcripts, when injected intradermally, resulted in marked, localized vascular extravasation. This response was rapid and transient, and could not be blocked by the administration of antihistamines. There was no evidence of vascular damage or inflammatory infiltrate around the sites of injection (128,140).

Unlike other growth factors such as bFGF, epidermal growth factor (EGF), and PDGF, the mitogenic activity of VEGF is specific to vascular endothelial cells. This has been demonstrated in veins, capillaries, and arteries of all calibers and from several anatomic sites. VEGF is also capable of stimulating angiogenesis in vivo both in chick chorioallantoic membrane and in the rat cornea (130,133,141). The angiogenic potency of VEGF is similar to that of bFGF in these assays.

In normal human and animal tissues, VEGF may play an important role in the development of blood vessels and in the maintenance of endothelial integrity and permeability. In situ hybridization studies have demonstrated elevated expression of VEGF in the developing brain and kidney. In adults, however, significant VEGF expression was found only in epithelial cells adjacent to fenestrated endothelium in these organs (139). Other studies have shown a more widespread distribution of VEGF expression in the adult, including the lung, adrenal gland, liver, spleen, heart, cardiac myocytes, and vascular smooth muscle cells (139,142–144).

VEGF also appears to be important in the maintenance of the normal ovarian cycle. Expression of the growth factor was found to be temporally and spatially related to the stage of the cycle and was most significant in tissues that acquired new capillary networks, such as lutein cells and the endometrial stroma (145). The stimulus for the increased expression of VEGF may be estrogen or progesterone. Injection of estradiol, estriol, and progesterone into rats induced a rapid increase in VEGF mRNA expression in the uterus (146).

In addition to its role in normal physiology, VEGF plays a critical role in a number of pathological conditions. Expression of VEGF has been documented in keratinocytes and the subepidermal granulation tissue adjacent to healing wounds. The expression was elevated for as long as 7 days following injury and was thought to be responsible for the increased vascular permeability and angiogenesis characteristic of healing wounds (147). Hypoxia has also been demonstrated to induce the expression of VEGF in a variety of cells. Levels of VEGF mRNA rise rapidly when cells are exposed to hypoxic conditions, and return to baseline in the presence of normal oxygen concentrations (148,149). VEGF also acts as an inducer of monocyte migration and activation, and may be expressed by these cells once they become activated (142,150). Thus, in many respects, VEGF functions much like a lymphokine and an inflammatory mediator.

VEGF RECEPTORS

A candidate receptor for VEGF was first identified by performing binding and crosslinking studies using radiolabeled, recombinant human VEGF (^{125}I-rhVEGF) in cultured endothelial cells. The apparent molecular weight of the receptor was approximately 180 to 185 kDa. Scatchard analysis demonstrated two high-affinity receptors having dissociation constants of 10^{-12} M and 10^{-11} M (151,152). Subsequent binding studies in whole rat tissues documented the presence of VEGF binding on the vascular endothelial cells lining arteries, veins, and microvessels in all organs. Levels were highest in brain, spinal cord, adrenal cortex, stomach, lung, spleen, and pancreas (153). In addition, binding sites for VEGF have also been identified on the cell surface of monocytes (154). The ubiquitous distribution of receptors for VEGF suggests that the factor is necessary for the maintenance and integrity of blood vessels throughout the body. In addition, VEGF binding was demonstrated along blood vessel lumens during embryogenesis and therefore may be essential for their normal development (155).

Similar to bFGF, the binding of VEGF to its receptor is dependent on the presence of heparin. Cells treated with heparinase were unable to bind VEGF; however, the addition of heparin to these cells restored their ability to bind the factor, and the specific binding of VEGF to the receptor was potentiated by increasing concentrations of heparin. In addition, it appears that the heparin-binding domain of the VEGF molecule is different from that responsible for receptor binding (151,156,157). The affinity of VEGF for heparin appears to be much lower than that of bFGF.

Three structurally related VEGF receptors have been identified thus far: *flt-1*, *flk-1*, and *flt-4* (158–161). All three have seven immunoglobulin-like loops in their extracellular domain, a single transmembrane sequence, and a tyrosine kinase region in their intracellular domain (162–164). Although their sequences are very homologous, they are each encoded by separate genes. These three receptors represent a subclass of the *fms* receptor tyrosine kinase family. In fact, for the *flt* receptors, their similarity to *fms* originally led to their name (*fms*-like tyrosine kinase). Other members of this family include the receptors for PDGF and the oncoprotein c-*kit*. These members of the *fms* family differ structurally from the VEGF receptors by having five immunoglobulin-like loops in their extracellular domains.

The expression of these three receptors varies among different tissues and different cells in culture, suggesting that each may play specific and unique physiologic roles. By Northern blot analysis, strong *flt-1* expression has been demonstrated in human placenta, with weaker expression in liver, muscle, and kidney. There was no detectable VEGF mRNA in tumor or normal cells in culture, except for those derived from vascular endothelium (160,162). The expression of *flt-4* can be detected in a variety of human embryonic tissues, with the highest levels being found in the spleen, brain, and lung. In the rat, *flt-4* mRNA could be found in these tissues, as well as in heart, kidney, liver, ovary, prostate, thymus, and testis (161). Using in situ hybridization, *flt-4* expression was found predominantly in bronchial epithelial cells, with none being found in pulmonary vascular endothelial cells (164). Interestingly, several cell lines not derived from vascular endothelium expressed this receptor, including those de rived from leukemia, Wilms' tumor, retinoblastoma, and teratocarcinoma cells (161,164). Thus, the role of the *flt-4* receptor in endothelial cell physiology and in angiogenesis may be minimal. In contrast, in situ hybridization studies demonstrated *flk-1* to be an endothelial-cell-specific receptor found in the endothelial cells of a variety of tissues throughout embryogenesis (158,159). By Northern blotting, *flk-1* expression was seen in adult brain, kidney, heart, and spleen (163).

The binding of VEGF to its cell surface receptors results in rapid and transient elevations of cytosolic [Ca^{2+}] in a variety of cultured endothelial cells. In contrast, the addition of VEGF to cultured human fibroblasts and vascular smooth muscle did not result in similar increases (165,166). The elevation of cytosolic [Ca^{2+}] appears to result from an influx of extracellular [Ca^{2+}] through non-voltage-gated channels, as it could not be blocked by treating endothelial cells with verapamil (166). The signal transduction pathway through which VEGF acts may be via activation of phospholipase C–mediated inositol lipid hydrolysis; treatment of cells with VEGF also resulted in increases in inositol triphosphate levels (137,165).

RELATION OF VEGF AND VEGF RECEPTORS TO TUMOR GROWTH AND TUMOR-DEPENDENT ANGIOGENESIS

The expression and production of VEGF have been demonstrated in numerous human tumors and tumor cell lines, including lymphomas, sarcomas (including AIDS-associated Kaposi's sarcoma), glioblastoma multiforme (GBM), adenocarcinomas, and melanoma (140,141,149, 167–170). In a study of CNS neoplasms, Berkman and coworkers found elevated expression of VEGF mRNA in 22 of 27 tumors that had a high degree of neovascularity and vasogenic cerebral edema. The most significant expression was in capillary hemangioblastomas, GBMs, and meningiomas. The average level of VEGF mRNA expression in these tumors was sixfold higher than in normal brain (168). Using a time-resolved immunofluorometric assay to quantitate VEGF in body fluids, Yeo and associates analyzed 67 human pleural and peritoneal effusions for elevated levels of VEGF. They found that their assay had a sensitivity of 66% and a specificity of 80% for diagnosing a malignancy when compared with cytologic evaluation of the fluid. If they omitted six cases with elevated VEGF whose cytologies were negative, but subsequently had cancer diagnosed by other means, the specificity of the assay increased to 97% (170).

VEGF, although clearly elevated in many tumors, does not act as a transforming growth factor or an oncoprotein. Nontumorigenic Chinese hamster ovary (CHO) cells transfected with an expression vector for $VEGF_{165}$ had no growth advantage over cells that were not transfected in either anchorage-dependent or anchorage-independent assays in vitro. In contrast, when transfected cells were injected into nude mice, they were able to grow small, nonmetastatic tumors. Microscopically, the lesions were well vascularized and benign in appearance (171). Similarly, treatment of cultured rhabdomyosarcoma cells and GBM cells with either exogenous VEGF or antibody specific for VEGF had no effect on cell growth or transformation (172). Together, these results suggest that VEGF does not function as a direct stimulator of tumor growth.

As mentioned earlier, a unique aspect of VEGF is that it is, almost exclusively, a mitogen for vascular endothelial cells. Several studies indicate that it is this property that may be responsible for promoting tumorigenesis. First, immunoreactive VEGF staining within tumor samples was most concentrated in and around blood vessels contained in the tumor mass. The level of VEGF decreased toward the periphery of the tumor, and could not be found in vessels more than 0.5 mm beyond the tumor margin (173). Second, in situ hybridization studies demonstrated that VEGF mRNA was produced by only a fraction of the tumor cells. In GBM samples, cells expressing VEGF were located around the periphery of necrotic regions, suggesting that cell hypoxia and necrosis may be the stimuli for the increased VEGF found in tumor samples (149). The ingrowth of vessels into these hypoxic regions would therefore promote tumorigenesis by providing oxygen, nutrients, and, possibly, mitogenic factors necessary for proliferating tumor cells. In addition, levels of mRNA for the VEGF receptors *flt-1* and *flk-1* were high in the GBM vascular endothelial cells. There was little or no expression of these receptors in the endothelial cells of normal brain (169,174). Finally, tumor growth in vivo could be significantly inhibited by treating animals with monoclonal antibodies specific for VEGF. Histologic examination of tumor samples demonstrated a decrease in the density of blood vessels in those taken from antibody-treated animals compared with those from saline-treated controls (172). Thus, these studies all suggest that the induction of angiogenesis is an important mechanism by which VEGF promotes tumorigenesis.

The increased vascular permeability induced by VEGF may also play a part in tumor progression and invasion. The extravasation of fibrinogen from the hyperpermeable tumor vessels leads to the formation of a fibrin gel that can serve as a substrate not only for new blood vessel growth, but also for the invasion of tumor cells (137,173). Furthermore, VEGF has been shown to induce the expression of both tissue-type and urokinase-type plasminogen activators. Activation of either of these proteases may contribute to tumor cell migration and invasion (175). Although acquisition of an angiogenic phenotype is associated with increased metastatic potential in cancer (4), there are no data yet that implicate VEGF in the genesis of metastases, nor have levels of VEGF in primary and metastatic tumors been compared.

The evidence implicating VEGF as an important factor in tumor angiogenesis and progression is compelling; however, it is important to determine whether the elevation of VEGF in tumors may be more representative of a reactive process than of an active one. In the study in which Plate and colleagues demonstrated elevated expression of VEGF in GBMs, only minimal expression of the factor could be found in less malignant gliomas (169). In addition, blood vessels within the lower grade gliomas failed to demonstrate VEGF immunoreactivity, while those in GBMs showed strong staining. Thus, angiogenesis was able to take place in these lower grade tumors in the absence of detectable VEGF. This might suggest that elevated VEGF is an end-stage response to the ongoing tumor progression and may be associated with hypoxia and necrosis. Hypoxic conditions encountered when tumors grow beyond the limits of metabolite diffusion would elevate VEGF, thus providing a signal to stimulate vascular recruitment to these areas. Hypoxia has been shown to be a potent inducer of VEGF expression (149). Similarly, if one were to implicate VEGF as an important factor for developing blood flow to a growing tumor, it would seem reasonable to expect expression of the factor at the growing or invading margin of the tumor mass. This has not been demonstrated in several studies (149,169,172). It may be that a number of distinct angiogenic factors are involved in promoting tumor growth and progression, and that they each play a seminal role at different stages of the process. This could have significance in designing therapies aimed at inhibiting angiogenesis; multiple angiogenic factors may need to be neutralized according to the stage of tumor development.

TARGETS FOR ANGIOGENIC INTERVENTION

The development and implementation of effective antiangiogenic therapies aimed at suppressing tumor growth are most likely to succeed if they are predicated upon a thorough understanding of the angiogenic process in tumors. A complete description of all the steps inherent in this process and their relation to intervention is beyond the scope of this chapter. Several excellent reviews have considered different aspects of this process and are recommended for further consideration (8–10,174).

An understanding of the multiple pathways used by tumor cells to promote neovascularization has suggested several strategies for impeding the angiogenic process. Many of these approaches are based on modulating the expression and action of angiogenic growth factors (Table 6.3). Specific strategies might include the following:

1. Inhibition of production and export of angiogenic proteins and cytokines. Antisense oligonucleotides have been used effectively to suppress expression of specific proteins, including growth factors (54,111,175). Recent studies have begun to identify relations between alterations in tumor suppressor genes and transcription factors in tumors with changes in the expression of growth factors. Introducing wild-type tumor suppressor genes, such as p53, back into transformed cells may successfully reduce growth factor expression. Because some tumor-derived growth factors are capable of directly stimulating tumor cell growth as well as promoting angiogenesis, this strategy could potentially reduce both tumor vascularity and, more directly, tumor cell growth.
2. Inactivation of angiogenic factors and cytokines following their release from tumor cells or macrophages or from the extracellular matrix. This could be accomplished by neutralizing antibodies such as those developed for bFGF (114,116) and vascular endothelial growth factor

Table 6.3: Modulating Growth Factor Activity

Pathways Involved in Growth Factor Action	*Currently Available Methods for Modulating Growth Factor Activity*
Synthesis	Antisense oligonucleotides, triplex tumor suppressors, transcription factors
Availability	Neutralizing antibodies, inhibitors of protease-mediated release from matrix
Receptor binding	Receptor antagonists
Receptor-mediated signal transduction	Dominant negative mutant receptors

(178). A second promising approach would involve the isolation of compounds that specifically bind and inactivate growth factors or prevent them from activating receptors on endothelial cells (see Fig. 6.1). This strategy is predicated on the requirement of some growth factors to bind heparan proteoglycans for biological activity. Interruption of this interaction would potentially inactivate growth factor activity. This may be effective for some potent endothelial cell mitogens such as bFGF and VEGF, which exhibit a strong affinity for heparan sulfates. Binding of bFGF to heparan is essential for promoting its biological activity. Some molecules that show promise in this regard are the carrageenans, which are polysulfated carbohydrates isolated from red algae (179), heparin derivatives (180,181), and the small molecular weight compound inositolhexakisphosphate (54,182,183).

3. Inhibition of vascular endothelial cells from responding to angiogenic stimuli. Because vascular endothelial cells are the primary target in this process, this area of research has received the most attention and has progressed the most. Several diverse chemical compounds have been identified recently that prevent endothelial cell proliferation in response to angiogenic stimuli. One of these, AGM-1470, a synthetic analogue of the fungal-derived antibiotic fumagillin, inhibits proliferation and migration of endothelial cells (7). It also inhibits the growth of a wide variety of tumors in nude mice (184–186). Because AGM-1470 inhibits tumor growth in vivo, but has little or no direct influence on tumor cell growth in vitro, it seems likely that AGM-1470 mediates its antitumor activity by inhibiting tumor angiogenesis.

Several other novel compounds directed toward endothelial cells may also hold promise as angiogenic inhibitors. For example, an antibody directed against the cell adhesion molecule E-selectin inhibited capillary formation in an in vitro capillary tube assay (107). The E-selectin adhesion molecule is postulated to participate in capillary morphogenesis by promoting endothelial cell interactions via its binding to endothelial cell surface oligosaccharides.

A 16-kDa amino-terminal fragment of rat prolactin (16K PRL) has been shown to inhibit capillary endothelial cell proliferation via a novel receptor (187,188). Also, concentrations of recombinant human 16K PRL in the nanogram range inhibit basal and bFGF- or VEGF-stimulated growth of bovine brain capillary endothelial cells in vitro (189). The 16K PRL does not appear to mediate its actions by binding to the PRL or FGF receptor. Both the human and rat 16K PRL were potent inhibitors of capillary formation when used in the in vivo chick embryo chorioallantoic membrane assay. Interestingly, the intact rat 23-kDa PRL had no effect on

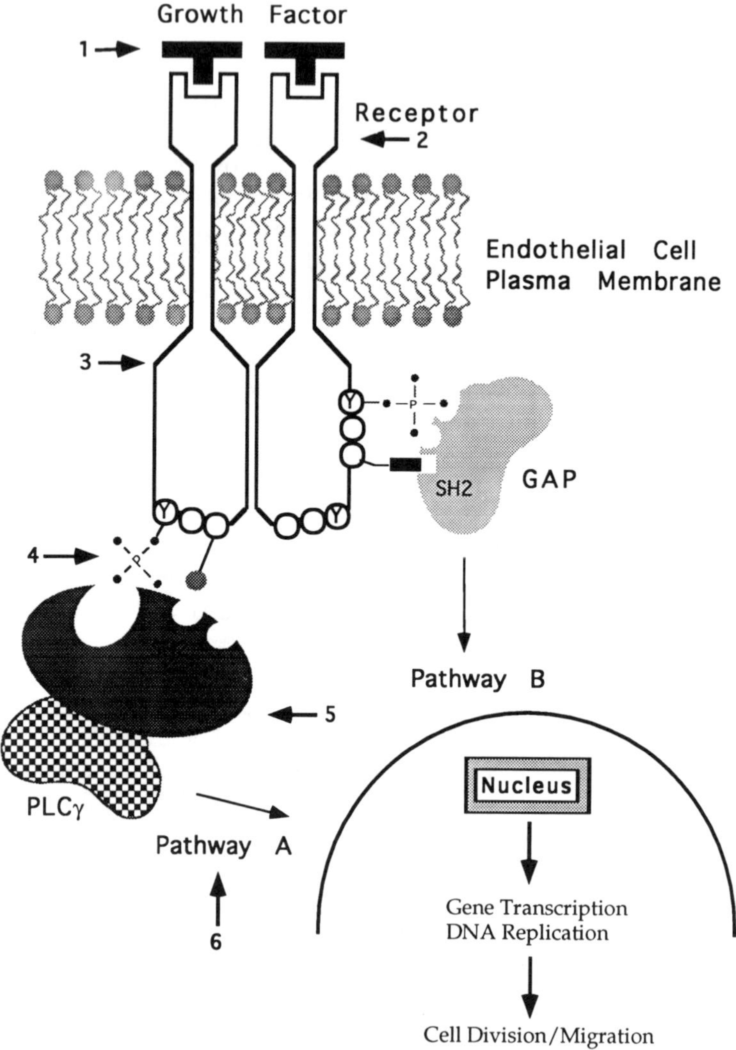
Growth Factor
1
Receptor
2
Endothelial Cell
Plasma Membrane
3
Y
P
GAP
SH2
4
5
Pathway B
Nucleus
PLCγ
Pathway A
6
Gene Transcription
DNA Replication
Cell Division/Migration

the basal or bFGF-stimulated growth of bovine capillary endothelial cells (187). The growth inhibitory actions of 16K PRL appeared to be specific for vascular endothelial cells, because it had no effect on bFGF-stimulated growth of BHK-21 cells and actually stimulated the growth of rat mammary epithelial cells. The 16K PRL is generated endogenously from the proteolysis of intact PRL in the pituitary gland (190), and its presence in the circulation suggests that 16K PRL could possess some important growth regulatory properties for endothelial cells. Further characterization will determine if this protein has therapeutic potential as an angiogenic inhibitor.

Peptides derived from the matrix protein thrombospondin-1 also have antiangiogenic activity (191). Thrombospondin-1 is a multifunctional protein present in platelet α granules and is secreted by a wide variety of cell types, including endothelial cells (192). Thrombospondin-1 is incorporated into the matrix surrounding mature vessels and is capable of inhibiting neovascularization (12). Thrombospondin-1 is a large molecule (180 kDa), and recently two separate domains of this protein were shown to be capable of inhibiting angiogenesis. More importantly, the identification of short peptides, derived from these two domains, that can block the angiogenic process may eventually be useful in the design of new antiangiogenic compounds.

←

Fig. 6.1: Mechanisms for suppressing growth-factor-mediated stimulation of endothelial cell growth. A number of potential avenues exist for modulating growth factor effects on endothelial cells: 1) Methods are available for suppressing growth factor expression or availability (see Table 6.3), thereby removing the stimulus. 2) Receptor antagonists would prevent growth factor–receptor interactions limiting signal transduction. Antagonists can be directed toward the high-affinity receptor or in some instances to companion molecules (such as heparan proteoglycans) that are required for growth factor binding. 3) Agents that prevent receptor dimerization would also limit the transduction pathway by inhibiting receptor *trans*-phosphorylation. The application of dominant negative receptors is also capable of inhibiting receptor transduction (199). 4) Peptides that prevent the binding of protein kinase receptors with intracellular second messenger complexes such as phospholipase-Cγ (PLCγ) would also limit the transduction cascade that ultimately impinges on the nucleus. Peptides would be directed toward phosphoamino acids located on receptors and SH2 domains located on secondary transduction complexes required for receptor-complex interactions. 5) Peptides or other agents that directly inactivate secondary signal transduction complexes might also provide a specific means of inhibiting endothelial cell responses to paracrine- and autocrine-mediated stimulation of blood vessel growth and remodeling; 6) as might peptides or other agents that directly inactivate more distal signal transduction pathways.

The growth of blood vessels in the chick chorioallantoic membrane was inhibited in vivo following the application of erbstatin, a specific tyrosine kinase inhibitor (193). Erbstatin also inhibited the growth of vascular endothelial cells in vitro in a dose-dependent manner. This is consistent with the concept that many angiogenic proteins, such as bFGF, induce biological actions after stimulating receptors possessing protein tyrosine kinase activity. However, growth factor receptors expressing protein tyrosine kinase activity are ubiquitous, and the application of a generalized inhibitor of tyrosine kinase activity could produce many nonspecific side effects. However, further work on these types of compounds is warranted because the mitotic rate of endothelial cells in developing tumors is very high relative to that in normal endothelial cells and in many parenchymal cells contained within the tissue of the developing tumor (194). This approach may become more fruitful as inhibitors of specific protein tyrosine kinase receptors become available.

Another potential class of inhibitor of endothelial cell proliferation is predicated upon the association of growth factor receptors with protein tyrosine kinase activity. Many kinase growth factor receptors are known to form dimers or oligomers after binding their ligands. This dimerization is essential for signal transduction and initiation of biological activity. Truncated forms of the kinase receptors, devoid of cytoplasmic kinase domains, form nonfunctional heterodimers with wild-type kinase receptors (195–197). This effectively prevents signal transduction from the receptor to the cytoplasm, so the truncated receptors are termed dominant negative mutations. Truncated receptors have been used successfully to limit growth responses to PDGF (195), EGF (196), FGF (197,198), and VEGF (199). One report has described the use of a retrovirus encoding a dominant negative mutant of the FLK-1/VEGF receptor to infect endothelial target cells in a glioblastoma in vivo (199). Tumor growth was prevented in nude mice following infection. These results further emphasize the important relation between angiogenic factors, angiogenesis, and solid tumor growth and point to the exciting possibility of using retroviruses encoding dominant negative receptor mutations as a potential modality to curtail tumor progression and metastases.

ACKNOWLEDGMENTS

We gratefully acknowledge Dr. Jordan Gutterman for critically reading the manuscript and Martha R. Morrison for preparing the manuscript.

REFERENCES

1. Folkman J. What is the evidence that tumours are angiogenesis dependent? J Natl Cancer Inst 1990;82:4–6.

2. Srivastava A, Laider P, Hughes L, Woodcock J, Shedden E. Neovascularization in human cutaneous melanoma: a quantitative morphological and Doppler ultrasound study. Eur J Cancer Clini Oncol 1986;22:1205–1209.

3. Sillman F, Boyce J, Fruchter R. The significance of atypical vessels and neovascularization in cervical neoplasia. Am J Obstet Gynecol 1981;139: 154–159.

4. Weidner N, Semple JP, Welch WR, Folkman J. Tumor angiogenesis correlates with metastasis in invasive breast carcinoma. N Engl J Med 1991; 324:1–8.

5. Weiss L, Orr FW, Honn KV. Interactions between cancer cells and the microvasculature: a rate-regulator for metastasis. Clin Exp Metastasis 1989;7:127–167.

6. Blood CH, Zetter BR. Tumor interactions with the vasculature: angiogenesis and tumor metastasis. Biochim Biophys Acta 1990;1032:89–118.

7. Ingber D, Fujita T, Kishimoto S, et al. Synthetic analogues of fumagillin that inhibit angiogenesis and suppress tumor growth. Nature 1990;348:555–557.

8. Folkman J, Hanahan D. Switch to the angiogenic phenotype during tumorigenesis. In: Harris CC, Hirohashi S, Ito N, et al., eds. Multistage carcinogenesis. Boca Raton, Florida: CRC Press, 1992:339–347.

9. Folkman J, Klagsbrun M. Angiogenic factors. Science 1987;235:442–447.

10. Folkman J, Shing Y. Angiogenesis. J Biol Chem 1992;267(16):10931-10934.

11. Bouck N. Angiogenesis: a mechanism by which oncogenes and tumor suppressor genes regulate tumorigenesis. In: Benz CC, Liu ET, eds. Oncogenes and tumor suppressor genes in human malignancies. Boston: Kluwer Academic Publishers, 1993:359–371.

12. Good D, Polverini PJ, Rastinejad F, et al. A tumor suppressor-dependent inhibitor of angiogenesis is immunologically and functionally indistinguishable from a fragment of thrombospondin. Proc Natl Acad Sci USA 1990;87:6624–6628.

13. Logan A, Frautschy SA, Gonzalez A, Baird A. A time course for the focal elevation of synthesis of basic fibroblast growth factor and one of its high-affinity receptors (flg) following a localized cortical brain injury. J Neurosci 1992; 12(10):3828–3837.

14. Plate KH, Greier G, Farrell CL, Risau W. Platelet-derived growth factor receptor β is induced during tumor development and upregulated during tumor progression in endothelial cells in human gliomas. Lab Invest 1992;67(4):529–534.

15. Burgess WH, Maciag T. The heparin-binding (fibroblast) growth factor family of proteins. Annu Rev Biochem 1989;58:575–606.

16. Rifkin DB, Moscatelli D. Recent developments in the cell biology of basic fibroblast growth factor. J Cell Biol 1989;109:1–6.

17. Jaye M, Howk R, Burgess W, et al. Human endothelial cell growth factor: cloning, nucleotide sequence, and chromosome localization. Science 1986;233:543–545.

18. Abraham JA, Whang JL, Tumolo A, et al. Human basic fibroblast growth factor: nucleotide sequence and genomic organization. EMBO J 1986;5:2523–2528.

19. Moore R, Casey G, Brookes S, Dixon M, Peters G, Dickson C. Sequence, topography and protein coding potential of mouse int-2: a putative oncogene activated by mouse mammary tumor virus. EMBO J 1986;5:919–924.

20. Delli-Bovi PD, Curatola AM, Kern FG, Greco A, Ittmann M, Basilico C. An oncogene isolated by transfection of Kaposi's sarcoma DNA encodes a growth factor that is a member of the FGF family. Cell 1987;50:729–737.

21. Sakamoto H, Mori M, Taira M, et al. Transforming gene from human stomach cancers and noncancerous portion of stomach mucosa. Proc Natl Acad Sci USA 1986;83:3997–4001.

22. Zhan X, Bates B, Hu X, Goldfarb M. The human FGF-5 oncogene encodes a novel protein related to fibroblast growth factors. Mol Cell Biol 1988;8:3487–4395.

23. Marics I, Adelaide J, Raybaud F, et al. Characterization of the HST-related FGF6 gene, a new member of the fibroblast growth factor gene family. Oncogene 1989;4:335–340.

24. Rubin JS, Osada H, Finch PW, Taylor WG, Rudikoff S, Aaronson SA. Purification and characterization of a newly identified growth factor specific for epithelial cells. Proc Natl Acad Sci USA 1989;86:802–806.

25. Finch PW, Rubin JS, Miki T, Ron D, Aaronson SA. Human KGF is FGF-related with properties of a paracrine effector of epithelial cell growth. Science 1989;245:752–755.

26. Tanaka A, Miyamoto K, Minamino N, et al. Cloning and characterization of an androgen-induced growth factor essential for the androgen-dependent growth of mouse mammary carcinoma cells. Proc Natl Acad Sci USA 1992; 89:8928–8932.

27. Naruo K, Seko C, Kuroshima K, et al. Novel secretory heparin-binding factors from human glioma cells (glia-activating factors) involved in glial cell growth. J Biol Chem 1993;268(4):2857–2864.

28. Miyamoto M, Naruo K, Seko C, Matsumoto S, Kondo T, Kurokawa T. Molecular cloning of a novel cytokine cDNA encoding the ninth member of the fibroblast growth factor family, which has a unique secretion property. Mol Cell Biol 1993;13(7):4251–4259.

29. Walicke P, Cowan WM, Ueno N, Baird A, Guillemin R. Fibroblast growth factor promotes survival of dissociated hippocampal neurons and enhances neurite extension. Proc Natl Acad Sci USA 1986;83:3012–3016.

30. Morrison RS, Sharma A, De Vellis J, Bradshaw R. Basic fibroblast growth factor supports the survival of cerebral cortical neurons in primary culture. Proc Natl Acad Sci USA 1986;83:7537–7541.

31. Unsicker K, Reichert-Preibsch H, Schmidt R, Pettmann B, Labourdette G, Sensenbrenner M. Astroglial and fibroblast growth factors have neurotrophic functions and cultured peripheral and central nervous system neurons. Proc Natl Acad Sci USA 1987;84:5459–5463.

32. Eckenstein FP, Shipley GD, Nishi R. Acidic and basic fibroblast growth factors in the nervous system: distribution and differential alteration of levels after injury of central versus peripheral nerve. J Neurosci 1991;11:412–419.

33. Kimelman D, Abraham A, Haaparanta T, Palisi TM, Kirschner M. The presence of fibroblast growth factor in the frog egg: its role as a natural mesoderm inducer. Science 1988;242:1053–1056.

34. Wilkinson DG, Bhatt S, McMahon AP. Expression pattern of the FGF-related proto-oncogene int-2 suggests multiple roles in fetal development. Development 1989;105:131–136.

35. De Lapeyriere O, Rosnet O, Benharroch D, et al. Structure, chromosome mapping and expression of the murine FGF-6 gene. Oncogene 1990;5:823–831.

36. Haub O, Goldfarb M. Expression of the fibroblast growth factor-5 gene in the mouse embryo. Development 1991;112:397–406.

37. Suzuki HR, Sakamoto H, Yoshida T, Sugimura T, Terada M, Solursh M. Localization of Hst1 transcripts to the apical ectodermal ridge in the mouse embryo. Dev Biol 1992;150:219–222.

38. Wilcox BJ, Unnerstall JR. Expression of acidic fibroblast growth factor mRNA in the developing and adult rat brain. Neuron 1991;6:397–409.

39. Elde R, Cao Y, Cintra A, et al. Prominent expression of acidic fibroblast growth factor in motor and sensory neurons. Neuron 1991;7:349–364.

40. Schnurch H, Risau W. Differentiating and mature neurons express the acidic fibroblast growth factor gene during chick neural development. Development 1991;111:1143–1154.

41. Fallon JH, Di Salvo J, Loughlin SE, et al. Localization of acidic fibroblast growth factor within the mouse brain using biochemical and immunocytochemical techniques. Growth Factors 1992;6:139–157.

42. Stock A, Kuzis K, Woodward WR, Nishi R, Eckenstein FP. Localization of acidic fibroblast growth factor in specific subcortical neuronal populations. J Neurosci 1992;12(12):4688–4700.

43. Abraham JA, Mergia A, Whang JL, et al. Nucleotide sequence of a bovine clone encoding the angiogenic protein, basic fibroblast growth factor. Science 1986;233:545–548.

44. Kurokawa T, Sasada R, Iwane M, Igaraski K. Cloning and expression of cDNA encoding human basic fibroblast growth factor. FEBS Lett 1987;213:189–194.

45. Florkiewicz RZ, Sommer A. Human basic fibroblast growth factor gene encodes four polypeptides: three initiate translation from non-AUG codons. Proc Natl Acad Sci USA 1989;86:3978–3981.

46. Prats H, Kaghad M, Prats AC, et al. High molecular mass forms of basic fibroblast growth factor are initiated by alternative CUG codons. Proc Natl Acad Sci USA 1989;86:1836–1840.

47. Kurokawa T, Seno M, Igarashi K. Nucleotide sequence of rat basic fibroblast growth factor cDNA. Nucleic Acids Res 1988;16:5201.

48. Renko M, Quarto N, Morimoto T, Rifkin DB. Nuclear and cytoplasmic localization of different basic fibroblast growth factor species. J Cell Physiol 1990;144:108–114.

49. Bugler B, Amalric F, Prats H. Alternative initiation of translation determines cytoplasmic or nuclear localization of basic fibroblast growth factor. Mol Cell Biol 1991;11:573–577.

50. Kandel J, Bossy-Wetzel E, Radvanyi F, Klagsbrun M, Folkman J, Hanahan D. Neovascularization is associated with a switch to the export of bFGF in the multistep development of fibrosarcoma. Cell 1991;66:1095–1104.

51. Mergia A, Eddy R, Abraham JA, Fiddes JC, Shows TB. The genes for basic and acidic fibroblast growth factors are on different human chromosomes. Biochem Biophys Res Commun 1986;138:644–651.

52. Shibata F, Baird A, Florkiewicz RZ. Functional characterization of the human basic fibroblast growth factor gene promoter. Growth Factors 1991; 4:277–287.

53. Murphy PR, Sato Y, Sato R, Friesen HG. Regulation of multiple basic fibroblast growth factor messenger ribonucleic acid transcripts by protein kinase C activators. Mol Endocrinol 1988;2:1196–1201.

54. Sherman L, Stocker KM, Morrison RS, Ciment G. Basic fibroblast growth factor (bFGF) acts intracellularly to cause the transdifferentiation of avian neural crest-derived Schwann cell precursors into melanocytes. Development 1993;118:1313–1326.

55. Houssaint E, Blanquet PR, Champion-Arnaud P, et al. Related fibroblast growth factor receptor genes exist in the human genome. Proc Natl Acad Sci USA 1990;87:8180–8184.

56. Johnson DE, Lu J, Chen H, Werner S, Williams LT. The human fibroblast growth factor receptor genes: a common structural arrangement underlies the mechanisms for generating receptor forms that differ in their third immunoglobulin domain. Mol Cell Biol 1991;11:4627–4634.

57. Keegan K, Johnson DE, Williams LT, Hayman MJ. Isolation of additional member of the fibroblast growth factor receptor family, FGFR-3. Proc Natl Acad Sci USA 1991;88:1095–1099.

58. Partanen J, Makela TP, Eerola E, et al. FGFR-4, a novel acidic fibroblast growth factor receptor with a distinct expression pattern. EMBO J 1991;10:1347–1354.

59. Ruta M, Burgess W, Givol D, et al. Receptor for acidic fibroblast growth factor is related to the tyrosine kinase encoded by the fms-like gene (FLG). Proc Natl Acad Sci USA 1989;86:8722–8726.

60. Kan M, DiSorbo D, Hou J, Hoshi H, Mansson PE, McKeehan WL. High and low affinity binding of heparin-binding growth factor to a 130-kDa receptor correlates with stimulation and inhibition of growth of a differentiated human hepatoma. Cell J Biol Chem 1988;263:11306–11313.

61. Moscatelli D. High and low affinity binding sites for basic fibroblast growth factor on cultured cells: absence of a role for low affinity binding in the

stimulation of plasminogen activator production by bovine capillary endothelial cells. J Cell Physiol 1987;131:123–130.

62. Vlodavsky I, Folkman J, Sullivan R, et al. Endothelial cell-derived basic fibroblast growth factor: synthesis and deposition into subendothelial extra cellular matrix. Proc Natl Acad Sci USA 1987;84:2292–2296.

63. Kan M, Shi E. Fibronectin, not laminin, mediates heparin-dependent heparin-binding growth factor type I binding to substrata and stimulation of endothelial cell growth. In Vitro Cell Dev Biol 1990;26:1151–1156.

64. Yayon A, Klagsbrun M, Esko JD, Leder P, Ornitz DM. Cell surface, heparin-like molecules are required for binding of basic fibroblast growth factor to its high affinity receptor. Cell 1991;64:841–848.

65. Rapraeger AC, Krufka A, Olwin BB. Requirement of heparan sulfate for bFGF-mediated fibroblast growth and myoblast differentiation. Science 1991; 252:1705–1708.

66. Halaban R, Ghosh S, Baird A. bFGF is the putative natural growth factor for human melanocytes. In Vitro Cell Dev Biol 1987;23:47–52.

67. Reid HH, Wilks AF, Bernard O. Two forms of basic fibroblast growth factor receptor-like mRNA are expressed in the developing mouse brain. Proc Natl Acad Sci USA 1990;87:1596–1600.

68. Heuer JG, Bartheld CS, Kinoshita Y, Evers PC, Bothwell M. Alternative phases of FGF receptor and NGF receptor expression in the developing chicken nervous system. Neuron 1990;5:283–296.

69. Wanaka A, Johnson EM Jr, Milbrandt J. Localization of FGF receptor mRNA in the adult rat central nervous system by in situ hybridization. Neuron 1990;5:267–281.

70. Wanaka A, Milbrandt J, Johnson EM Jr. Expression of FGF receptor gene in rat development. Development 1991;111:455–468.

71. Lai C, Lemke G. An extended family of protein-tyrosine kinase genes differentially expressed in the vertebrate nervous system. Neuron 1991;6:691–704.

72. Stark K, McMahon JA, McMahon AP. FGFR-4, a new member of the fibroblast growth factor receptor family, expressed in the definitive endoderm and skeletal muscle lineages of the mouse. Development 1991;113:641–651.

73. Peters KG, Werner S, Chen G, Williams LT. Two FGF receptor genes are differentially expressed in epithelial and mesenchymal tissues during limb formation and organogenesis in the mouse. Development 1992;114:233–243.

74. Werner S, Duan D-S R, de Vries C, Peters KG, Johnson DE, Williams LT. Differential splicing in the extracellular region of fibroblast growth factor receptor 1 generates receptor variants with different ligand-binding specificities. Mol Cell Biol 1992;12:82–88.

75. Peters K, Ornitz D, Werner S, Williams L. Unique expression pattern of the FGF receptor 3 gene during mouse organogenesis. Dev Biol 1993;155:423–430.

76. Dionne CA, Crumley G, Bellot F, et al. Cloning and expression of two distinct high-affinity receptors cross-reacting with acidic and basic fibroblast growth factors. EMBO J 1990;9:2685–2692.

77. Johnson DE, Lee PL, Lu J, Williams LT. Diverse forms of a receptor for acidic and basic fibroblast growth factors. Mol Cell Biol 1990;10:4728–4736.

78. Hou J, Kan M, McKeehan K, McBride G, Adams P, McKeehan WL. Fibroblast growth factors from liver vary in three structural domains. Science 1991;251:665–668.

79. Dell KR, Williams LT. A novel form of fibroblast growth factor receptor 2. J Biol Chem 1992;267(29):21225–21229.

80. Miki T, Bottaro DP, Fleming TP, et al. Determination of ligand-binding specificity by alternative splicing: two distinct growth factor receptors encoded by a single gene. Proc Natl Acad Sci USA 1992;89:246–250.

81. Yayon A, Zimmer Y, Guo-Hong S, Avivi A, Yarden Y, Givol D. A confined variable region confers ligand specificity on fibroblast growth factor receptors: implications for the origin of immunoglobulin fold. EMBO J 1992; 11:1885–1890.

82. Shi E, Kan M, Xu J, Wang F, Hou J, McKeehan WL. Control of FGF receptor kinase signal transduction by heterodimerization of combinatorial splice variants. Mol Cell Biol 1993;13:3907–3918.

83. Johnson DE, Williams LT. Structural and functional diversity in FGF receptor multigene family. Adv Cancer Res 1993;60:1–41.

84. Eisemann A, Ahn JA, Graziani G, Tronick SR, Ron D. Alternative splicing generates at least five different isoforms of the human basic-FGF receptor. Oncogene 1991;6:1195–1202.

85. Hou J, Kan M, Wang F, et al. Substitution of putative half-cysteine residues in heparin-binding fibroblast growth factor receptors. J Biol Chem 1992; 267(25):17804–17808.

86. Yan G, Fukabori Y, McBride G, Nikolaropolous S, McKeehan W. Exon switching and activation of stromal and embryonic fibroblast growth factor (FGF)-FGF receptor genes in prostate epithelial cells accompany stromal independence and malignancy. Mol Cell Biol 1993;13:4513–4522.

87. Yamaguchi F, Hideyuki S, Bruner JM, Morrison RS. Differential expression of two fibroblast growth factor-receptor genes is associated with malignant progression in human astrocytomas. Proc Natl Acad Sci USA 1994;91:484–488.

88. Blam SB, Mitchell R, Tischer E, et al. Addition of growth hormone secretion signal to basic fibroblast growth factor results in cell transformation and secretion of aberrant forms of the protein. Oncogene 1988;3:129–136.

89. Neufeld G, Mitchell R, Ponte P, Gospodarowicz D. Expression of human basic fibroblast growth factor cDNA in baby hamster kidney-derived cells results in autonomous cell growth. J Cell Biol 1988;106:1385–1394.

90. Rogelj S, Weinberg RA, Fanning P, Klagsbrun M. Basic fibroblast growth factor fused to a signal peptide transforms cells. Nature 1988;331:173–175.

91. Sasada R, Kurokawa T, Iwane M, Igarashi K. Transformation of mouse BALB/c 3T3 cells with human basic fibroblast growth factor cDNA. Mol Cell Biol 1988;8:588–594.

92. Dotto GP, Moellmann G, Ghosh S, Edwards M, Halaban R. Transformation of murine melanocytes by basic fibroblast growth factor cDNA and

oncogenes and selective suppression of the transformed phenotype in a reconstituted cutaneous environment. J Cell Biol 1989;109:3115–3128.

93. Taira M, Yoshida T, Miyagawa K, Sakamoto H, Terada M, Sugimura T. cDNA sequence of human transforming gene hst and identification of the coding sequence required for transforming activity. Proc Natl Acad Sci USA 1987; 84:2980–2984.

94. Smith R, Peters G, Dickson C. Multiple RNAs expressed from the int-2 gene in mouse embryonal carcinoma cell lines encode a protein with homology to fibroblast growth factors. EMBO J 1988;7:1013–1022.

95. Libermann TA, Friesel R, Jaye M, et al. An angiogenic growth factor is expressed in human glioma cells. EMBO J 1987;6:1627–1632.

96. Maxwell M, Nabor SP, Wolfe HJ, et al. Expression of angiogenic growth factor genes in primary human astrocytomas may contribute to their growth and progression. Cancer Res 1991;51:1345–1351.

97. Takahashi JA, Mori H, Fukumoto M, et al. Gene expression of fibroblast growth factors in human gliomas and meningiomas: demonstration of cellular source of basic fibroblast growth factor mRNA and peptide in tumor tissues. Proc Natl Acad Sci USA 1990;87:5710–5714.

98. Murphy PR, Sato Y, Sato R, Friesen HG. Regulation of multiple basic fibroblast growth factor messenger ribonucleic acid transcripts by protein kinase C activators. Mol Endocrinol 1988;2:1196–1201.

99. Morrison RS, Gross JL, Herblin WF, et al. Basic fibroblast growth factor-like activity and receptors are expressed in a human glioma cell line. Cancer Res 1990;50:2524–2529.

100. Paulus W, Grothe C, Sensenbrenner M, et al. Localization of basic fibroblast growth factor, a mitogen and angiogenic factor, in human brain tumors. Acta Neuropathol 1990;79:418–423.

101. Zagzag D, Miller DC, Sato Y, Rifkin DB, Burstein DE. Immunohistochemical localization of basic fibroblast growth factor in astrocytomas. Cancer Res 1990;50:7393–7398.

102. Stefanik DF, Rizakalla LR, Soi A, Goldblatt SA, Rizkalla WM. Acidic and basic fibroblast growth factors are present in glioblastoma multiforme. Cancer Res 1991;51:5760–5765.

103. Tanimoto H, Yoshida K, Yokozaki H, et al. Expression of basic fibroblast growth factor in human gastric carcinomas. Virchows Arch B Cell Pathol 1991;61:263–267.

104. Ohtani H, Nakamura S, Watanabe Y, Mizoi T, Saku T, Nagura H. Immunocytochemical localization of basic fibroblast growth factor in carcinomas and inflammatory lesions of the human digestive tract. Lab Invest 1993; 68(5):520–527.

105. Eguchi J, Nomata K, Kanda S, et al. Gene expression and immunohistochemical localization of basic fibroblast growth factor in renal cell carcinoma. Biochem Biophys Res Commun 1992;183(3):937–944.

106. Yamanaka Y, Friess H, Buchler M, et al. Overexpression of acidic and basic fibroblast growth factors in human pancreatic cancer correlates with advanced tumor stage. Cancer Res 1993;53:5289–5296.

107. Nguyen M, Strubel NA, Bischoff J. A role for sialyl Lewis-X/A glycoconjugates in capillary morphogenesis. Nature 1993;365:267–269.

108. Lidereau R, Callahan R, Dickson C, Peters G, Escot C, Ali IU. Amplification of the int-2 gene in primary human breast tumors. Oncogene Res 1988; 2:285–291.

109. Adnane J, Guadray P, Dionne CA, et al. BEK and FLG, two receptors to members of the FGF family, are amplified in subsets of human breast cancers. Oncogene 1991;6:659–663.

110. Becker D, Meier CB, Herlyn M. Proliferation of human malignant melanomas is inhibited by antisense oligodeoxynucleotides targeted against basic fibroblast growth factor. EMBO J 1989;8(12):3685–3691.

111. Morrison RS. Suppression of basic fibroblast growth factor expression by antisense oligodeoxynucleotides inhibits the growth of transformed human astrocytes. J Biol Chem 1991;266:728–734.

112. Murphy PR, Sato Y, Knee RS. Phosphorothioate antisense oligonucleotides against basic fibroblast growth factor inhibit anchorage-dependent and anchorage-independent growth of a malignant glioblastoma cell line. Mol Endocrinol 1992;6:877–884.

113. Allen LE, Maher PA. Expression of basic fibroblast growth factor and its receptor in an invasive bladder carcinoma cell line. J Cell Physiol 1993;155:368–375.

114. Takahashi JA, Fukumoto M, Kozai Y, et al. Inhibition of cell growth and tumorigenesis of human glioblastoma cells by a neutralizing antibody against human basic fibroblast growth factor. FEBS Lett 1991;288(1,2):65–71.

115. Gross JL, Herblin WF, Dusak BA, et al. Effects of modulation of basic fibroblast growth factor on tumor growth in vivo. J Natl Cancer Inst 1993; 85(2):121–131.

116. Hori A, Sasada R, Matsutani E, et al. Suppression of solid tumor growth by immunoneutralizing monoclonal antibody against human basic fibroblast growth factor. Cancer Res 1991;51:6180–6184.

117. Dennis PA, Rifkin DB. Studies on the role of basic fibroblast growth factor in vivo: inability of neutralizing antibodies to block tumor growth. J Cell Physiol 1990;144:84–98.

118. Matsuzaki K, Yoshitake Y, Matuo Y, Saskai H, Nishikawa K. Monoclonal antibodies against heparin-binding growth factor II/basic fibroblast growth factor that block its biological activity: invalidity of the antibodies for tumor angiogenesis. Proc Natl Acad Sci USA 1989;86:9911–9915.

119. Asai T, Wanaka A, Kato H, Masana Y, Seo M, Tohyama M. Differential expression of two members of FGF receptor gene family, FGFR-1 and FGFR-2 mRNA, in the adult rat central nervous system. Mol Brain Res 1993;17:174–178.

120. Kato H, Wanaka A, Tohyama M. Co-localization of basic fibroblast growth factor-like immunoreactivity and its receptor mRNA in the rat spinal cord and the dorsal root ganglion. Brain Res 1992;576:351–354.

121. Orr-Urtreger A, Givol D, Yayon A, Yarden Y, Lonai P. Developmental expression of two murine fibroblast growth factor receptors, flg and bek. Development 1991;113:1419–1434.

122. Friesel R, Brown SAN. Spatially restricted expression of fibroblast growth factor receptor-2 during Xenopus development. Development 1992; 116:1051–1058.

123. Patstone G, Pasquale EB, Maher PA. Different members of the fibroblast growth factor family are specific to distinct cell types in the developing chicken embryo. Dev Biol 1993;155:107–123.

124. Morrison RS, Yamaguchi F, Bruner J, et al. Fibroblast growth factor receptor gene expression and immunoreactivity are elevated in human glioblastoma multiforme. Cancer Research 1994;54:2794–2799.

125. Lyons MK, Anderson RE, Meyer FB. Basic fibroblast growth factor promotes in vivo cerebral angiogenesis in chronic forebrain ischemia. Brain Res 1991;558:315–320.

126. Cuevas P, Gimenez-Gallego G, Carceller F, Cuevas B, Crespo A. Single topical application of human recombinant basic fibroblast growth factor (rbFGF) promotes neovascularization in rat cerebral cortex. Surg Neurol 1993;39:380–384.

127. Nabel EG, Yang Z, Plautz G, et al. Recombinant fibroblast growth factor-1 promotes intimal hyperplasia and angiogenesis in arteries in vivo. Nature 1993;362:844–846.

128. Senger D, Galli S, Dvorak A, et al. Tumor cells secrete a vascular permeability factor that promotes accumulation of ascites fluid. Science 1983; 219:983–985.

129. Keck P, Hauser S, Krivi G, et al. Vascular permeability factor, a endothelial cell mitogen related to PDGF. Science 1989;246:1309–1312.

130. Leung D, Cachianes G, Kuang W, et al. Vascular endothelial growth factor is a secreted angiogenic mitogen. Science 1989;246:1306–1309.

131. Houck K, Ferrara N, Winer J, et al. The vascular endothelial growth factor family: identification of a fourth molecular species and characterization of alternative splicing of RNA. Mol Endocrinol 1991;5(12):1806–1814.

132. Tischer E, Mitchell R, Hartman T, et al. The human gene for vascular endothelial growth factor: multiple protein forms are encoded through alternative exon splicing. J Biol Chem 1991;266(18):11947–11954.

133. Ferrara N, Henzwl W. Pituitary follicular cells secrete a novel heparin-binding growth factor specific for vascular endothelial cells. Biochem Biophys Res Commun 1989;161(2):851–858.

134. Plouet J, Schilling J, Gospodarowicz D. Isolation and characterization of a newly identified endothelial cell mitogen produced by AtT-20 cells. EMBO J 1989;8:3801–3806.

135. Gospodarowicz D, Abraham J, Schilling J. Isolation and characterization of a vascular endothelial cell mitogen produced by pituitary-derived folliculo stellate cells. Proc Natl Acad Sci USA 1989;86:7311–7315.

136. Conn G, Bayne M, Soderman D, et al. Amino acid and cDNA sequences of a vascular endothelial cell mitogen that is homologous to platelet-derived growth factor. Proc Natl Acad Sci USA 1990;87:2628–2632.

137. Ferrara N, Houck K, Jakeman L, et al. Molecular and biological properties of the vascular endothelial growth factor family of proteins. Endocr Rev

1992;13(1):18–32.

138. Houck K, Leung D, Rowland A, et al. Dual regulation of vascular endothelial growth factor: bioavailability by genetic and proteolytic mechanisms. J Biol Chem 1992;267(36):26031–26037.

139. Breier G, Albrecht U, Sterrer S, et al. Expression of vascular endothelial growth factor during embryonic angiogenesis and endothelial cell differentiation. Development 1992;114:521–532.

140. Senger D, Perruzzi C, Feder J, et al. A highly conserved vascular permeability factor secreted by a variety of human and rodent tumor cell lines. Cancer Res 1986;46:5629–5632.

141. Connolly D, Olander J, Heuvelman D, et al. Human vascular permeability factor. J Biol Chem 1989;265(33):20017–20024.

142. Berse B, Brown L, Van De Water L, et al. Vascular permeabillity factor (vascular endothelial growth factor) gene is expressed differentially in normal tissues, macrophages and tumors. Mol Biol Cell 1992;3:211–220.

143. Monacci W, Merrill M, Oldfield E. Expression of vascular permeability factor/vascular endothelial growth factor in normal rat tissues. Am J Physiol 1993;264:C995–C1002.

144. Ferrara N, Winer J, Burton T. Aortic smooth muscle cells express and secrete vascular endothelial growth factor. Growth Factors 1991;5:141–148.

145. Shweiki D, Itin A, Neufeld G, et al. Patterns of expression of vascular endothelial growth factor (VEGF) and VEGF receptors in mice suggest a role in hormonally regulated angiogenesis. J Clin Invest 1993;91:2235–2243.

146. Cullinan-Bove K, Koos R. Vascular endothelial growth factor/vascular permeability factor expression in the rat uterus: rapid stimulation by estrogen correlates with estrogen-induced increases in uterine capillary permeability and growth. Endocrinology 1993;133:829–837.

147. Brown L, Yeo K-T, Berse B, et al. Expression of vascular permeability factor (vascular endothelial growth factor) by epidermal keratinocytes during wound healing. J Exp Med 1992;176:1375–1379.

148. Ladoux A, Frelin C. Hypoxia is a strong inducer of vascular endothelial growth factor mRNA expression in the heart. Biochem Biophys Res Commun 1993;195(2):1005–1010.

149. Shweiki D, Itin A, Soffer D, et al. Vascular endothelial growth factor induced by hypoxia may mediate hypoxia-initiated angiogenesis. Nature 1992; 359:843–845.

150. Clauss M, Gerlach M, Gerlach H, et al. Vascular permeability factor: a tumor-derived polypeptide that induces endothelial cell and monocyte procoagulant activity, and promotes monocyte migration. J Exp Med 1990; 172:1535–1545.

151. Vaisman N, Gospodarowicz D, Neufeld G. Characterization of the receptors for vascular endothelial growth factor. J Biol Chem 1990; 265(32):19461–19466.

152. Plouet J, Moukadiri H. Characterization of the receptor to vasculotropin on bovine adrenal cortex-derived capillary endothelial cells. J Biol Chem 1990; 265(36):22071–22074.

153. Jakeman L, Winer J, Bennett G, et al. Binding sites for vascular endothelial growth factor are localized on endothelial cells in adult rat tissues. J Clin Invest 1992;89:244–253.

154. Shen H, Clauss M, Ryan J, et al. Characterization of vascular permeability/vascular endothelial growth factor receptors on mononuclear phagocytes. Blood 1993;81(10):2767–2773.

155. Jakeman L, Armanini M, Phillips H, et al. Developmental expression of binding sites and messenger ribonucleic acid for vascular endothelial growth factor suggests a role for this protein in vasculogenesis and angiogenesis. Endocrinology 1993;133(2):848–858.

156. Gitay-Goren H, Soker S, Vlodavsky I, et al. The binding of vascular endothelial growth factor to its receptor is dependent on cell surface-associated heparin-like molecules. J Biol Chem 1992;267(9):6093–6098.

157. Soker S, Svahn C, Neufeld G. Vascular endothelial growth factor is inactivated by binding to α_2-macroglobulin and the binding is inhibited by heparin. J Biol Chem 1993;268(11):7685–7691.

158. Millauer B, Wizigmann-Voos S, Schnurch A, et al. High affinity VEGF binding and developmental expression suggest Flk1 as a major regulator of vasculogenesis and angiogenesis. Cell 1993;72:835–846.

159. Quinn T, Peters K, De Vries C, et al. Fetal liver kinase 1 is a receptor for vascular endothelial growth factor and is selectively expressed in vascular endothelium. Proc Natl Acad Sci USA 1993;90:7533–7537.

160. De Vries C, Escobedo J, Ueno H, et al. The fms-like tyrosine kinase, a receptor for vascular endothelial growth factor. Science 1992;255:989–991.

161. Galland F, Karamysheva A, Pebusque M-J, et al. The FLT4 gene encodes a transmembrane tyrosine kinase related to the vascular endothelial growth factor receptor. Oncogene 1993;8:1233–1240.

162. Shibuya M, Yamaguchi S, Yamane A, et al. Nucleotide sequence and expression of a novel human receptor-type tyrosine kinase gene (flt) closely related to the fms family. Oncogene 1990;5:519–524.

163. Matthews W, Jordan C, Gavin M, et al. A receptor tyrosine kinase cDNA isolated from a population of enriched primitive hematopoietic cells and exhibiting close genetic linkage to c-kit. Proc Natl Acad Sci USA 1991;88:9026–9030.

164. Pajusola K, Aprelikova O, Korhonen J, et al. FLT4 receptor tyrosine kinase contains seven immunoglobulin-like loops and is expressed in multiple human tissues and cell lines. Cancer Res 1992;52:5738–5743.

165. Brock T, Dvorak H, Senger D. Tumor-secreted vascular permeability factor increases cytosolic Ca^{2+} and von Willebrand factor release in human endothelial cells. Am J Pathol 1991;138:213–221.

166. Criscuolo G, Lelkes P, Rotrosen D, et al. Cytosolic calcium changes in endothelial cells induced by a protein product of human gliomas containing vascular permeability factor activity. J Neurosurg 1989;71:884–891.

167. Weindel K, Weich HA. AIDS-associated Kaposi's sarcoma cells in culture express vascular endothelial growth factor. Biochem Biophys Res Commun 1992;183(3):1167–1174.

168. Berkman R, Merrill M, Reinhold W, et al. Expression of the vascular permeability factor/vascular endothelial growth factor gene in central nervous system neoplasms. J Clin Invest 1993;91:153–159.

169. Plate K, Breier G, Weich H, et al. Vascular endothelial growth factor is a potential tumour angiogenesis factor in human gliomas in vivo. Nature 1992;359:845–848.

170. Yeo K-T, Wang H, Nagy J, et al. Vascular permeability factor (vascular endothelial growth factor) in guinea pig and human tumor and inflammatory effusions. Cancer Res 1993;53:2912–2918.

171. Ferrara N, Winer J, Burton T, et al. Expression of vascular endothelial growth factor does not promote transformation but confers a growth advantage in vivo to Chinese hamster ovary cells. J Clin Invest 1993;91:160–170.

172. Kim K, Bing L, Winer J, et al. Inhibition of vascular endothelial growth factor–induced angiogenesis suppresses tumour growth in vivo. Nature 1993; 362:841–844.

173. Dvorak H, Sioussat T, Brown L, et al. Distribution of vascular permeability factor (vascular endothelial growth factor) in tumors: concentration in tumor blood vessels. J Exp Med 1991;174:1275–1278.

174. Plate KH, Breier G, Millauer B, Ullrich A, Risau W. Up-regulation of vascular endothelial growth factor and its cognate receptors in a rat glioma model of tumor angiogenesis. Cancer Res 1993;53:5822–5827.

175. Pepper M, Ferrara N, Orci L, et al. Vascular endothelial growth factor (VEGF) induces plasminogen activators and plasminogen activator inhibitor-1 in microvascular endothelial cells. Biochem Biophys Res Commun 1991; 181(2):902–906.

176. Bicknell R, Harris AL. Anticancer strategies involving the vasculature: vascular targeting and the inhibition of angiogenesis. Cancer Biol 1992;3:399–407.

177. Becker D, Meier CB, Herlyn M. Proliferation of human malignant melanomas is inhibited by antisense oligodeoxynucleotides targeted against basic fibroblast growth factor. EMBO J 1989;8(12):3685–3691.

178. Jin Kim K, Li B, Winer J, et al. Inhibition of vascular endothelial growth factor–induced angiogenesis suppresses tumour growth in vivo. Nature 1993; 362:841–844.

179. Hoffman R. Carrageenans inhibit growth-factor binding. Biochem J 1993;289:331–334.

180. Ishihara M, Tyrrell DJ, Stauber GB, Brown S, Cousens LS, Stack RJ. Preparation of affinity-fractionated, heparin-derived oligosaccharides and their effects on selected biological activities mediated by basic fibroblast growth factor. J Biol Chem 1993;268(7):4675–4683.

181. Tyrrell DJ, Ishihara M, Rao N, et al. Structure and biological activities of a heparin-derived hexasaccharide with high affinity for basic fibroblast growth factor. J Biol Chem 1993;268(7):4684–4689.

182. Morrison RS, Giordano S, Yamaguchi F, Hendrickson S, Berger MS, Palczewski K. Basic fibroblast growth factor expression is required for clonogenic growth of human glioma cells. J Neurosci Res 1993;34:502–509.

183. Morrison RS, Shi E, Kan M, et al. Inositol hexakisphosphate ($InsP_6$): An antagonist of fibroblast growth factor receptor binding and activity, in press.

184. Yanase T, Tamura M, Fujita K, Kodama S, Tanaka K. Inhibitory effect of angiogenesis inhibitor TNP-470 on tumor growth and metastasis of human cell line in vitro and in vivo. Cancer Res 1993;53:2566–2570.

185. Yamaoka M, Yamamoto T, Masaki T, Ikeyama S, Sudo K, Fujita T. Inhibition of tumor growth and metastasis of rodent tumors by the angiogenesis inhibitor *O*-(chloroacetyl-carbamolyl)fumagillol (TNP-470; AGM-1470). Cancer Res 1993;53:4262–4267.

186. Yamaoka M, Yamamoto T, Ikeyama S, Sudo K, Fujita T. Angiogenesis inhibitor TNP-470 (AGM-1470) potently inhibits the tumor growth of hormone-independent human breast and prostate carcinoma cell lines. Cancer Res 1993; 53:5233–5236.

187. Ferrara N, Clapp C, Weiner RI. The 16K fragment of prolactin specifically inhibits basal or fibroblast growth factor stimulated growth of capillary endothelial cells. Endocrinology 1991;129:896–900.

188. Clapp C, Weiner RI. A specific, high affinity, saturable binding site for the 16-kilodalton fragment of prolactin on capillary endothelial cells. Endocrinology 1992;130:1380–1386.

189. Clapp C, Martial JA, Guzman RC, Rentier-Delrue F, Weiner RI. The 16-kilodalton N-terminal fragment of human prolactin is a potent inhibitor of angiogenesis. Enocrinology 1993;133(3):1292–1299.

190. Sinha YN, Gilligan TA, Lee DW, Hollingsworth D, Markoff E. Cleaved prolactin: evidence for its occurrence in human pituitary gland and plasma. J Clin Endocrinol Metab 1985;60:239–243.

191. Tolsma SS, Volpert OV, Good DJ, Frazier WA, Polverini PJ, Bouck N. Peptides derived from two separate domains of the matrix protein thrombospondin-1 have anti-angiogenic activity. J Cell Biol 1993;122(2):497–511.

192. Lawler J. The structural and functional properties of thrombospondin. Blood 1986;67:1197–1209.

193. Oikawa M, Morita I, Shimamura M. Inhibition of angiogenesis by erbstatin, an inhibitor of tyrosine kinase. J Antibiot (Tokyo) 1993;46(5):785–790.

194. Denekamp J, Hobson B. Endothelial cell proliferation in experimental tumours. Br J Cancer 1982;46:711–720.

195. Ueno H, Colbert H, Escobedo JA, Williams LT. Inhibition of PDGF beta receptor signal transduction by coexpression of a truncated receptor. Science 1991;252:844–848.

196. Kashles O, Yarden Y, Fischer R, Ullrich A, Schlessinger J. A dominant negative mutation suppresses the function of normal epidermal growth factor receptors by heterodimerization. Mol Cell Biol 1991;11:1454–1463.

197. Werner S, Weinberg W, Liao X, et al. Targeted expression of a dominant-negative FGF receptor mutant in the epidermis of transgenic mice reveals a role of FGF in keratinocyte organization and differentiation. EMBO J 1993; 12:2635–2643.

198. Ueno H, Gunn M, Dell K, Tseng AJ, Williams L. A truncated form of

fibroblast growth factor receptor 1 inhibits signal transduction by multiple types of fibroblast growth factor receptor. J Biol Chem 1992;267:1470–1476.

199. Millauer B, Shawver LK, Plate KH, Risau W, Ullrich A. Glioblastoma growth inhibited in vivo by a dominant-negative Flk-1 mutant. Nature 1994;367:576–579.

CHAPTER 7

Molecular Genetics of Chronic Myelogenous Leukemia

Razelle Kurzrock
Meir Wetzler
Zeev Estrov
Moshe Talpaz

The Philadelphia (Ph) chromosome is found in chronic myelogenous leukemia (CML) in 90% of patients, in acute lymphoblastic leukemia (ALL) in 20% of adult patients and 5% of pediatric patients, and in acute myelogenous leukemia (AML) in 2% of patients (1,2) (Table 7.1). At the cytogenetic level, the Ph chromosome is indistinguishable in these distinct disorders; at the molecular level, subtle differences are present (3–10). Herein, we will present the current state of knowledge regarding the Ph-translocation-mediated molecular events in CML, related processes in Ph-positive acute leukemia, and events driving progression in CML.

BACKGROUND

Defining the Philadelphia Chromosome

The Ph chromosome refers to a chromosome 22 with a truncated long arm (2,11). This shortened chromosome results from a reciprocal exchange of material between the long (or "q") arms of chromosomes 22 and 9. Chromosome 22 is broken at band q11; chromosome 9, at band q34. The distal part of the long arm of chromosome 22 is attached to the broken end of chromosome 9, and the distal part of the long arm of chromosome 9 is attached to the broken end of chromosome 22. Standard nomenclature designates this configuration a t(9;22)(q34;q11) translocation.

Table 7.1: The Philadelphia Translocation in Human Leukemia

Disease	*Percentage Ph-Positive*
CML	90
Adult ALL	15–20
Childhood ALL	5
AML	2

In Ph+ patients with CML, the Ph translocation is generally found in 100% of bone marrow metaphases. Patients with AML or ALL are more likely to show a mosaic cytogenetic pattern, with both normal diploid and Ph+ metaphases.

Subgroups of patients with CML and acute leukemia carry translocations that bear similarities to, but are not identical with, the classic t(9;22)(q34;q11). For instance, some CML patients have complex translocations involving three, four, or five chromosomes (including chromosomes 9 and 22), or variant translocations in which cytogenetic analysis demonstrates involvement of chromosome 22 and a chromosome other than 9, or vice versa (12,13). Further, a subset of "Ph+" ALL patients exhibit a shortened chromosome 22, with the break in band q11, but without cytogenetic evidence of a 9q34 anomaly (14). The opposite also occurs; patients with CML and acute leukemia may have an isolated 9q34 abnormality. There are also individuals with AML who carry t(6;9)(p23;q34), and individuals with T cell ALL who carry t(7;9)(q36;q34) aberrations. Recent molecular studies have begun to dissect the genetic correlates of these disparate karyotypes. Current evidence indicates that, regardless of cytogenetic findings, patients with the BCR-ABL molecular abnormality have a clinical outcome indistinguishable from that of their Ph+ counterparts, whereas those patients with distinct molecular events follow a different course.

Chronic Myelogenous Leukemia: A Clinical Model for Malignant Evolution

Chronic myelogenous leukemia was originally described in 1845 by Craigie (15), Bennett (16), and Virchow (17) as a purulence of the blood with splenomegaly. From the classification standpoint, CML is a myeloproliferative disorder and is nosologically related to polycythemia vera, agnogenic myeloid metaplasia, myelofibrosis, and essential thrombocythemia. Despite the myeloid phenotype of the early stages of CML, the clonal CML progenitor actually arises from neoplastic transformation of a hematopoietic stem cell (18,19). CML is known to account for about 25% of adult leukemias and has an incidence of one case per 100,000 persons per year (20). Although CML can occur at any age, it is rare in childhood and peaks in the middle 40s. There is no marked sex preponderance.

The clinical course of CML is triphasic (20,21) (Table 7.2). This course exemplifies the multistep process of tumor progression, a phenomenon that is probably characteristic of most malignancies but is less well delineated in other cancers. The disease ineluctably evolves from an indolent benign or chronic phase to a more aggressive accelerated phase, and terminates in blast crisis. The chronic phase is marked by a greatly increased number of committed myeloid precursor cells (22); this situation is postulated to be a consequence of unregulated growth or discordant maturation (23). Because terminal differentiation of cells is maintained,

Table 7.2: Clinical Evolution of Chronic Myelogenous Leukemia

Stage of Disease	*Characteristics*
Chronic	*Early*
	Interferon-α responsive
	Increase in mature myeloid cells
	Ph chromosome present
	Hematologic remission achieved easily
	Late
	Interferon-α resistant
	Hematologic remission achieved with hydroxyurea or other single-agent chemotherapy
Accelerated	Increasing resistance to therapy
	Thrombocytosis or thrombocytopenia
	Clonal evolution
	Increasing bone marrow or peripheral blood blasts
	Increasing basophils
	Increasing spleen size
	Occasional weight loss, fever, night sweats
Blast transformation	Bone marrow blasts $\geqslant$ 30%
	Blasts usually myeloid (60% of cases) or lymphoid (30% of cases)
	Remission induction requires AML-like high-dose chemotherapy
	Remission duration is short (though may last one year in lymphoid blast crisis)

the salient clinical feature of this stage is a significantly increased number of circulating mature neutrophils. It is not known why the myeloid compartment is preferentially affected in the chronic stage if CML is a stem cell disorder. During this phase, the high white blood cell counts can be readily managed by any one of a variety of nontoxic therapies, and the vast majority of patients maintain a near-normal lifestyle (24,25). Unfortunately, after a median interval of about 4.0 years (range, several weeks to more than 27 years) (26–28), a more aggressive hematologic state of myeloproliferative acceleration develops, wherein the myeloid cell loses its capacity for terminal differentiation. In about 80% of cases, new cytogenetic abnormalities also appear (11). An increase in immature myeloid precursors and clonal evolution are grave signs, for they are precursors of the treatment-resistant blast transformation phase, a terminal stage closely resembling Ph+ acute leukemia.

Blast crisis may have a slow onset following an accelerated phase that lasts about 6 months. Alternatively, the aggressive blastic clone may emerge abruptly. The variables dictating the length of time to blast crisis are unknown. Although clear-cut criteria separating the chronic, accelerated, and blast crisis stages are difficult to establish, most clinicians consider the detection of 30% or more blasts in the bone marrow specimen as indicative of blast transformation. The blastic clone morphology can reflect any of the hematopoietic lineages arising from the Ph chromosome: myeloid (blasts or, rarely, promyelocytes), T- and B-lymphoid, erythroid, and megakaryocytic (26,19–36). Accurate classification can be accomplished by the combined use of morphological, cytochemical, enzymatic, immunologic, and molecular techniques. Myeloid and lymphoid blast crises occur most frequently, and constitute about 65% and 25% of cases, respectively. It is important to be aware of these subtypes because they respond differently to treatment.

Patients rarely die during the chronic phase of CML. Conventional therapy of this disorder includes well-tolerated drugs such as busulfan and hydroxyurea (37,38). With these agents, the proliferative thrust of the disease can be controlled and hematologic remission can be achieved with few side effects. Unfortunately, the malignant clone remains fully predominant, as evidenced by the persistence of 100% Ph+ metaphases in bone marrow specimens from CML patients in hematologic remission. There is no cure fraction, and canonical treatment regimens have minimal impact on survival; progression to blast crisis occurs regardless of hematologic response. These observations underscore the importance of developing therapeutic approaches that prevent this terminal event. It is clear that impeding disease evolution requires eradication of the Ph+ clone or blocking of the processes that lead to blastic conversion.

Eradication of the Ph+ clone can only be accomplished in select individuals. For instance, interferon-α therapy can induce a prolonged complete suppression of the Ph chromosome in a portion of patients with newly diagnosed CML; even so, a significant number of patients remain resistant (25,39,40). Supralethal chemoradiotherapy followed by allogeneic bone marrow transplantation obliterates the malignant clone, and 50% to 70% of CML patients receiving bone marrow transplants during the chronic phase are long-term, disease-free survivors (reviewed in 25). However, patient age (usually >50) and a shortage of histocompatible sibling donors make many CML patients ineligible for bone marrow transplantation. The alternative strategy of preventing disease progression, without elimination of the Ph chromosome, is not clinically feasible because the factors causing blast crisis are unknown.

One of the most fascinating enigmas concerning the malignant evolution of CML is the extraordinary range of time intervals to blast crisis (27,28). Most patients reach this end-stage in about 4 years; however, blast crisis can occur within several weeks of diagnosis, and a case of ongoing benign phase lasting more than 27 years has been reported (27). Although a host of prognostic variables have been calculated by multivariate regression analyses of large numbers of CML patients, the actual determinants of blast crisis remain elusive. Further, a debate as to whether progression is caused by stochastic or by preordained genetic events is ongoing. The reports showing that patients with very long chronic phases (10 to 27 years) also have very long first remissions (27,41) support the school of thought that the length of time to blast crisis is predetermined.

CML Blast Crisis and Ph+ Acute Leukemia—One or Two Diseases? The Clinical Evidence The Ph chromosome was initially believed to be manifested exclusively in CML. However, it has become apparent that a t(9;22) anomaly is present in a small percentage of patients with ALL and AML (Table 7.1) (11,42,43), and is cytogenetically identical with the t(9;22) of CML. In general, its presence in acute leukemia is a poor prognostic factor (44).

A question that remains is whether Ph+ acute and chronic leukemia are actually manifestations of the same malignancy or, alternatively, are different disease states with molecularly distinct derangements, despite a cytogenetically similar appearance (14). These are not merely academic points, as understanding the relation between chronic and acute Ph+ leukemia may help uncover the pathogenetic events responsible for the conversion of CML from a chronic phenotype (the benign phase) to an acute phenotype (blast crisis). In this regard, several lines of evidence support a single underlying process for Ph+ acute and chronic leukemia (14,45). The most cogent of these are the well-documented reports of patients with Ph+ ALL or AML in whom therapy-induced transition to CML occurred, and the clinical and morphological similarity of Ph+ AML and ALL to the myeloid and lymphoid blast crises of CML, respectively. Even so, the possibility that Ph+ acute and chronic leukemia are distinct diseases is inferred from their different clinical presentations.

ONCOGENES IN MURINE LEUKEMIA

Virally induced murine leukemias presented investigators with the first clues as to the importance of altered cellular oncogenes. Because, in humans, these alterations could be caused by transfer of cytogenetic material, it followed that translocation breakpoints might yield critical information regarding the residence of cancer-related genes.

The ABL oncogene was initially discovered through analysis of the viral-ABL (v-ABL) gene of Abelson murine leukemia virus (Ab-MuLV) (46). Ab-MuLV is a replication-defective virus that transforms a wide range of lymphoid and myeloid cells in vitro (reviewed in 47). In vivo, Ab-MuLV primarily induces nonthymic pre-B lymphomas in susceptible mice (48,49); however, a variety of other murine hematopoietic neoplasms including plasmacytomas, T cell lymphomas, myelomonocytic leukemias, and mastocytomas can also be induced (50–54). Ab-MuLV may therefore cause deregulated proliferation in many hematopoietic lineages in addition to its most frequent target—the immature B cell.

The Ab-MuLV transforming gene is a fusion product of viral GAG (core) sequences encoded by Moloney murine leukemia virus (M-MuLV), and a component derived from a normal mouse gene—cellular-ABL (c-ABL, reviewed in 47). In other words, v-ABL originated from c-ABL, and the Ab-MuLV arose from recombination between M-MuLV and the mouse c-ABL gene, an event resulting in the formation of a chimeric GAG-ABL gene.

There are two main prototypic strains of Ab-MuLV, and they produce v-Abl proteins of 120 and 160 kDa, respectively (reviewed in 47 and 55). The two viruses differ in that the strain producing p120 has an internal in-frame deletion. Transformation by Ab-MuLV is mediated by the v-Abl protein. Both $p120^{GAG\text{-}ABL}$ and $p160^{GAG\text{-}ABL}$ consist of a truncated c-Abl polypeptide attached in its amino-terminus to the viral GAG (core) protein. In infected cells, the Gag-Abl proteins have cytoplasmic and plasma membrane localization (56). In addition, Gag-Abl proteins are constitutively active as tyrosine protein kinase enzymes.

Protein kinases catalyze the transfer of the terminal phosphate of adenosine triphosphate (ATP) to the amino acid residue of the substrate protein. Autophosphorylation can also occur, in which case the target protein is the enzyme itself. Phosphorylation is a basic cellular mechanism of regulating protein function. Protein kinases usually add phosphate to the amino acids serine or threonine that have a hydroxyl (OH) group to which the phosphate can be linked. Another amino acid that has a hydroxyl group and thus can be phosphorylated is tyrosine; however, phosphotyrosine accounts for less than 1% of cellular phosphoamino acids, indicating that tyrosine is phosphorylated infrequently.

The importance of the tyrosine phosphokinase function is assumed because of the association of this enzymatic property with several growth factor receptors (57) and because loss of the phosphokinase activity of the v-Abl protein abrogates its transforming capability (58). The manner in which the enzymatically enhanced v-Abl product modulates growth regulatory circuits has not been clearly defined. However, various in vitro models have supported a spectrum of mechanisms including circumvention of a growth factor–growth factor receptor pathway (59–61) and induction of new growth factors (62).

Despite the critical part played by the ABL gene in both in vitro cellular transformation and in vivo tumor production by Ab-MuLV, the complexity of these processes mandates that other variables must also be involved. Indeed, the phenotype of Ab-MuLV-induced malignancies varies with several factors: age and strain of mice, site of virus inoculation, and the helper virus used to package the Ab-MuLV genome. Further, cells transformed by viral-ABL in vitro often exhibit low tumorigenic properties at first, but evolve to highly malignant cells with time (63). For lymphocytes to exhibit a fully oncogenic growth phenotype, changes must occur in cellular genes subsequent to expression of v-ABL (63,64). In some instances, Ab-MuLV may only initiate these changes, after which it is no longer needed.

These observations on Ab-MuLV-induced murine malignancies may provide a paradigm for understanding the role of the human cellular-ABL gene—the cellular counterpart of the v-ABL gene—in the initiation and phenotypic evolution of CML. In ABL-related diseases of humans, viral mechanisms are presumably not operative. However, the Ph chromosomal translocation may cause alterations in the c-ABL gene analogous to those evoked by viral transduction. A process paralleling that in Ab-MuLV-infected mice is then postulated to occur—changes in ABL probably initiate the CML process, but secondary changes are needed to drive the progression of the disease.

THE PHILADELPHIA TRANSLOCATION AND HUMAN LEUKEMIAS

The BCR Gene

It is now known that the chromosome 22 breakpoints in Ph+ CML are restricted to a DNA segment of 5.8 kilobase pairs (kb)—the breakpoint cluster region (bcr) (6,7). Rearrangements within the 5.8-kb bcr have been confirmed in hundreds of cases of Ph+ CML (65–72). The specificity of this molecular test is underscored by the observation that bcr rearrangement is never observed in normal individuals or in victims of other diseases.

The precise point of breakage within bcr is different in each patient. Bcr sequence deletions (especially in the 3′ end) occur in 10% to 20% of patients (71,73), and mandate the use of probes corresponding to 5′ (proximal) as well as 3′ (distal) bcr regions, to avoid false-negative results for bcr rearrangement.

The 5.8-kb bcr is part of a gene (7). This gene is called the PHL gene in some of the older literature, but is currently designated as the BCR gene. The BCR gene spans 130 kb, contains 21 exons, and is oriented with its 5′ end toward the centromere (74–77) (Table 7.3). The 5.8-kb bcr segment occupies the central portion of the gene and includes five

Table 7.3: Molecular Characteristics of BCR and ABL Genes

Gene	*Characteristics*
BCR	Gene spans 130 kb and contains 21 exons
	Transcripts of 4.5 and 6.7 kb
	Protein of 160 kDa
	Protein has an associated serine/threonine kinase activity
	Protein has GTPase activity and serves as a modulator of $p21^{rac}$ on the cytoskeleton*
	Bcr expression decreases with myeloid maturation
ABL	Gene spans > 230 kb
	Transcripts of 6.0 and 7.0 kb
	Two alternative 145-kDa proteins (type 1a and 1b) that differ in their amino-terminal 26 and 44 amino acids
	Belongs to family of tyrosine kinase enzymes but not constitutively active
	Abl protein has DNA-binding properties
	Abl expression decreases with myeloid maturation

* Diekmann et al. (78)

small exons, termed bcr exons 1 to 5 (7). Three other chromosome 22 DNA segments, which are highly homologous to the seven 3′ exons of the original BCR gene, have also been cloned (79). Because transcripts have not yet been found for these other DNA segments, they may be pseudogenes.

The function of the BCR gene in normal hematopoiesis is not known. However, it is known that the BCR gene codes for messenger RNAs of 4.5 and 6.7 kb (Table 7.3); some 5′ probes also recognize a 1.2-kb mRNA (8,75,80). The 4.5-kb and 6.7-kb transcripts are expressed ubiquitously, having been detected in fibroblasts and in myeloid, lymphoid, and epithelial cells (8,75,80).

The major BCR translation product is a 160-kDa phosphoprotein ($p160^{BCR}$) with an associated kinase activity (81,82). Phosphorylation of $p160^{BCR}$ occurs on serine or threonine. Because the Bcr sequence does not bear significant homology to any known protein kinases, it is possible that $p160^{BCR}$ is not itself enzymatically active, but rather that another protein with this activity coprecipitates with it.

The ABL Gene

The other gene affected by the Ph translocation is the c-ABL gene (Table 7.3). In Ph+ CML patients, this gene is transposed from its normal residence on the long arm of chromosome 9 (band q34) to the long arm of chromosome 22 (band q11) (4). Cloning of the normal ABL gene has revealed that it spans at least 230 kb and contains at least 11 exons (7,8,83,84). It is oriented with its 5′ end toward the centromere. Two alternative first exons (exons 1a and 1b) exist and are governed by independent promoters. Exon 1a is 19 kb proximal to exon 2; exon 1b is a formidable 200 kb or more proximal to exon 2 (8,83). As a result of this genomic configuration, two major ABL transcripts are produced: 6.0-kb and 7.0-kb mRNAs (85) (Table 7.3). The 6.0-kb transcript consists of exons 1a through 11; the 7.0-kb transcript begins with exon 1b, skips exon 1a and the immense distance (in molecular terms) to exon 2, and continues with exons 2 through 11 (reviewed in 9). A similar situation occurs in the mouse, in which four types of ABL message with distinct first exons (designated exons I through IV) exist (86). All the ABL messages share a common set of 3′ exons. Therefore, the splice acceptor site of exon 2 is unusual in its promiscuous accommodation of multiple donors, and in its propensity to jump nearby donor splice sites in favor of those much farther away. The ability of ABL exon 2 to permit this facile interchange of attaching ABL DNA sequences may be a critical factor in its malignant potential, for it also allows ABL to be fused with non-ABL sequences and thus become oncogenic.

The normal human Abl proteins (designated as 1a and 1b) have a molecular mass of 145 kDa, and are termed p145^{ABL} (87). The Abl polypeptide belongs to a family of tyrosine protein kinases. Analysis of the predicted amino acid sequence of the human ABL product reveals neither a transmembrane region nor a signal peptide, indicating that ABL is not a receptor for extracellular factors (8). Investigations suggest that Abl has DNA-binding properties (56).

The BCR-ABL Gene: A Unique Cancer-Specific Marker

Chronic Myelogenous Leukemia (Chronic Phase) Interpatient variability exists in the breakpoints on both chromosomes 22 and 9 in Ph+ CML. On chromosome 22, the breakpoints span a region of 5.8 kb within bcr (6,7); on chromosome 9, they extend over a region of more than 200 kb at the 5′ (proximal) end of the ABL gene (83,84). However, ABL exon 2 is always included in the segment transposed to chromosome 22, and exons 1a and 1b are often included as well. As a consequence of the reciprocal exchange of cytogenetic material between chromosomes 9 and 22, proximal BCR sequences (ending within the central 5.8-kb bcr segment) are juxtaposed to ABL sequences in a head-to-tail fashion, and thus form a chimeric BCR-ABL gene on chromosome 22 (8). This aberrant genomic configuration occurs even in CML patients with variant Ph translocations (88). Therefore, at the molecular level, chromosome 9 is universally involved in CML, regardless of cytogenetic findings. When fused to BCR, the splice acceptor site associated with ABL exon 2 skips splice donor sites in ABL exon 1a, and frequently in ABL exon 1b, to fuse with the splice donor sites of bcr exons. ABL exons 1a and 1b do not have a splice acceptor site and, hence, cannot fuse with the donor sites within bcr. The result is that ABL exons 1a and 1b are not included in the BCR-ABL mRNA, even though they may be part of the BCR-ABL gene on chromosome 22q− (reviewed in 9). The 8.5-kb transcript consists of 5′ BCR exons ending in bcr exon 2 or 3 fused to ABL exons 2 through 11 (8) (Table 7.4). The point of fusion includes a chimeric codon with one nucleotide from bcr exon 2 or 3 and two nucleotides from ABL exon 2. It is also possible for bcr exon 1 to be fused with ABL exon 2 in this manner, but fusion of bcr exons 4 or 5 to ABL exon 2 is not consonant with formation of a BCR-ABL message because this configuration would result in a reading frameshift (7).

The translation product is a chimeric protein of 210 kDa, p210$^{BCR\text{-}ABL}$ (91–93). At least two molecular variants of the Bcr-Abl protein are generated: those with and without bcr exon 3 (94). The electrophoretic mobilities of these variants are extremely similar, probably because bcr exon 3 is small, containing only 25 amino acids.

Protein p210$^{BCR\text{-}ABL}$ is not detected in hematologic malignancies other

Table 7.4: Molecular Characteristics of BCR-ABL Genes

Gene	*Characteristics*
BCR-ABL encoding $p210^{BCR-ABL}$	Gene has amino-terminal BCR sequences and carboxy-terminal ABL sequences Break is in central region of BCR gene known as 5.8-kb breakpoint cluster region (bcr) BCR-ABL transcript of 8.5 kb (An ABL-BCR fusion transcript of uncertain significance is also expressed)* $p210^{BCR-ABL}$ is localized in the cytoplasm and can bind cytoskeletal actin microfilaments $p210^{BCR-ABL}$ is constitutively active as a tyrosine kinase enzyme $p210^{BCR-ABL}$ expression decreases with myeloid maturation†
BCR-ABL encoding $p190^{BCR-ABL}$	Gene has amino-terminal BCR sequences and carboxy-terminal ABL sequences Break is in proximal 5′ region of BCR gene (in the first intron) Transcript of 7.0 kb $p190^{BCR-ABL}$ is localized in the cytoplasm $p190^{BCR-ABL}$ is constitutively active as a tyrosine kinase enzyme; activity is more potent than that of $p210^{BCR-ABL}$

* Melo et al. (89)
† Wetzler et al. (90)

than Ph+ leukemia (95). The high specificity of $p210^{BCR-ABL}$, its altered tyrosine phosphokinase activity, and recent evidence that it can transform hematopoietic cells in vitro (96,97) constitute powerful support for the belief that this protein is directly involved in the malignant process.

CML Blast Crisis and Ph+ Acute Leukemia—One or Two diseases? The Molecular Evidence The bcr status of hundreds of cases of Ph+ CML has been documented in the literature, and at least 98% of these patients, regardless of disease stage, have bcr rearrangement. In contrast, only about 50% of adult Ph+ ALL or AML patients, and 10% of pediatric Ph+ ALL patients demonstrate bcr rearrangement (Table 7.5). Therefore, Ph+ ALL and AML include both bcr rearrangement-positive (bcr+) and bcr rearrangement-negative (bcr−) subgroups. In general, bcr rearrangement-positivity is associated with expression of $p210^{BCR-ABL}$. Patients who do not show a rearrangement within the bcr region of the BCR gene, nevertheless have a break within this gene, albeit at a more proximal position—the first BCR intron (74,98–100). It is now known that the first intron of the BCR gene, like that of the ABL gene, is large—68 kb

Table 7.5: p190$^{BCR-ABL}$ versus p210$^{BCR-ABL}$ in Philadelphia-Positive Leukemias

Type of Leukemia	*Percentage of Patients Expressing p190$^{BCR-ABL}$**
CML	<1%
Adult ALL	50%
Childhood ALL	90%
AML	50%

* The remainder of the Ph-positive patients express p210$^{BCR-ABL}$

(77). The breakpoints in Ph+ acute leukemia appear to cluster in the 3′ or distal 35 kb of this intron (77,101). In these patients, BCR exon 1 is juxtaposed to the ABL gene. The resultant chimeric gene encodes an abnormal 7.0-kb BCR-ABL mRNA consisting of BCR exon 1 fused to ABL exons 2 through 11 (Table 7.4) (reviewed in 9). The translation product is a 190-kDa Bcr-Abl protein (p190$^{BCR-ABL}$) (72,102–104). There is also a third, small (about 10%) subgroup of Ph+ acute leukemia patients who exhibit a rearrangement localizing immediately proximal to the 5.8-kb bcr segment (105); the impact of this breakpoint on protein expression is not known.

Initially, p190$^{BCR-ABL}$ was discerned in ALL cells; hence, it was thought to participate in lymphoid leukemogenesis. More recently, this protein has also been detected in Ph+ AML and in acute leukemia with a mixed myeloid-lymphoid phenotype; these data oppose the concept of lineage specificity for p190$^{BCR-ABL}$ (106,107).

Ph+ bcr+ (p210+) acute leukemia is strongly reminiscent of CML blast crisis, and both entities carry the same molecular aberration; therefore, it seems reasonable to postulate that they are a single disease process, with the acute leukemia representing the blast crisis stage of CML in patients who failed to be diagnosed while in the asymptomatic, early benign phase. The corollary to this notion is that Ph+ bcr− (p190+) ALL and AML represent bona fide acute leukemias. Evidence for such assumptions includes the following findings: 1) transition to a chronic-phase CML phenotype has been seen in several cases of p210+, but not p190+, acute leukemia (108, and R. Kurzrock, unpublished data), suggesting that the disease was really blast crisis of a previously undiagnosed CML; 2) bcr positivity and expression of p210$^{BCR-ABL}$ are found in nearly all cases of CML, while p190$^{BCR-ABL}$ has been reported almost exclusively in acute leukemia rather than in CML; and 3) the absence of a preexisting CML state has been documented in the time interval immediately preceding the onset of Ph+ bcr− (p190+) AML (106). Therefore, expression of p190$^{BCR\text{-}ABL}$ versus p210$^{BCR\text{-}ABL}$ may play a critical role in the manifestation of an acute, rather than a chronic, Ph+ leukemia phenotype. The above hypothesis is especially appealing in light of recent data indicating that, in an in vitro system using retroviral constructs (109), p190$^{BCR\text{-}ABL}$ is a more potent stimulator of hematopoietic transformation and proliferation than p210$^{BCR\text{-}ABL}$.

It still remains possible that the significant event in Ph+ leukemogenesis lies solely in the truncation and resultant activation of ABL. According to this alternative hypothesis, leukemia phenotype is determined by the target cell affected by the t(9;22) translocation, regardless of whether it is p190$^{BCR\text{-}ABL}$ or p210$^{BCR\text{-}ABL}$ that is expressed. CML results because of the acquisition of the Ph translocation by a pluripotent stem cell; ALL or AML, by a lymphoid or myeloid lineage progenitor, respectively. Against this proposition are research data generated by Turhan and colleagues (110), who have shown that Ph+ bcr+ ALL can originate in either a pluripotent stem cell or a lineage-restricted lymphoid progenitor. In addition, our analysis of individually plucked granulocyte-macrophage colony-forming units (CFU-GM) and erythrocyte blast-forming units (BFU-E) colonies for BCR-ABL has shown that both p210- and p190-positive ALL can be either lymphoid-lineage-restricted or multilineage in origin (111).

The discovery of two molecularly distinct subgroups of Ph+ acute leukemia prompted a search for clinical distinguishing features. Among 32 adults whom we studied (112), no striking differences between bcr+ (p210+) and bcr− (p190+) Ph+ acute leukemia patients were seen, a result consistent with smaller studies in the literature (105, 106, 113–115, reviewed in 9). As mentioned above, because both p210-positive and p190-positive disease may originate either in a pluripotent stem cell or in a committed progenitor, it may be that differences between p210- and p190-positive leukemias will be distinguishable only if the cell of origin is taken into account (111).

Transformation by Bcr-Abl May Involve Enzymatic Activation or a Change in Subcellular Localization

The possible mechanisms governing oncogenic activation of the ABL gene through usurpation of its amino-terminus by the BCR gene include a change in enzymatic activity, subcellular localization, or substrate affinity. In this regard, p210$^{BCR\text{-}ABL}$ (regardless of the presence of bcr exon 3) has a much higher tyrosine kinase activity than its normal counterpart p145^{ABL} in vitro (91–94); p210$^{BCR\text{-}ABL}$ and p145^{ABL} also have dissimilar patterns of autophosphorylation and may interact differently with substrates (87).

However, though several candidate substrate phosphoproteins have been identified for $p210^{BCR-ABL}$ (95,96), work in this area remains rather limited in scope. $p190^{BCR-ABL}$ is also enzymatically active as a tyrosine protein kinase (72,116), and its enzymatic activity is more potent than that of $p210^{BCR-ABL}$ (116); this may explain the association of $p190^{BCR-ABL}$ with the more aggressive acute leukemia phenotype.

Concerning subcellular localization, studies of the normal mouse Abl proteins, which, as mentioned above, differ in their amino-termini, may provide insights. Simple overexpression of these proteins is insufficient for transformation of mouse NIH3T3 cells. However, deletion of a putative regulatory domain (53 amino-terminal amino acids) of the mouse ABL IV transcript activates its transforming potential fully, with respect both to fibroblast and B-lymphocyte transformation in vitro and to leukemogenicity in vivo (56). Apparently, deletion of this sequence is also associated with transfer of the Abl protein from its normal residence in the nucleus to the cytoplasm and abrogation of the DNA-binding properties of normal Abl (56). In both normal and leukemic human hematopoietic cells, however, c-Abl is discerned predominantly in the cytoplasm (90). The reason for the difference in localization between mouse and human Abl is not known but could relate to the predominant Abl subtype expressed, with human type 1a corresponding to mouse Abl I, and human Abl 1b corresponding to mouse Abl IV. In support of this concept are data demonstrating that the Abl protein encoded by the 7.0-kb ABL mRNA (type 1a) has a glycine at its amino-terminus and that such a structure in several other proteins (the SRC oncogene product and the catalytic unit of cyclic adenosine-monophosphate-dependent kinase [117,118]) is linked to myristic acid, which is thought to stabilize the interaction of proteins with the lipid bilayer of cell membranes.

The normal endogenous Bcr protein and the aberrant $p210^{BCR-ABL}$ and $p190^{BCR-ABL}$ proteins also consistently localize to the cytoplasm in both cell lines and fresh cells (90). Whether or not a difference in subcellular localization between the normal human Abl protein and the aberrant Bcr-Abl protein occurs and is operative in the emergence of human leukemia therefore remains speculative. Cytoplasmic localization of $p210^{BCR-ABL}$ is consistent with the experiments of Dikstein and coworkers (119), who showed that $p210^{BCR-ABL}$ lacks the ability to bind DNA, and with those of McWhirter and Wang (120), who showed that this protein binds cytoplasmic actin microfilaments, perhaps because BCR sequences promote serine-threonine phosphorylation of the c-Abl carboxy-terminus, an event associated with attenuation of Abl DNA-binding properties.

MOLECULAR MECHANISMS OF EVOLUTION TO BLAST CRISIS AND PREDICTION OF CHRONIC PHASE DURATION

Altered Oncogenes

The Ph chromosome is found in the early chronic phase; therefore it seems likely that bcr rearrangement is necessary, but not sufficient, for disease evolution, and that superimposed secondary genetic driving forces are required (11). Cytogenetic and clinical heterogeneity characterizes the blast crisis stage. For example, blasts can have a myeloid, lymphoid, or, less commonly, an erythroid or megakaryocytic morphology. Secondary cytogenetic abnormalities can include an additional Ph chromosome, trisomy of chromosome 8, isochromosome 17(q), trisomy of chromosome 19, and loss of chromosome 7 or the Y chromosome. This heterogeneity suggests that several different subsets of blast crisis exist and that the genetic events in each of these subsets are distinct. It is therefore highly unlikely that any single molecular abnormality will account for blast crisis in all patients. The known cytogenetic anomalies may provide localization clues in the search for relevant molecular perturbations.

To date, several aberrant molecular events have been implicated in the development of blast crisis. In this regard, one of the most common manifestations of clonal evolution is the appearance of an additional Ph chromosome. The presence of this aberration can be used to argue for the association of quantitative changes in Abl-related expression with disease progression. A role for increased Bcr-Abl expression in driving metamorphosis to the blastic phenotype is also supported by the observation of elevated levels of BCR-ABL mRNA in some patients with advanced rather than early CML, though this finding has been disputed by at least one group of investigators (121–124).

Alternatively, a role for a qualitative change in Bcr-Abl expression has also been advanced. Several observations support this concept. First, in some cases, a new bcr rearrangement (albeit with retention of the original rearrangement) can be detected during the course of disease evolution (71,125). Further, even in those instances in which only a single rearranged bcr band is discerned, the presence of a second BCR gene rearrangement, which is outside the region recognized by the probe for the 5.8-kb bcr, cannot be ruled out. It is therefore conceivable that another recombination event coincides with the development of a second Ph chromosome. Second, all the sequences needed to code for $p190^{BCR\text{-}ABL}$ are included in the BCR-ABL gene of CML (74,98,99). In the event that a switch from the exclusive expression of this gene as $p210^{BCR\text{-}ABL}$ to the expression of $p190^{BCR\text{-}ABL}$ in addition to $p210^{BCR\text{-}ABL}$ was possible, would a change in phenotype occur? We studied 37 patients with CML and found that 3 (9%) expressed both $p190^{BCR\text{-}ABL}$ and $p210^{BCR\text{-}ABL}$ in the advanced (blast crisis or accelerated) stage, but only $p210^{BCR\text{-}ABL}$ in the chronic phase (126). In addition, a CML-type breakpoint (within the 5.8-kb bcr) coexisting with an ALL-type breakpoint (within the first BCR exon) has been described in a child with ALL and two Ph chromosomes (127).

Another controversial theory is that patients with 5′ (proximal) breakpoints within the 5.8-kb bcr survive longer than those with 3′ (distal) breakpoints (66,71,128–130, reviewed in 131). The most cogent evidence against this theory is that two alternate 210-kDa Bcr-Abl products (those with and those without exon 3 [94]) often coexist in CML patients (132,133), thus obviating the influence of distinct breakpoints within the 5.8-kb bcr. In addition, most of these studies suffer from a fundamental defect in that they are confined to analyzing DNA breakpoints, which do not necessarily correlate with the structure of the RNA and protein produced. Some investigators who segregated patients by the type of protein produced (with or without bcr exon 3) suggested that those patients whose Bcr-Abl protein includes bcr exon 3 are more prone to thrombocytosis (134). These authors could not find other clinical differences between the two molecular subgroups.

Lastly, several lines of evidence suggest that the role of $p210^{BCR\text{-}ABL}$ in advanced CML is supplanted by other factors. Convincing arguments against the participation of Bcr-Abl expression in disease evolution are as follows. First, we found equivalent expression of $p210^{BCR\text{-}ABL}$ enzymatic activity in myeloid, lymphoid, and undifferentiated blast crisis (95); yet the distinct phenotypes of these disease manifestations should reflect disparate genetic events. Second, Bartram and colleagues (135) documented deletion of BCR-ABL sequences in cells derived from a CML patient entering blast crisis. Third, we have observed that the growth rate of CML blast crisis cells in vitro can be maintained even in the presence of severely reduced levels of Bcr-Abl protein and enzymatic activity (136). Finally, Young and Witte have demonstrated that although $p210^{BCR\text{-}ABL}$ can transform hematopoietic cells in vitro, these cells do not display an aggressive oncogenic phenotype (137). This result is consistent with a role for BCR-ABL in the indolent, chronic phase of CML, but not in the virulent blastic phase.

Other oncogenes have also been analyzed in CML blast transformation (Table 7.6). LeMaistre and associates (138) studied 22 samples derived from CML blast crisis patients and 29 samples from chronic-phase CML patients for mutations of codons 12, 13, or 61 of the N-RAS, H-RAS, and K-RAS genes; two blast crisis patients (9%) exhibited mutations,

Table 7.6: Altered Oncogenes in Advanced Chronic Granulocytic Leukemia

Gene Studied	*Chronic Phase (% of Patients with Alteration)*	*Accelerated/Blast Crisis (% of Patients with Alteration)*
p53	2%	13%
Rb gene	1%	12%
RAS gene	1%	7%
p190$^{BCR\text{-}ABL}$*	0%	9%

The data in this table are summarized from a survey of the literature.
* Refers to the development of the acute-leukemia-associated p190$^{BCR\text{-}ABL}$ in addition to the CML-associated p210$^{BCR\text{-}ABL}$ (126).

both in the K-RAS gene. Collins et al. (139) could not, however, identify mutations in any of the 21 cases of CML blast crisis that they analyzed. These data indicate that RAS mutations can occur in advanced disease, albeit rarely. Alteration in p53, a gene whose protein product is associated with tumorigenicity, has also been described (140,141). Ahuja and coworkers found rearrangement in the p53 gene in 7 (30%) of 23 patients with CML blast crisis, but not in patients with other types of leukemia (140). The MYC protooncogene has also been analyzed in advanced CML, with investigators reporting both elevated MYC RNA expression (142) and MYC amplification (143).

Autocrine and Paracrine Phenomena

Signaling pathways that mediate the normal functions of growth factors or, alternatively, growth factors themselves are often subverted in cancer. We have published a series of papers suggesting that there is evidence for aberrant growth factor production in the more advanced stages of CML (144,145).

Our experiments indicate that higher levels of interleukin-1β (IL-1β) protein can be found in leukocyte lysates from patients in the accelerated or blast crisis phase of the disease as compared with those in the chronic phase and may be predictive of duration of survival (M. Wetzler, manuscript submitted). In addition, adherent layers derived from patients in myeloid and undifferentiated blast crises exhibit constitutive growth factor expression (IL-1β, IL-6, and GM-CSF) not seen in the chronic phase of the disease (144) (Table 7.7). Support for the notion of autocrine and paracrine effects comes from experiments showing that progenitor cells from the interferon-resistant phase display less response to exogenous growth factors and a relatively autonomous growth pattern in colony

Table 7.7: Altered Cytokine Expression in Cultured Stromal Cells from Patients with Advanced Chronic Myelogenous Leukemia

Diagnosis	*Percentage of Patients Positive for*			*Total No. of Patients Studied*
	IL-1β	*IL-6*	*GM-CSF*	
Normal	0	0	0	11
Chronic-phase CML	0	0	0	8
Blast crisis CML	50	50	10	10

Data are from Wetzler M, Kurzrock R, Lowe DG, Kantarjiaw H, Gutterman JU, Talpaz M. Alteration in bone marrow adherent layer growth factor expression: a novel mechanism of chronic myelogenous leukemia progression. Blood 1991;78:2400–2406.

assays (145). Interestingly, soluble IL-1 receptors and IL-1 receptor antagonist, both of which inhibit IL-1 actions, suppress CML colony growth in a dose-dependent fashion; a similar effect has been noted with the use of anti-IL-1β neutralizing antibodies (145).

Our observations suggest the possibility that factors, such as IL-1β, generated by the leukemic cells may influence the activity of the microenvironment as well as that of accessory cells, and, hence, direct interactions between the regulatory mesenchymal cells and the leukemic cells that determine the proliferative and differentiation state of the latter. Because constitutive activation of growth-promoting pathways is amenable to diagnostic testing, these findings offer opportunities for improving our ability to predict prognosis (unpublished data). In addition, on the basis of our in vitro results showing suppression of CML clonogenic growth by IL-1 inhibitors, we have recently initiated a clinical trial using IL-1 receptor antagonist in patients with CML.

Molecular Studies and Classification of Philadelphia Chromosome-Negative Leukemias

The Ph chromosome is not discernible in about 5% to 10% of patients with a presumptive diagnosis of CML (146) or in the majority of patients with acute leukemia. Nevertheless, approximately 50% of Ph-negative CML patients and just under 10% of ALL patients who are Ph-negative will show molecular evidence of the BCR-ABL abnormality (Table 7.8). In the case of Ph-negative, BCR-ABL-positive CML, the clinical course has been shown to be identical with that of Ph-positive disease (147,148).

Ph−, bcr+ CML In 1985, Bartram described a Ph-negative CML patient who exhibited a typical bcr rearrangement and juxtaposition of the BCR and ABL genes (149). This observation has since been confirmed by

Table 7.8: BCR-ABL-Positivity in Philadelphia-Negative Leukemia

Disease	*Percentage of Patients Who Are Ph-Negative*	*Percentage of Ph-Negative Patients Who Are BCR-ABL-Positive*
CML	5–10	50
Adult ALL	80–85	9
AML	98	0*

* Only a small number of patients have been studied; therefore, the accuracy of this percentage is not totally clear.

several investigators (147,150–157); it has also been demonstrated that Ph− bcr+ CML patients produce the 8.5-kb BCR-ABL mRNA (147) and the 210-kDa Bcr-Abl protein (150) (Fig. 7.1). In Ph− CML, as in Ph+ disease (71,73), deletions in the 3′ region of bcr can occur and result in false-negative results for bcr rearrangement (147); therefore, accurate assessment of bcr status in these patients requires the use of probes encompassing both the 5′ and 3′ segments of the 5.8-kb bcr region.

Usually Ph− bcr+ CML patients have either a normal cytogenetic profile or an abnormality involving chromosome 9 band q34. The mechanism by which the BCR and ABL genes are juxtaposed in these patients has been investigated by in situ hybridization. These studies suggest that

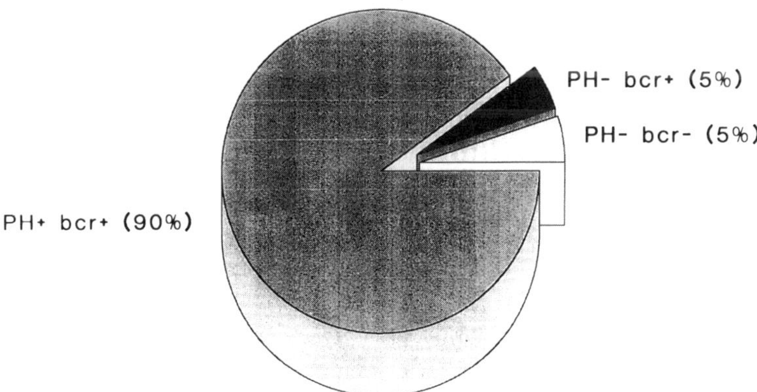

Fig. 7.1: Diagrammatic representation of the percentage of CML patients with Ph-positive and BCR-ABL-positive ($p210^{BCR\text{-}ABL}$) disease.

in cytogenetically diploid cells, interstitial insertion of the ABL gene next to the bcr segment of the BCR gene on chromosome 22 occurs without reciprocal translocation of genomic material to chromosome 9 (154); in patients with a translocation between 9q34 and another chromosome, a complex three-way set of recombination events is needed to produce the juxtaposition of the BCR and ABL genes (149).

Patients with Ph− bcr+ CML have clinical and laboratory features that are indistinguishable from those of Ph+ CML patients, and both groups respond to hydroxyurea, busulfan, and interferon-α therapy in an identical manner (147,150). Importantly, during hematologic remission, CML patients with a diploid karyotype exhibit a normal morphological and cytogenetic blood and bone marrow profile. Even so, molecular studies will reveal bcr rearrangement, reflecting the persistence of the malignant clone (147).

Overall, several important concepts have emerged from the molecular investigations of Ph− CML. First, the ability of molecular techniques to define subchromosomal genomic aberrations in Ph− CML indicates that these methods are a crucial adjunct to clinical and morphological evaluation of disease nosology. Second, Ph+ and Ph− CML with bcr rearrangement have similar clinical outcomes and an identical molecular emblem, suggesting that they represent a single disease. Ph− bcr+ CML patients should therefore be eligible for therapeutic protocols aimed at Ph+ CML patients, including bone marrow transplantation and interferon therapy. Finally, Ph− bcr+ patients should be followed by molecular techniques, as bcr rearrangement may be the only evidence of residual malignancy at the time of hematologic remission.

Ph−, bcr+ Acute Leukemia Of our adult ALL patients, 14% demonstrate the Ph chromosome (158) (Fig. 7.2). We investigated 59 adults with Ph-negative acute leukemia (24 with AML and 35 with ALL) for the presence of BCR-ABL transcripts (158). Three patients, all of whom had ALL, had molecular evidence of BCR-ABL-positivity (one had p190 and two had p210). Each of these three patients had insufficient metaphases on cytogenetic analysis, and since the total number of patients with insufficient metaphases was 16, the incidence of Ph-related molecular abnormalities in this group was 19%. Based on a 32% incidence of insufficient metaphases in our ALL population, we projected an additional detection rate of 6% for Ph-positivity in ALL patients if molecular techniques are used.

Ph−, bcr− CML The question arises whether all Ph-negative patients with classic CML will exhibit bcr rearrangement. Intuitively, we have felt

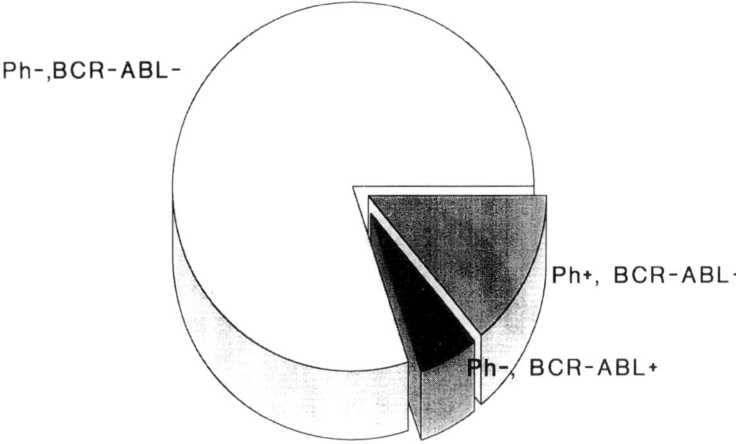

Fig. 7.2: Diagrammatic representation of the estimated percentage of adult ALL patients with Ph-positive and BCR-ABL-positive ($p190^{BCR\text{-}ABL}$ or $p210^{BCR\text{-}ABL}$) disease. About 80% of adult ALL patients have neither the Ph chromosome nor a BCR-ABL protein. About 14% of ALL patients are both Ph-positive and BCR-ABL-positive, while about 6% have a BCR-ABL protein in the absence of cytogenetic evidence of the Ph chromosome.

that this should be true. Yet, we have now encountered 11 patients with classic chronic-phase CML who have neither a discernible bcr rearrangement nor a BCR-ABL message (148). At the time of diagnosis, these patients had splenomegaly, neutrophilia, occasional basophilia and thrombocytosis, a generally low leukocyte alkaline phosphatase score, and a hypercellular bone marrow specimen with an increased myeloid-to-erythroid ratio, a shift to the left in myeloid maturity, and a pronounced granulocytic hyperplasia. Monocytosis, polycythemia, dysplasia, and disproportionate thrombocytosis were not observed. Cytogenetic analysis revealed that six of these patients had a normal diploid karyotype; five patients had karyotypic abnormalities in a fraction of their metaphases: trisomy 21 in two patients; trisomy 8 in one patient; trisomy 19 in one patient; 20q− in one patient; t(1p−;13q+) in one patient; and monosomy 17 with an i(17q) in one patient. Individuals with Ph−, bcr− CML responded to hydroxyurea, busulfan, and interferon-α, although the small number of cases precluded a statistically valid comparison of their reponse rates to those of bcr+ patients. Their predicted 50% survival was 37 months by Kaplan and Meir analysis. Distinguishing characteristics appeared to be limited to a higher median age (60 years) at diagnosis than that of bcr+ CML patients (46 years), and the occasional presence of B symptoms (fever, weight loss, or night sweats), a finding that is rare in early bcr+ CML.

Despite the overall striking resemblance between chronic-phase bcr+ and bcr− CML, disease progression manifests differently in these two groups. In bcr+ disease, the chronic phase ineluctably yields to a blast transformation phase typified by growing numbers of rapidly proliferating immature cells. In contrast, blast crisis did not occur in the bcr− CML patients; rather, progression was characterized by increasing leukemia burden with profound leukocytosis, organomegaly, extramedullary infiltrates, and eventual bone marrow failure (anemia and thrombocytopenia, with blast counts <20%).

The molecular forces driving Ph−, bcr− CML are not known. However, it may be that $p210^{BCR\text{-}ABL}$ is part of a growth factor–growth factor receptor signal transduction pathway. Perturbation of any part of this pathway could conceivably result in a disease phenotype similar to bcr+ CML. In this regard, $p210^{BCR\text{-}ABL}$ can transform IL-3-dependent hematopoietic cells to factor independence without autocrine secretion of IL-3 (159). It is conceivable that IL-3, or another growth factor, forms part of the BCR-ABL pathway, and that an aberration of this yet-to-be-determined factor or its receptor may result in leukemic proliferation and, hence, Ph− bcr− CML.

Distinguishing the Ph Translocation from Other Anomalies at 22q11 and 9q34

Ph− Acute Leukemias with 9q34 Chromosomal Abnormalities As mentioned above, patients with a CML phenotype may have several types of karyotypic alterations that are variants of the classic t(9;22)(q34;q11) Ph anomalies: complex translocations between at least three chromosomes including 9 and 22; variant translocations between chromosome 22q11 and a chromosome other than 9; and variant translocations between chromosome 9q34 and a chromosome other than 22. At the molecular level, these patients exhibit the BCR-ABL genotype that is pathognomonic for CML, and their clinical outcome mirrors that of their counterparts carrying the standard t(9;22).

Even so, cases exist of acute leukemias with a 9q34 chromosomal abnormality that does not involve the ABL gene. For instance, a subgroup of AML patients have a t(6;9)(p23;q34) aberration (160). Westbrook and coworkers (161) studied these patients and found that at the molecular level, the break on chromosome 9q34 occurs distal to the ABL gene and no anomalous Abl protein is produced. Similarly, in an acute T cell leukemia line carrying a t(7;9)(q36;q34) abnormality, Westbrook and coworkers (162) demonstrated that the ABL gene is not involved. The breakpoint in these cells occurs at chromosome 9q34, but is at least 250 kb proximal to the ABL gene.

ALL with a 22q11 Variant Chromosomal Abnormality A subgroup exists of "Ph+" ALL patients who display a deletion of the long arm of chromosome 22, with the breakpoint occurring at 22q11, but without cytogenetic evidence of involvement of 9q34. In contrast to the situation in CML, in which variant Ph translocations are always associated with production of a Bcr-Abl product (163), the 22q11 variant of ALL has been shown to encode only the normal $p145^{ABL}$ in some patients (164). It follows that the disruption of chromosome 22 in these patients does not involve the BCR gene. It is theoretically possible, but experimentally unproven, that the break occurs in the immunoglobulin λ light chain, a region on 22q11 known to be rearranged in some lymphoid malignancies. Although systematic correlation of the outcome of BCR-ABL-positive versus BCR-ABL-negative, 22q− patients has not been performed, the poor prognosis associated with the t(9;22) in acute leukemias suggests that the two subgroups of patients should behave differently. Therefore, even though BCR-ABL status (p210-positivity versus p190-positivity) in patients with acute leukemia carrying a classic Ph chromosome is not currently useful as a predictive test, further molecular assessment of Ph variants in this disease is warranted, and may yield important prognostic information.

Detection of Minimal Residual Disease: Advantages and Limitations of Current Technology

Several different techniques have been used to measure minimal residual disease in CML: cytogenetic analysis, Southern blotting of DNA, polymerase chain reaction (PCR), and fluorescent in situ hybridization (FISH). Table 7.9 presents a comparison of these assays.

The polymerase chain reaction is a tremendously potent technique that allows amplification of specific DNA or RNA (using complementary DNA) sequences, and the detection of small numbers of cells containing these sequences (one malignant cell among up to 1×10^6 normal cells). Currently, this technology can be applied to the detection of minimal residual disease only if a known nucleotide sequence is unique to the malignant cells. Ph+ leukemias are therefore the perfect prototypes on which to use PCR. Because of its exquisite sensitivity, this technique

Table 7.9: Methods for Detecting Minimal Residual Disease in Chronic Myelogenous Leukemia

Method	*Advantages*	*Disadvantages*
Cytogenetics	Very specific for Ph chromosome; false-positive results rare.	Only 25 metaphases usually counted; therefore, cannot detect < 5% residual leukemia cells. Time-consuming and labor-intensive.
Southern blotting of DNA	Very specific for bcr rearrangement; false-positive results rare.	Cannot detect < 1% to 5% residual Ph-positive cells. Time-consuming and labor-intensive.
Polymerase chain reaction	Can detect one leukemic cell among 10^4 to 10^6 normal cells.	Between 5% and 15% of samples may represent false-positive results because of danger of contamination with such an exquisitely sensitive technique. Not quantitative (but adaptations to make this assay quantitative are under development). Relies on detection of mRNA and may therefore be inaccurate if Ph-positive cells that produce little or no message exist.
Fluorescent in situ hybridization (FISH)	Allows rapid detection of Ph translocation among hundreds of cells.	Up to 5% of cells may show false-positive reaction, and this may be the limiting factor in the use of this assay for detection of minimal residual disease.

provides a significantly better rate of detection than older techniques such as cytogenetics and Southern blotting, both of which have a lower limit of sensitivity of about 5%. Amplification of cDNA sequences corresponding to the BCR-ABL mRNA is easily exploited for determining the presence of residual malignant cells in CML and Ph+ acute leukemia (165,166). However, extreme caution in preventing false-positive results must be exercised by laboratories utilizing PCR, as the powerful amplification process makes sample contamination a constant hazard.

Using this technology, several groups have demonstrated BCR-ABL transcripts in some CML patients in clinical and cytogenetic remission after bone marrow transplantation (133,167) and interferon-α therapy (168). However, most studies have reported a high incidence of PCR-positivity early after allogeneic transplant that does not correlate with relapse (169). Indeed, more than 75% of transplant recipients will have at least one PCR-positive test within one year of transplant, and 40% to 50% will be positive at one year. Results of PCR studies late after transplant are also highly variable, and do not appear to indicate a significant risk of relapse. In contrast to the results of single determinations, serial PCR determinations have allowed the identification of three patient categories—persistently PCR-positive, intermittently positive or negative, and persistently PCR-negative, which correlate with high, intermediate, and low risks of relapse, respectively (169). Finally, as most PCR assays are currently not considered quantitative, more information may be derived when a refinement of the assay for quantitation becomes routinely available.

Another technique that has recently been applied to the detection of Ph-positive cells is fluorescent in situ hybridization (FISH). This technique is a form of "molecular cytogenetics" that allows the detection of chromosomal abnormalities in large numbers of interphases or metaphases, rather than the small number (about 25) usually counted during classic cytogenetic analysis. In contrast to PCR determinations in Ph-positive CML, which are performed using BCR-ABL transcripts, FISH detects DNA-based genetic abnormalities and would therefore not be subject to the hypothetical possibility of variability due to differential transcription. However, the application of FISH to the detection of residual disease may be limited by false-positive rates of up to 5% of cells in individual patient samples.

MOLECULAR THERAPY

Exploitation of BCR-ABL for diagnostic purposes and to measure residual disease is common. The possibility of targeting this defect for therapeutic purposes is beginning to be pursued. Two characteristics distinguish Bcr-Abl from other tumor markers and make it an attractive target for treatment: it is found exclusively in the tumor cells and is never expressed in normal cells, and it is directly involved in the pathogenesis of the leukemic process. Even so, since Bcr-Abl is not a cell-surface protein, mitigating its actions is not simple, and this field is still in its infancy.

Antisense Purging for Autologous Transplantation

In both early and advanced CML, high-dose therapy followed by autologous transplantation can produce transient cytogenetic responses. To increase response rates, ex vivo purging techniques to eliminate residual

Ph-positive cells are being pioneered. One gene-directed approach would be the use of antisense oligodeoxynucleotides complementary to BCR-ABL sequences (170). Current limitations of this approach include rapid degradation of unmodified oligomers, and nonspecific inhibition or solubility problems in oligomers that have been modified. An exciting new strategy to overcome this problem involves liposomal encapsulation of antisense oligomers.

Tyrosine Kinase Inhibitors

BCR-ABL is active as a tyrosine kinase enzyme, and this activity is believed to be critical to its transforming capability. Several tyrosine kinase inhibitors are available (reviewed in 171). They include tyrphostins (a group of low molecular weight dihydroxybenzylidene derivatives that inhibit tyrosine kinases in a noncompetitive manner), genistein (an isoflavone derivative), herbimycin A (a benzenoid ansamycin antibiotic), and erbstatin (another antibiotic compound). Because tyrosine kinase activity plays a fundamental role in the growth and development of normal tissues, the major concern with tyrosine kinase inhibitors is their nonspecificity and hence their potential for toxicity. However, some of these inhibitors, such as the tyrphostins, may be more specific for certain kinases including $p210^{BCR\text{-}ABL}$.

Interleukin-1 Inhibitors

It is generally believed that the BCR-ABL molecular perturbation is critical in the initiation and promotion of early disease; however, secondary genetic forces drive disease progression. As mentioned above, both altered oncogenes and autocrine or paracrine production of growth factors (Fig. 7.3) such as IL-1β may be involved in various subsets of patients. The latter leads to the possibility of using growth factor inhibitors in the clinic.

Over the last few years, numerous ways to inhibit the production or action of cytokines have been discovered. The cloning of a series of novel molecules—IL-1 receptor antagonist, soluble IL-1 receptors, and IL-1 converting enzyme inhibitor—that specifically interfere with IL-1 function may be of particular relevance (Table 7.10). There molecules can suppress CML clonogenic growth in vitro (145). They are currently undergoing preclinical and clinical testing in a variety of settings including autoimmune disorders, infection, and graft-versus-host disease, and most recently in neoplastic diseases such as CML.

CONCLUSION

Cytogenetic aberrations are a hallmark of malignancy and provide a pointer for the molecular localization of activated cancer genes

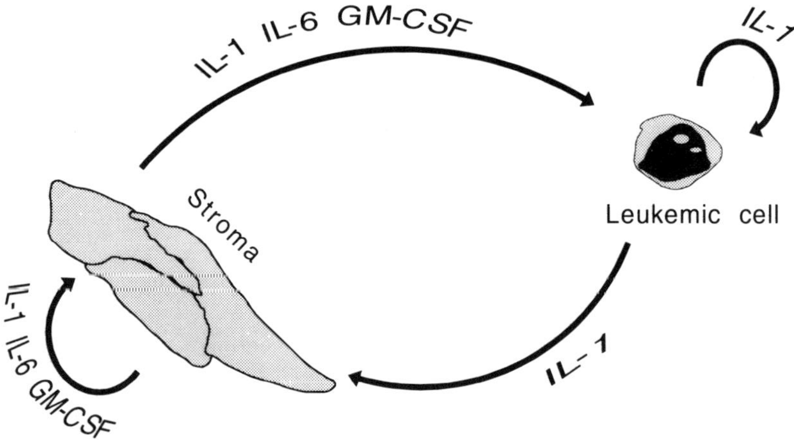

Fig. 7.3: Postulated autocrine and paracrine network in CML.

(oncogenes). The first and best-studied consistent karyotypic abnormality found to be associated with neoplastic disease was the Philadelphia (Ph) chromosome, an anomaly identified in 1960 by Nowell and Hungerford (2). In leukemia patients bearing this abnormality, the Ph translocation t(9;22)(q34;q11) results in transposition of the ABL protooncogene from chromosome 9q34 to 22q11, where it is fused with part of the BCR gene. The translation product is a chimeric Bcr-Abl protein with markedly enhanced enzymatic properties and possibly with a displaced subcellular

Table 7.10: Characteristics of Interleukin-1 Inhibitors

Inhibitor	*Characteristics*
Neutralizing antibodies	Antibody that neutralizes IL-1 action
IL-1 receptor antagonist	Naturally occurring molecule that competes with IL-1 for its receptor without agonist activity
Soluble IL-1 receptors	Truncated IL-1 receptor that binds 1 mol IL-1/mol receptor
IL-1 converting enzyme inhibitors	Inhibits the converting enzyme that mediates the cleavage of the relatively inactive pro-IL-1β to its active form
IL-4	Pleiotropic lymphokine that inhibits IL-1 transcription
IL-10	Inhibits IL-1 production by blood mononuclear cells and up-regulates the synthesis of IL-1RA
IL-1 antisense	Oligomer that inhibits translation of IL-1

Reviewed by Estrov et al. (172)

localization. Cognizance of these molecular events is crucial to understanding the processes promoting initiation and progression of Ph+ hematologic malignancies and is providing fundamental tools for the diagnosis and detection of minimal residual disease. This knowledge may also eventually be exploitable as the basis of gene-directed therapy of Ph-positive leukemias.

REFERENCES

1. Mitelman F, ed. Catalog of chromosome aberrations in cancer. New York: Alan R. Liss, 1988.

2. Nowell PC, Hungerford DA. A minute chromosome in human chronic granulocytic leukemia. Science 1960;132:1197.

3. Rowley JD. A new consistent chromosomal abnormality in chronic myelogenous leukemia identified by quinacrine fluorescence and Giemsa staining. Nature 1973;243:290–293.

4. Bartram CR, de Klein A, Hagemeijer A, et al. Translocation of the ABL oncogene correlates with the presence of the Philadelphia chromosome in chronic myelocytic leukaemia. Nature 1983;306:277–280.

5. Heisterkamp N, Stephenson JR, Groffen J, et al. Localization of the *c-abl* oncogene adjacent to a translocation breakpoint in CML. Nature 1983;306:239–242.

6. Groffen J, Stephenson JR, Heisterkamp N, de Klein A, Bartram CR, Grosveld G. Philadelphia chromosomal breakpoints are clustered within a limited region, bcr, on chromosome 22. Cell 1984;36:93–99.

7. Heisterkamp N, Stam K, Groffen J, de Klein A, Grosveld G. Structural organization of the *bcr* gene and its role in the Ph^1 translocation. Nature 1985; 316:758–761.

8. Shtivelman E, Lifshitz B, Gale RP, Canaani E. Fused transcript of *abl* and *bcr* genes in chronic myelogenous leukemia. Nature 1985;315:550–554.

9. Kurzrock R, Gutterman JU, Talpaz M. The molecular genetics of Philadelphia chromosome-positive leukemias. N Engl J Med 1988;319:990–998.

10. Kurzrock R, Shtalrid M, Gutterman JU, Talpaz M. Molecular diagnostics of chronic myelogenous leukemia and Philadelphia-positive acute leukemia. Cancer Cells 1989;7:9–13.

11. Rowley JD. Chromosome abnormalities in human leukemia. Annu Rev Genet 1980;14:17–39.

12. Sandberg AA. Chromosomes and causation of human cancer and leukemia: XL. The Ph^1 and other translocations in CML. Cancer 1980;46:2221–2226.

13. Sonta S, Sandberg AA. Chromosomes and causation of human cancer and leukemia. XXIV. Unusual and complex Ph^1 translocations and their clinical significance. Blood 1977;50:691–697.

14. Catovsky K. Ph[1]-positive acute leukaemia and chronic granulocytic leukaemia: one or two diseases? Br J Haematol 1979;42:493–498.

15. Craigie D. Case of disease of the spleen in which death took place in consequence of the presence of purulent matter in the blood. Edinburgh Med Surg J 1845;64:400–412.

16. Bennett JH. Case of hypertrophy of the spleen and liver in which death took place from suppuration of the blood. Edinburgh Med Surg J 1845;64:413–418.

17. Virchow R. Weisses blut. Froiep Notizen 1845;36:151–154.

18. Fialkow PJ, Jacobson RJ, Papayannopoulou T. Chronic myelocytic leukemia: clonal origin in a stem cell common to the granulocyte, erythrocyte, platelet and monocyte/macrophage. Am J Med 1977;63:125–130.

19. Barr RD, Fialkow PJ. Clonal origin of chronic myelocytic leukemia. N Engl J Med 1973;289:307–309.

20. Sokal JE. Evaluation of survival data for chronic myelocytic leukemia. Am J Hematol 1976;1:493–500.

21. Cervantes F, Rozman C. A multivariate analysis of prognostic factors in chronic myeloid leukemia. Blood 1982;60:1298–1304.

22. Galbraith PR, Abu-Zahra HT. Granulopoiesis in chronic granulocytic leukaemia. Br J Haematol 1972;22:135–143.

23. Strife A, Clarkson B. Biology of chronic myelogenous leukemia: is discordant maturation the primary defect? Semin Hematol 1988;25:1–19.

24. Koeffler HP, Golde DW. Chronic myelogenous leukemia—new concepts. N Engl J Med 1981;304:1201–1209, 1269–1274.

25. Talpaz M, Kurzrock R, Kantarjian HM, Gutterman JU. Recent advances in the therapy of chronic myelogenous leukemia. In: DeVita V, Hellman S, Rosenberg S, eds. Important advances in oncology. Philadelphia: JB Lippincott Co, 1988:297–321.

26. Champlin RE, Golde DW. Chronic myelogenous leukemia: recent advances. Blood 1985;65:1039–1047.

27. Nowell PC, Jackson L, Weiss A, Kurzrock R. Historical communication: Philadelphia-positive chronic myelogenous leukemia followed for 27 years. Cancer Genet Cytogenet 1988;34:57–61.

28. Kantarjian HM, Smith TL, McCredie KB, et al. Chronic myelogenous leukemia: a multivariate analysis of the associations of patient characteristics and therapy with survival. Blood 1985;66:1326–1335.

29. Griffin JD, Todd RF, Ritz J, et al. Differentiation patterns in the blastic phase of chronic myeloid leukemia. Blood 1983;61:85–91.

30. Ekblom M, Borgström G, von Willebrand E, et al. Erythroid blast crisis in chronic myelogenous leukemia. Blood 1983;62:591–596.

31. Bain B, Catovsky D, O'Brien M, Spiers ASD, Richards HGH. Megakaryoblastic transformation of chronic granulocytic leukaemia. An electron microscopy and cytochemical study. J Clin Pathol 1977;30:235–242.

32. Bakshi A, Minowada J, Arnold A, et al. Lymphoid blast crisis of chronic myelogenous leukemia represent stages in the development of B-cell precursors. N Engl J Med 1983;309:826–831.

33. Hernandez P, Carrot J, Cruz C. Chronic myeloid leukemia blast crisis with T-cell features. Br J Haematol 1982;51:175–180.

34. Sarin PS, Anderson PN, Gallo RC. Terminal deoxynucleotidyl-transferase activities in human blood leukocytes and lymphoblast cell lines. High levels in lymphoblast cell lines and blast cells of some patients with chronic myelogenous leukemia in acute phase. Blood 1976;47:11–20.

35. Rosenthal S, Canellos GP, Gralnick H. Erythroblastic transformation of chronic granulocytic leukemia. Am J Med 1977;63:116–124.

36. Kuriyama K, Gale RP, Tomonaga M, et al. CML-T1: a cell line derived from T-lymphocyte acute phase of chronic myelogenous leukemia. Blood 1989; 74:1381–1387.

37. Rushing D, Goldman A, Gibbs G, Howe R, Kennedy BJ. Hydroxyurea versus busulfan in the treatment of chronic myelogenous leukemia. Am J Clin Oncol 1982;5:307–313.

38. Talpaz M, Kantarjian HM, Kurzrock R, Gutterman J. Therapy of chronic myelogenous leukemia: chemotherapy and interferons. Semin Hematol 1988; 25:62–73.

39. Talpaz M, Kantarjian HM, McCredie K, Trujillo JM, Keating MJ, Gutterman JU. Hematologic remission and cytogenetic improvement induced by recombinant human interferon α_A in chronic myelogenous leukemia. N Engl J Med 1986;314:1065–1069.

40. Kurzrock R, Gutterman JU, Kantarjian H, Talpaz M. Therapy of chronic myelogenous leukemia with interferon. Cancer Invest 1989;7:83–91.

41. Prischl FC, Haas OA, Lion T, Eyb R, Schwarzmeier JD. Duration of first remission as an indicator of long-term survival in chronic myelogenous leukaemia. Br J Haematol 1989;71:337–342.

42. Look TA. The emerging genetics of acute lymphoblastic leukemia: clinical and biologic implications. Semin Oncol 1985;12:92–104.

43. Poplack DG. Acute lymphoblastic leukemia in childhood. Pediatr Clin North Am 1985;32:669–697.

44. Bloomfield CD, Goldman AI, Alimena G, et al. Chromosomal abnormalities identify high-risk and low-risk patients with acute lymphoblastic leukemia. Blood 1986;67:415–420.

45. Jain K, Arlin A, Mertelsmann R, et al. Philadelphia chromosome and terminal transferase positive acute leukemia: similarity of terminal phase of chronic myelogenous leukemia and *de novo* acute presentation. J Clin Oncol 1983;1:669–676.

46. Abelson HT, Rabstein LS. Lymphosarcoma: virus-induced thymic-independent disease in mice. Cancer Res 1970;30:2213–2222.

47. Whitlock CA, Witte ON. The complexity of virus-cell interactions in Abelson virus infection of lymphoid and other hematopoietic cells. Adv Immunol 1985;37:73–98.

48. Siden EJ, Baltimore D, Clark D, Rosenber NE. Immunoglobulin synthesis by lymphoid cells transformed in vitro by Abelson murine leukemia virus. Cell 1979;16:389–396.

49. Boss M, Greaves M, Teich N. Abelson virus-transformed haematopoietic cell lines with pre-B cell characteristics. Nature 1979;278:551–553.

50. Cook W. Rapid thymomas induced by Abelson murine leukemia virus. Proc Natl Acad Sci USA 1982;79:2917–2921.

51. Raschke WC, Baird S, Ralph P, Nakoinz I. Functional macrophage cell lines transformed by Abelson leukemia virus. Cell 1978;15:261–267.

52. Mendoza GR, Metzger H. Disparity of IgE binding between normal and tumor mouse mast cells. J Immunol 1976;117:1573–1578.

53. Risser R, Potter M, Rowe WP. Abelson virus-induced lymphomagenesis in mice. J Exp Med 1978;148:714–726.

54. Potter M, Sklar MD, Rowe WP. Rapid viral induction of plasmacytomas in pristane-primed BALB/c mice. Science 1973;182:592–594.

55. Konopka JB, Witte ON. Activation of the *abl* onocogene in murine and human leukemias. Biochim Biophys Acta 1985;823:1–17.

56. Van Etten RA, Jackson P, Baltimore D. The mouse type IV *c-abl* gene product is a nuclear protein, and activation of transforming ability is associated with cytoplasmic localization. Cell 1989;58:669–678.

57. Hunter T, Cooper JA. Protein-tyrosine kinases. Annu Rev Biochem 1985;4:897–930.

58. Prywes R, Foulkes JG, Baltimore D. The minimum transforming region of *v-abl* is the segment encoding protein-tyrosine kinase. J Virol 1985;54:114–122.

59. Pierce JH, Di Fiore PP, Aaronson SA, et al. Neoplastic transformation of mast cells by Abelson-MuLV: abrogation of IL-3 dependence by a nonautocrine mechanism. Cell 1985;41:685–693.

60. Cook WD, Metcalf D, Nicola NA, Burgess AW, Walker F. Malignant transformation of a growth factor-dependent myeloid cell line by Abelson virus without evidence of an autocrine mechanism. Cell 1985;41:677–683.

61. Broxmeyer HE, Ralph P, Gilbertson S, Margolis VB. Induction of leukemia-associated inhibitory activity and bone marrow granulocyte-macrophage progenitor cell alterations during infection with Abelson virus. Cancer Res 1980; 40:3928–3933.

62. Twardzik DR, Todaro GJ, Marquardt H. Transformation induced by Abelson murine leukemia virus involves production of a polypeptide growth factor. Science 1982;216:894–897.

63. Whitlock CA, Witte ON. Abelson virus-infected cells can exhibit restricted in vitro growth and low oncogenic potential. J Virol 1981; 40:577–584.

64. Lane MA, Neary C, Cooper GM. Activation of a cellular transforming gene in tumors induced by Abelson leukemia virus. Nature 1982;300:659–661.

65. Collins SJ. Breakpoints on chromosomes 9 and 22 in Philadelphia chromosome-positive chronic myelogenous leukemia (CML). Amplification of rearranged *c-abl* oncogenes in CML blast crisis. J Clin Invest 1986;78:1392–1396.

66. Schaefer-Rego K, Dudek H, Popenoe D, et al. CML patients in blast crisis have breakpoints localized to a specific region of the *bcr*. Blood 1987; 70:448–455.

67. Boehm TLJ, Drahovsky D. Application of a *bcr*-specific probe in the classification of human leukemia. J Cancer Res Clin Oncol 1987;113:267–272.

68. Benn P, Soper L, Eisenberg A, et al. Utility of molecular genetic analysis of *bcr* rearrangement in the diagnosis of chronic myeloid leukemia. Cancer Genet Cytogenet 1987;29:1–7.

69. Selleri L, Narni F, Emilia G, et al. Philadelphia-positive chronic myeloid leukemia with a chromosome 22 breakpoint outside the breakpoint cluster region. Blood 1987;70:1659–1664.

70. Hirosawa S, Aoki N, Shibuya M, Onozawa Y. Breakpoints in Philadelphia chromosome (Ph^1)-positive leukemias. Jpn J Cancer Res 1987;78:590–595.

71. Shtalrid M, Talpaz M, Kurzrock R, et al. Analysis of breakpoints with the *bcr* gene and their correlation with the clinical course of Philadelphia-positive chronic myelogenous leukemia. Blood 1988;72:485–490.

72. Kurzrock R, Shtalrid M, Romero P, et al. A novel *c-abl* protein in Philadelphia-positive acute lymphoblastic leukemia. Nature 1987;325:631–635.

73. Popenoe DW, Schaefer-Rego K, Mears JG, Bank A, Leibowitz D. Frequent and extensive deletion during the 9,22 translocation in CML. Blood 1986;68:1123–1128.

74. Hermans A, Heisterkamp N, von Lindern M, et al. Unique fusion of *bcr* and *c-abl* genes in Philadelphia chromosome positive acute lymphoblastic leukemia. Cell 1987;51:33–40.

75. Hariharan IK, Adams JM. cDNA sequences for human *bcr*, the gene that translocates to the *abl* oncogene in chronic myeloid leukemia. EMBO J. 1987;6:115–119.

76. Mes-Masson AM, McLaughlin J, Daley GQ, Paskind M, Witte ON. Overlapping cDNA clones define the complete coding region for the $p210^{c\text{-}abl}$ gene product associated with chronic myelogenous leukemia cells containing the Philadelphia chromosome. Proc Natl Acad Sci USA 1986;83:9768–9772.

77. Heisterkamp N, Knoppel E, Groffen J. The first BCR gene intron contains breakpoints in Philadelphia chromosome positive leukemia. Nucleic Acids Res 1988;16:10069–10181.

78. Diekmann K, Brill S, Garrett MD, et al. *Bcr* encodes a GTPase-activating protein for $p21^{rac}$. Nature 1991;351:400–402.

79. Croce CM, Huebner K, Isobe M, et al. Mapping of four distinct *BCR*-related loci to chromosome region 22q11: order of *BCR* loci relative to chronic myelogenous leukemia and acute lymphoblastic leukemia breakpoints. Proc Natl Acad Sci USA 1987;84:7174–7178.

80. Collins S, Coleman H, Groudine M. Expression of *bcr* and *bcr-abl* fusion transcripts in normal and leukemic cells. Mol Cell Biol 1987;7:2870–2876.

81. Stam K, Heisterkamp N, Reynolds FH, Groffen J. Evidence that the *phl* gene encodes a 160,000-Dalton phosphoprotein with associated kinase activity. Mol Cell Biol 1987;7:1955–1960.

82. Timmons MS, Witte ON. Structural characterization of the BCR gene product. Oncogene 1989;4:559–567.

83. Bernards A, Rubin CM, Westbrook CA, Paskind M, Baltimore D. The first intron in the human *c-abl* gene is at least 200 kilobases long and is a target for translocations in chronic myelogenous leukemia. Mol Cell Biol 1987;7:3231–3236.

84. Grosveld G, Verwoerd T, van Agthoven T, et al. The chronic myelocytic cell line K562 contains a breakpoint in *bcr* and produces a chimeric *bcr/c-abl* transcript. Mol Cell Biol 1986;6:607–616.

85. Shtivelman E, Lifshitz B, Gale RP, Roe BA, Canaani E. Alternative splicing of RNAs transcribed from the human *abl* gene from the *bcr-abl* fused gene. Cell 1986;47:277–284.

86. Ben-Neriah Y, Bernards A, Paskind M, Daley GQ, Balitmore D. Alternative 5' exons in *c-abl* mRNA. Cell 1986;44:577–586.

87. Konopka JB, Witte ON. Detection of *c-abl* tyrosine kinase activity in vitro permits direct comparison of normal and altered *abl* gene products. Mol Cell Biol 1985;5:3116–3123.

88. Browett PJ, Cooke HMG, Secker-Walker LM, Norton JD. Chromosome 22 breakpoints in variant Philadelphia translocations and Philadelphia-negative chronic myeloid leukemia. Cancer Genet Cytogenet 1989;37:169–177.

89. Melo JV, Gordon DE, Cross NCP, Goldman JM. The ABL-BCR fusion gene is expressed in chronic myeloid leukemia. Blood 1993;81:158–165.

90. Wetzler M, Talpaz M, Van Etten RA, Hirsh-Ginsberg C, Beran M, Kurzrock R. Subcellular localization of Bcr, Abl, and *Bcr-Abl* proteins in normal and leukemic cells and correlation of expression with myeloid differentiation. J Clin Invest 1993;92:1925–1939.

91. Konopka JB, Watanabe SM, Singer JW, Collins SJ. Cell lines and clinical isolates derived from Ph[1]-positive chronic myelogenous leukemia patients express *c-abl* proteins with a common structural alteration. Proc Natl Acad Sci USA 1985;82:1810–1814.

92. Kloetzer W, Kurzrock R, Smith L, et al. The human cellular *abl* gene product in the chronic myelogenous leukemia cell line K562 has an associated tyrosine protein kinase activity. Virology 1985;140:230–238.

93. Konopka JB, Watanabe SM, Witte ON. An alteration of the human *c-abl* protein in K562 leukemia cells unmasks associated tyrosine kinase activity. Cell 1984;37:1035–1042.

94. Kurzrock R, Kloetzer WS, Talpaz M, et al. Identification of molecular variants of p210 $^{BCR\text{-}ABL}$ in chronic myelogenous leukemia. Blood 1987;70:233–236.

95. Maxwell SA, Kurzrock R, Parsons SJ, et al. Analysis of P210$^{BCR\text{-}ABL}$ tyrosine protein kinase activity in various subtypes of Philadelphia chromosome-positive cells from chronic myelogenous leukemia patients. Cancer Res 1987; 47:1731–1739.

96. Huhn RD, Posner MR, Rayter SI, Foulkes JG, Frackelton AR. Cell lines and peripheral blood leukocytes derived from individuals with chronic myelogenous leukemia display virtually identical proteins phosphorylated on tyrosine residues. Proc Natl Acad Sci USA 1987;84:4408–4412.

97. McLaughlin J, Chianese E, Witte ON. In vitro transformation of immature hematopoietic cells by the p210 *BCR-ABL* oncogene product of the Philadelphia chromosome. Proc Natl Acad Sci USA 1987;84:6558–6562.

98. ar-Rushdi A, Negrini M, Kurzrock R, Huebner K, Croce CM. Fusion of the *bcr* and the *c-abl* genes in Ph[1]-positive acute lymphocytic leukemia with no

rearrangement in the breakpoint cluster region. Oncogene 1988;2:353–357.

99. Fainstein E, Marcelle C, Rosner A, et al. A new fused transcript in Philadelphia chromosome positive acute lymphocytic leukemia. Nature 1987; 330:386–388.

100. Van der Feltz MJM, Shivji MKK, Grosveld G, Wiedemann LM. Characterization of the translocation breakpoint in a patient with Philadelphia-positive, *bcr*-negative acute lymphoblastic leukaemia. Oncogene 1988;3:215–219.

101. Chen SJ, Chen Z, Hillion J, et al. Ph1-positive, bcr-negative acute leukemias: clustering of breakpoints on chromosome 22 in the 3' end of the BCR gene first intron. Blood 1989;73:1312–1315.

102. Clark SS, McLaughlin J, Crist WM, Champlin R, Witte ON. Unique forms of the *abl* tyrosine kinase distinguish Ph^1-positive CML from Ph^1-positive ALL. Science 1987;235:85–88.

103. Chan LC, Karhi KK, Rayter SI. A novel abl protein expressed in Philadelphia chromosome positive acute lymphoblastic leukemia. Nature 1987; 323:635–637.

104. Walker LC, Ganesan TS, Dhut S, et al. Novel chimaeric protein expressed in Philadelphia positive acute lymphoblastic leukaemia. Nature 1987; 329:851–853.

105. Kurzrock R, Shtalrid M, Gutterman JU, et al. Molecular analysis of chromosome 22 breakpoints in adult Philadelphia-positive acute lymphoblastic leukemia. Br J Hematol 1987;67:55–59.

106. Kurzrock R, Shtalrid M, Talpaz M, Kloetzer WS, Gutterman JU. Expression of *c-abl* in Philadelphia-positive acute myelogenous leukemia. Blood 1987;70:1584–1588.

107. Okamura J, Yamada S, Ishii E, et al. A novel leukemia cell line, MR-87, with positive Philadelphia chromosome and negative breakpoint cluster region rearrangement coexpressing myeloid and early B-cell markers. Blood 1988; 72:1261–1268.

108. Morgan GJ, Wiedemann LM, Chan LC, Price CM, Kanfer EJ, Galton DAG. A case of M-bcr-rearranged, Philadelphia-positive AML that relapsed as chronic phase CML. Blood 1990;75:317–323.

109. McLaughlin J, Chianese E, Witte ON. Alternative forms of the *bcr-abl* oncogene have quantitatively different potencies for stimulation of immature lymphoid cells. Mol Cell Biol 1989;9:1866–1874.

110. Turhan AG, Eaves CJ, Eaves J, Kalousek DK, Eaves AC, Humphries RK. Molecular analysis of clonality and *bcr* rearrangements in Philadelphia chromosome-positive acute lymphoblastic leukemia. Blood 1988;71:1495–1498.

111. Estrov Z, Talpaz M, Kantarjian HM, Zipf TF, McClain KL, Kurzrock R. Heterogeneity in lineage derivation of Philadelphia-positive acute lymphoblastic leukemia expressing $p190^{BCR-ABL}$ or $p210^{BCR-ABL}$: determination by analysis of individual colonies with the polymerase chain reaction. Cancer Res 1993; 53:3289–3293.

112. Kantarjian HM, Talpaz M, Dhingra K, et al. Significance of the p210 versus p190 molecular abnormalities in adults with Philadelphia chromosome-positive acute leukemia. Blood 1991;78:2411–2418.

113. Rodenhuis S, Smets LA, Slater RM, Behrendt H, Veerman AJP. Distinguishing the Philadelphia chromosome of acute lymphoblastic leukemia from its counterpart in chronic myelogenous leukemia. N Engl J Med 1985;313:51–52.

114. Erikson J, Griffin CA, ar-Rushdi A, et al. Heterogeneity of chromosome 22 breakpoints in Philadelphia (Ph^+) acute lymphoblastic leukemia. Proc Natl Acad Sci USA 1986;82:1807–1811.

115. De Klein A, Hagemeijer A, Bartram CR, et al. *bcr* rearrangement and translocation of the *c-abl* oncogene in Philadelphia positive acute lymphoblastic leukemia. Blood 1986;68:1369–1375.

116. Lugo TG, Pendergast AM, Muller AJ, Witte ON. Tyrosine kinase activity and transformation potency of *bcr-abl* oncogene products. Science 1990;247:1079–1082.

117. Carr SA, Biemann K, Shoji S, Parmelee DC, Titani K. *n*-Tetradecanoyl is the NH_2-terminal blocking group of the catalytic subunit of cyclic AMP-dependent protein kinase from bovine cardiac muscle. Proc Natl Acad Sci USA 1982;79:6128–6131.

118. Buss JE, Sefton BM. Myristic acid, a rare fatty acid, is the lipid attached to the transforming protein of rous sarcoma virus and its cellular homolog. J Virol 1985;53:7–12.

119. Dikstein R, Heffetz D, Ben-Neriah Y, Shaul Y. c-abl has a sequence-specific enhancer binding activity. Cell 1992;69:751–757.

120. McWhirter JR, Wang JYJ. Activation of tyrosine kinase and microfilament-binding function of *c-abl* by *bcr* sequences in *bcr/abl* fusion proteins. Mol Cell Biol 1991;11:1553-1565.

121. Romero P, Blick M, Talpaz M, Murphy E, Hester J, Gutterman J. C-*sis* and *c-abl* expression in chronic myelogenous leukemia and other hematologic malignancies. Blood 1986;67:839–841.

122. Collins SJ, Kubonishi I, Miyoshi I, Groukine M. Altered transcription of the *c-abl* oncogene in K562 and other chronic myelogenous leukemia cells. Science 1984;225:72–74.

123. Konopka JB, Clark S, McLaughlin J, et al. Variable expression of the translocated *c-abl* oncogene in Philadelphia-chromosome-positive B-lymphoid cell lines from chronic myelogenous leukemia patients. Proc Natl Acad Sci USA 1986;83:4049–4052.

124. Shtivelman E, Gale RP, Dreazen O, et al. *bcr-abl* RNA in patients with chronic myelogenous leukemia. Blood 1987;69:971–973.

125. Bartram CR, De Klein A, Hagemeijer A, Carbonell F, Kleihauer E, Grosveld G. Additional *c-abl/bcr* rearrangements in a CML patient exhibiting two Ph^1 chromosomes during blast crisis. Leuk Res 1986;10:221–225.

126. Dhingra K, Talpaz M, Kantarjian H, Ku S, Rothberg J, Gutterman JU, Kurzrock R. Appearance of acute leukemia-associated $p190^{BCR\text{-}ABL}$ in chronic myelogenous leukemia may correlate with disease progression. Leukemia 1991; 5:191–195.

127. Heisterkamp N, Jenkins R, Thibodeau S, Testa JR, Weinberg K, Groffen J. The *bcr* gene in Philadelphia chromosome positive acute lymphoblastic leukemia. Blood 1989;73:1307–1311.

128. Mills KI, MacKenzie ED, Birnie GD. The site of the breakpoint within the bcr is a prognostic factor in Philadelphia-positive CML patients. Blood 1988;72:1237–1241.

129. Ogawa H, Sugiyama H, Soma T, Masaoka T, Kishimoto S. No correlation between locations of *bcr* breakpoints and clinical states in Ph[1]-positive CML patients. Leukemia 1989;3:492–496.

130. Przepiorka D. Breakpoint zone of *bcr* in chronic myelogenous leukemia does not correlate with disease phase or prognosis. Cancer Genet Cytogenet 1988; 36:117–122.

131. Birnie GD, Mills KI, Benn P. Does the site of the breakpoint on chromosome 22 influence the duration of the chronic phase in chronic myeloid leukemia? Leukemia 1989;3:545–547.

132. Lee MS, LeMaistre A, Kantarjian HM, et al. Detection of two alternative *bcr/abl* mRNA junctions and minimal residual disease in Philadelphia chromosome positive chronic myelogenous leukemia by polymerase chain reaction. Blood 1989;73:2165–2170.

133. Lange W, Synder DS, Castro R, Rossi JJ, Blume KG. Detection by enzymatic amplification of *bcr-abl* mRNA in peripheral blood and bone marrow cells of patients with chronic myelogenous leukemia. Blood 1989;73:1735–1741.

134. Inokuchi K, Inoue T, Tojo A, et al. A possible correlation between the type of *bcr-abl* hybrid messenger RNA and platelet count in Philadelphia-positive chronic myelogenous leukemia. Blood 1991;78:3125–3127.

135. Bartram CR, Janssen JWG, Becher R, De Klein A, Grosveld G. Persistence of chronic myelocytic leukemia despite deletion of rearranged *bcr/c-abl* sequences in blast crisis. J Exp Med 1986;164:1389–1396.

136. Eisbruch A, Blick M, Evinger-Hodges MJ, et al. Effect of differentiation-inducing agents on oncogene expression in a chronic myelogenous leukemia cell line. Cancer 1988;62:1171–1178.

137. Young JC, Witte ON. Selective transformation of primitive lymphoid cells by the *bcr/abl* oncogene expressed in long-term lymphoid or myeloid cultures. Mol Cell Biol 1988;8:4079–4087.

138. LeMaistre A, Lee MS, Talpaz M, et al. RAS oncogene mutations are rare late stage events in chronic myelogenous leukemia. Blood 1989;73:889–891.

139. Collins SJ, Howard M, Andrews DF, Agura E, Radich J. Rare occurrence of N-*ras* point mutations in Philadelphia chromosome positive chronic myeloid leukemia. Blood. 1989;73:1028–1032.

140. Ahuja H, Bar-Eli M, Advani SH, Benchimol S, Cline MJ. Alteration in the p53 gene and the clonal evolution of the blast crisis of the chronic myelocytic leukemia. Proc Natl Acad Sci USA 1989;86:6783–6787.

141. Mashal R, Shtalrid M, Talpaz M, et al. Rearrangement and expression of p53 in the chronic phase and blast crisis of chronic myelogenous leukemia. Blood 1990;75:180–189.

142. Blick M, Romero P, Talpaz M, et al. Molecular characteristics of chronic myelogenous leukemia in blast crisis. Cancer Genet Cytogenet 1987; 27:349–356.

143. McCarthy DM, Goldman JM, Rassool FV, Graham SV, Birnie GD.

Genomic alterations involving the c-myc proto-oncogene locus during the evolution of a case of chronic granulocytic leukemia. Lancet 1984;2:1362–1365.

144. Wetzler M, Kurzrock R, Lowe DG, Kantarjian H, Gutterman JU, Talpaz M. Alteration in bone marrow adherent layer growth factor expression: a novel mechanism of chronic myelogenous leukemia progression. Blood 1991; 78:2400–2406.

145. Estrov Z, Kurzrock R, Wetzler M, et al. Suppression of chronic myelogenous leukemia colony growth by interleukin-l (IL-1) receptor antagonist and soluble IL-1 receptors: a novel application for inhibitors of IL-1 activity. Blood 1991;78:1476–1484.

146. Ezdinli EZ, Sokal JE, Crosswhite L, Sandberg AA. Philadelphia-chromosome-positive and -negative chronic myelocytic leukemia. Ann Intern Med 1970;72:175–182.

147. Shtalrid M, Talpaz M, Blick M, Kurzrock R. Philadelphia-negative chronic myelogenous leukemia with breakpoint cluster region rearrangement: molecular analysis, clinical characteristics, and response to therapy. J Clin Oncol 1988;6:1569-1575.

148. Kurzrock R, Kantarjian HM, Shtalrid M, Gutterman JU, Talpaz M. Philadelphia chromosome-negative chronic myelogenous leukemia without breakpoint cluster region rearrangement: a chronic myeloid leukemia with a distinct clinical course. Blood 1990;75:445–452.

149. Bartram CR. *bcr* rearrangement without juxtaposition of *c-abl* in chronic myelocytic leukemia. J Exp Med 1985;162:2175–2179.

150. Kurzrock R, Blick MB, Talpaz M, et al. Rearrangement in the breakpoint cluster region and the clinical course in Philadelphia-negative chronic myelogenous leukemia. Ann Intern Med 1986;105:673–679.

151. Morris CM, Reeve AE, Fitzgerald PH, Hollings PE, Beard MEJ, Heaton DC. Genomic diversity correlates with clinical variation in Ph'-negative chronic myeloid leukaemia. Nature 1986;320:281–283.

152. Stoll DB, Peterson P, Exten R, et al. Clinical presentation and natural history of patients with essential thrombocythemia and the Philadelphia chromosome. Am J Hematol 1988;27:77–83.

153. Dreazen O, Rassol F, Sparkes RS, Klisak I, Goldman JM, Gale RP. Do oncogenes determine clinical features in chronic myeloid leukemia? Lancet 1987; 1:1402–1405.

154. Palumbo AP, Boccadoro M, Battaglio S, et al. Philadelphia-positive thrombocythemia with a complex translocation involving chromosomes 9, 15, and 22. Cancer Genet Cytogenet 1989;39:77–80.

155. Wiedemann LM, Karhi KK, Shivji MKK, et al. The correlation of breakpoint cluster region rearrangement and p210 *phl/abl* expression with morphological analysis of Ph-negative chronic myeloid leukemia and other myeloproliferative disorders. Blood 1988;71:349–355.

156. Morris CM, Fitzgerald PH, Hollings PE, et al. Essential thrombocythaemia and the Philadelphia chromosome. Br J Haematol 1988; 70:13–19.

157. Melani C, Canepa L, Sessarego M, Miglino M, Ferraris AM, Gaetani GF. Molecular analysis of the bcr rearrangement in a case of Ph'-negative blastic

crisis of Ph'-positive chronic myelogenous leukemia. Eur J Haematol 1989;42:32–37.

158. Kantarjian H, Talpaz M, Estey E, Ku S, Kurzrock R. What is the contribution of molecular studies to the diagnosis of BCR-ABL positive disease in adult acute leukemia? Am J Med, in press.

159. Daley GQ, Baltimore D. Transformation of an interleukin 3–dependent hematopoietic cell line by the chronic myelogenous leukemia-specific $p210^{BCR-ABL}$ protein. Proc Natl Acad Sci USA 1988;85:9312.

160. Pearson MG, Vardiman JW, LeBeau MM, et al. Increased numbers of marrow basophils may be associated with a t(6;9)in ANLL. Am J Hematol 1985; 18:393–403.

161. Westbrook CA, LeBeau MM, Diaz MO, Groffen J. Chromosomal localization and characterization of *c-abl* in the t(6;9) of acute nonlymphocytic leukemia. Proc Natl Acad Sci USA 1985;82:8742–8746.

162. Westbrook CA, Rubin CM, LeBeau MM, et al. Molecular analysis of *TCR-B* and *ABL* in a t(7;9)-containing cell line (SUP-T3) from a human T-cell leukemia. Proc Natl Acad Sci USA 1987;84:251-255.

163. Morris CM, Rosman I, Archer SA, Cochrane JM, Fitzgerald PH. A cytogenetic and molecular analysis of five variant Philadelphia translocations in chronic myeloid leukemia. Cancer Genet Cytogenet 1988;35:179–197.

164. Dow LW, Tachibana N, Raimondi SC, Lauer SJ, Witte ON, Clark SS. Comparative biochemical and cytogenetic studies of childhood acute lymphoblastic leukemia with the Philadelphia chromosome and other 22q11 variants. Blood 1989;73:1291–1297.

165. Kawasaki ES, Clark SS, Coyne MY, et al. Diagnosis of chronic myeloid and acute lymphocytic leukemias by detection of leukemia-specific mRNA sequences amplified in vitro. Proc Natl Acad Sci USA 1988;85:5698–5702.

166. Dobrovic A, Trainor KJ, Morley AA. Detection of the molecular abnormality in chronic myeloid leukemia by use of the polymerase chain reaction. Blood 1988;72:2063–2065.

167. Gabert J, Lafage M, Maraninchi D, Thuret I, Carcassonne Y, Mannoni P. Detection of residual *bcr-abl* translocation by polymerase chain reaction in chronic myeloid leukaemia patients after bone-marrow transplantation. Lancet 1989;2:1125–1128.

168. Dhingra K, Kurzrock R, Kantarjian H, et al. Minimal residual disease in interferon-treated chronic myelogenous leukemia: results and pitfalls of analysis based on polymerase chain reaction. Leukemia 1992;6:754–760.

169. Pichert G, Ritz J. Clinical significance of *bcr/abl* gene rearrangement detected by the polymerase chain reaction after allogeneic bone marrow transplantation in chronic myelogenous leukemia. Leuk Lymphoma 1993;10:1–8.

170. Szczylik C, Skorski T, Nicolaides NC, et al. Selective inhibition of leukemia cell proliferation by BCR-ABL antisense oligodeoxynucleotides. Science 1992;253:562–565.

171. Van Etten RA. The molecular pathogenesis of the Philadelphia-positive leukemias: implications for diagnosis and therapy. In: Freireich EJ, Kantarjian H,

eds. Leukemia: advances in research and treatment. Boston: Kluwer Academic Publishers, 1993:295–325.

172. Estrov Z, Kurzrock R, Talpaz M. Role of interleukin-1 inhibitory molecules in therapy of acute and chronic myelogenous leukemia. Leuk Lymphoma 1992;10:407–418.

CHAPTER 8

Molecular Genetics of Acute Myeloid Leukemia

Kun-Sang Chang
Michael J. Siciliano

It is now clear that cancer is a genetic disease; genes involved in cell differentiation, proliferation, signal transduction, cell growth, and transcription regulation of gene activities are often disrupted or overexpressed in cancer cells as a result of chromosomal translocation, deletion, mutation, or gene amplification. These chromosome abnormalities can result in activation of oncogenes, loss of function of a tumor suppressor, creation of new oncogenic fusion proteins, loss of control of cellular regulatory genes, etc. All these molecular events result in a common goal that eventually leads to a malignant transformation or carcinogenesis. Carcinogenesis is a multistep event, as is evident in many types of malignancies including human leukemia. Chromosomal abnormalities may be the primary events of carcinogenesis. Secondary events are necessary for tumor progression and could include mutations within oncogenes, such as N-*ras* or *p53*, or other yet unknown events. Multistep carcinogenesis has been demonstrated to occur in hematologic neoplasia, for example, Burkitt's lymphoma and chronic myelogenous leukemia (CML).

Strong evidence indicates that abnormalities in both growth and differentiation are required for full leukemic transformation (1). It appears that a loss of control of growth as well as a differentiation block in the hematopoietic cells are required for leukemogenesis. Recent advances in the technology of molecular genetics and resources made available by the Human Genome Project have made significant contributations in the cloning and characterization of chromosome abnormalities in patients with acute myeloid leukemia (AML) (Table 8.1). Studies on the genes involved at the sites of these abnormalities have provided important information to help understand the molecular mechanism of leukemogenesis of AML.

Table 8.1: Chromosome Abnormalities in Acute Myeloid Leukemia

Type	*Abnormality*	*Genes Involved*	*Functions*
AML-M2	t(8;21)(q22;q23)	*AML1/ETO*	Transcription factors, AML1-ETO fusion protein
AML-M2 AML-M4 AUL	t(6;9)(p23;q34)	*DEK/SET/ CAN*	Transcription factors, DEK-CAN, SET-CAN fusion proteins
AML-M3	t(15;17)(q22;q12)	*PML/RARα*	Transcription factors, PML-RARα, RARα-PML fusion proteins
AML-M4Eo	inv(16)(p13.1;q22) t(16;16)(p13.1;q22)	*MYH11/CBFB*	Transcription factor
AML-M4 AML-M5	t(9;11)(q22;q23) t(6;11)(q27;q23) t(11;19)(q23;p13)	*MLL (ALL1)*	Homeobox regulator, *trithorax* homology, fusion proteins
AML	5q−	Unknown	Unknown

NONRANDOM CHROMOSOMAL TRANSLOCATION IN ACUTE MYELOID LEUKEMIA

Acute Myeloid Leukemia with Maturation

Acute myeloid leukemia with maturation (AML-M2 according to the French-American-British classification [FAB]) is a distinct subtype of AML. A nonrandom chromosomal translocation t(8;21)(q22;q22) is found in about 8% to 18% of the patients with AML-M2 by cytogenetic analysis. The t(8;21)-positivity is associated with good prognosis and high long-term remission or survival rates. A high incidence of this type of AML occurs in young patients.

The t(8;21) translocation breakpoint was initially characterized by using a NotI linking clone proximal to the chromosome 21q22 locus as the hybridization probe in a pulsed-field gel Southern blot (2–4). Rearrangement of a DNA fragment was detected in t(8;21)-positive AMLs. A yeast artificial chromosome (YAC) clone that spanned the breakpoint on chromosome 21 was isolated and partially characterized. Genomic DNA fragments from the breakpoint region of the translocation were used to identify complementary DNA clones of the normal and fusion transcripts. These studies concluded that the t(8;21) translocation occurs within the *AML1* gene from chromosome 21 and the *ETO* or the *MTG8* gene from chromosome 8. Southern blot analysis of the t(8;21)-positive AML patients using the *AML1* cDNA probe concluded that the translocation breakpoints occurred within a single intron of the *AML1* and the *ETO* genes. Northern blot analysis of the RNAs isolated from various t(8;21)-positive AMLs identified abnormal messenger RNAs of 7 and 10 kb, representing the fusion transcripts of *AML1-ETO* made at the breakpoint junction. Multiple species of normal *AML1* mRNAs of 8.5, 6.2, 3.9, and 2.2 kb and a single 5.5-kb *ETO* mRNA were detected. Some of these mRNA species appear to be predominantly expressed in certain cell types. These different mRNA species are believed to be the result of alternative splicing of the primary RNA transcript. The *AML1* gene is ubiquitously expressed, but the *ETO* gene is not normally expressed in the hematopoietic cells or in any other leukemia samples tested. Therefore, expression of the *ETO* gene appears to be activated by the t(8;21) translocation. Unexpectedly, an unusually large noncoding region was found in both the *AML1* and the *ETO* mRNAs, which encode normal proteins of 250 and 604 amino acids, respectively. Using reverse transcription–polymerase chain reaction (RT-PCR), fusion transcript *AML1-ETO* was detected in all the t(8;21)-positive AML patients. Sequence analysis of the amplified DNA fragment confirmed that all the t(8;21) translocations occurred within a single intron of the *AML1* and the *ETO* genes.

Rearrangements of the *AML1* or the *ETO* genes can be identified by standard agarose gel electrophoresis and Southern blotting in only a small percentage of patients with t(8;21), thus limiting its use as a diagnostic tool for the detection of this type of AML. However, pulse-field gel electrophoresis (which separates larger fragments) and Southern blotting can detect rearrangement in all t(8;21)-positive AML patients. The fusion transcript *AML1-ETO* transcribed from the t(8;21) breakpoint is a specific tumor marker for this type of AML. DNA sequences derived from the fusion junction can be used to design specific oligonucleotide primers and to amplify this unique DNA segment by RT-PCR. In this way an extremely small number of residual leukemia cells (e.g., one in a million) can be detected in patients in complete remission. Therefore, RT-PCR is a powerful technique for monitoring disease progression during patient treatment and remission. RT-PCR detected the *AML1-ETO* fusion transcript in all the t(8;21)-positive AML patients during long-term complete remission. This finding suggests that the t(8;21) translocation is the primary event in the leukemogenesis of AML, but a second event may be necessary for disease progression.

Variant forms of t(8;V;21) translocation and masked translocation were found in 8 of the 33 patients with AML-M2. Rearrangement of the *AML1* gene was detected in these patients. Using RT-PCR, the fusion transcript *AML1-ETO* was found in t(8;21)-negative AML patients with normal karyotypes. Sequence analysis of the amplified DNA fragment demonstrated that the fusion junction of the two genes is identical to the one transcribed from the t(8;21)-positive AML patients. In these patients activation of the *ETO* gene expression was also similar to that in t(8;21)-positive AML patients. These results confirm that a masked translocation of t(8;21) that is not detectable by the conventional cytogenetic technique exists in AML. This type of masked translocation represents about 16% of the total AML-M2 cases. From these studies it appears that the incidence of AML-M2 cases involving the t(8;21) translocation is significantly higher than previouly estimated.

Analysis of the amino acid sequences of the AML1 protein showed that 132 amino acids have strong sequence homology with the *Drosophila* segmentation gene, *runt*. In *Drosophila*, *runt* is required to establish the segmentation pattern during development. The t(8;21) translocation fused the AML1 protein immediately downstream of the *runt* homologous region to the amino-terminal end of the ETO protein (Fig. 8.1). The *runt* homology sequence was also found in the core binding factor-α (CBFα) or polyomavirus enhancer binding protein 2 (PEBP2) family of transcription factors. When the cDNA of the mouse *CBFα* gene was isolated and characterized, it was found that CBFα protein has 99% amino acid sequence homology with AML1 in the first 241 amino acids, indicating that AML1 is the human counterpart of the mouse CBFα protein

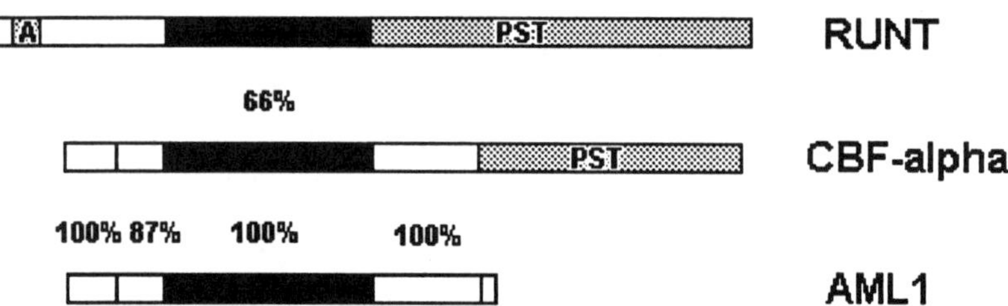

Fig. 8.1: Amino acid sequence homology between AML1 (human), CBFα (mouse), and RUNT (*Drosophila*). A dark box indicates the *runt* homology domain of the three proteins. A, regions of alanine stretch; PST, the proline, serine, and threonine-rich domains. The overall amino acid sequence homology between mouse CBFα and AML1 proteins is more than 98%, indicating that AML1 is the human homologue of mouse CBFα.

(Fig. 8.1). Expression of the *CBFα* (or *AML1*) gene was detected in almost all cell lines tested except for F9 cells.

Almost the entire ETO protein is included in the AML1-ETO fusion protein as a result of the t(8;21) translocation (Fig. 8.2). Several stretches of proline-rich domains, which resemble the transcriptional activating motif of some transcription factors, were found in the ETO protein. A cysteine-rich zinc-finger motif able to act as a DNA-binding domain was found close to the carboxy-terminus of the ETO protein. These structural features imply that ETO is most likely a transcription factor.

It has been reported that AML1 and AML1-ETO are sequence-specific DNA-binding proteins (5). The amino acid sequence with homology to *runt* is required for the DNA-binding and protein-protein interaction. This result strongly suggests that AML1 is a transcription factor. The fusion protein AML1-ETO is capable of binding the AML1 recognition sequences, and its transcription regulatory properties may have been altered as a result of fusing with the ETO protein. No information about whether the fusion of AML1 with ETO will alter its normal

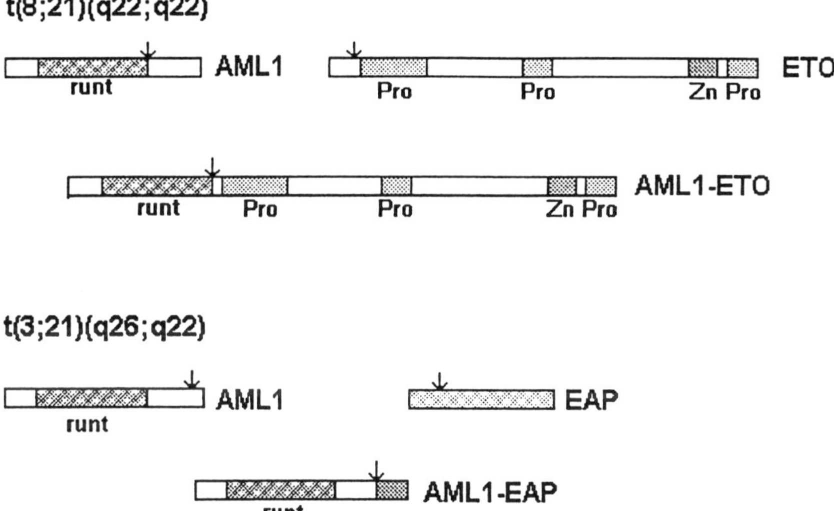

Fig. 8.2: Structural features of AML1, ETO, and EAP proteins involved in the t(8;21) and t(3;21) translocations. Fusion proteins AML1-ETO and AML1-EAP are encoded from the breakpoint sites. Pro represents the proline-rich domains, and Zn the putative zinc-finger motif of the ETO protein; runt indicates *Drosophila runt* homology of the AML1 protein. Arrowheads indicate the sites of translocation breakpoints.

cellular localization is available at this stage. There is evidence that AML1 binds DNA by association with other forms of DNA-binding proteins (see below, section on acute myelomonocytic leukemia with bone marrow eosinophilia). Thus, AML1-ETO may be a dominant negative inhibitor for an unknown transcription factor by forming a nonfunctional heterocomplex. Whether AML1 can form a homodimer by itself or a nonfunctional heterodimer with the AML1-ETO fusion protein, which will result in sequestration of the normal AML1 in the leukemia cells, is also unknown. As the biological function of ETO remains unknown, activation of this gene may also play a role in the pathogenesis of AML.

A different form of chromosome translocation, t(3;21)(q26;q22), was also cloned, and genes involved at the breakpoint site were characterized (6). The breakpoint of this translocation occurs within the same intron of the *AML1* gene found in the t(8;21) translocation. The gene from chromosome 3 was identified to be the *EAP* gene, which is similar to a small nuclear protein gene previously called EBER1. However, in this case fusion of *AML1* with the *EAP* gene resulted in a frameshift of the *EAP* portion of the DNA sequence (Fig. 8.2). This result indicates that the *EAP* gene is not important to the pathogenesis of this type of leukemia. Therefore, *AML1* may be the critical gene for leukemogenesis in AML.

Acute Promyelocytic Leukemia

Acute promyelocytic leukemia (APL or AML-M3) arises as a result of the clonal proliferation of abnormal promyelocytes. During the acute phase of the disease, the abnormal promyelocytes account for more than 90% of the total white cells in patient bone marrow specimens. A nonrandom chromosome translocation, t(15;17)(q22;q12), can be found in more than 90% of the patients with APL and is a cytogenetic hallmark for this type of leukemia. The prognosis for this group of patients is good once remission is induced. Frequently, however, bleeding problems arise as a result of disseminated intravascular coagulation (DIC). One major problem is that chemotherapy potentiates the bleeding diathesis as a result of cell killing, and the consequences can be fatal.

Huang and coworkers (7) found that APL can be treated with all-*trans* retinoic acid (ATRA), and more than 95% of their patients achieved complete clinical remission by induced maturation of the leukemic blasts. This finding was confirmed by several laboratories in a large number of patients. When patients were treated with ATRA, coagulopathy caused by DIC or fibrinolysis disappeared rapidly, and only minor side effects were found. Cytochemical and electron-microscopic analysis of the patients' samples after ATRA treatment confirmed that ATRA had induced abnormal promyelocytes to differentiate into mature granulocytes. Cell surface immunophenotyping, premature chromosome condensation, and fluorescent in situ hybridization (FISH) studies demonstrated that complete clinical remission as a result of ATRA treatment was the result of induced maturation of the abnormal promyelocytes.

Unfortunately, APL patients quickly developed resistance to ATRA treatment after an average of 6 months, even under continuous treatment. Thus, using ATRA as a maturation-induction therapy alone is not a curative treatment for APL. A more successful treatment can be achieved by a combination of ATRA-induced maturation followed by intensive chemotherapy (8,9).

Using somatic cell hybrids and chromosome-mediated gene transfer, a hybrid cell that carried a human chromosome fragment close to the t(15;17) breakpoint site was identified. A NotI linking library was constructed from DNA isolated from this clone. A linking clone that encompasses the breakpoint region was identified. Characterization of this clone revealed that the retinoic acid receptor-α (*RAR*α) gene is involved at the breakpoint. Armed with the knowledge that ATRA induces maturation of APL and that the *RAR*α gene is localized to the region proximal to the t(15;17) translocation, other laboratories confirmed *RAR*α gene rearrangement in APL patients by Southern blot analysis. All of these studies concluded that the t(15;17) translocation breakpoint in APL occurs within the 17-kb second intron of the *RAR*α gene (10,11) (Fig. 8.3).

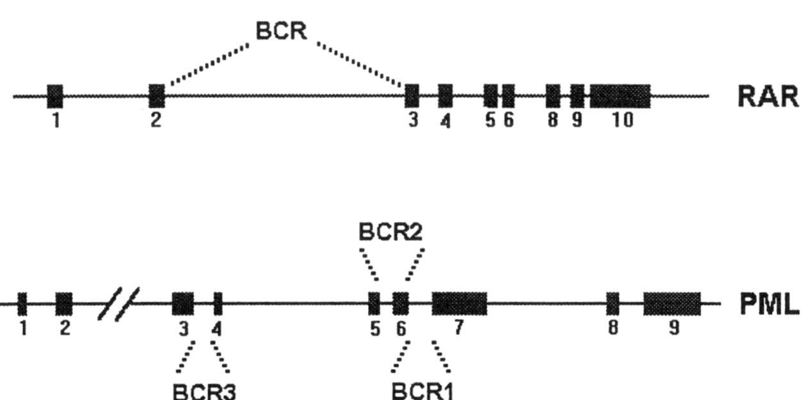

Fig. 8.3: The breakpoint cluster regions on *PML* and *RAR*α genes involved in the t(15;17) translocation. The t(15;17) translocation occurs within the 17-kb second intron of *RAR*α and at two major sites (BCR1 and BCR3) of *PML* genes. A breakpoint occurring at the BCR2 site of the *PML* gene is uncommon.

The *RARα* cDNA was used as a hybridization probe to screen the cDNA library made from the RNA of APL cells, and cDNA clones representing the fusion transcripts PML-RARα and RARα-PML were identified and characterized. These results indicate that both translocated chromosomes, the 17q− and the 15q+, transcribe fusion mRNAs.

The full-length cDNA of the normal *PML* gene was isolated and characterized by several groups. The predicted amino acid sequence of PML consisted of several interesting features (Fig. 8.4). A proline-rich domain found close to the amino-terminal end resembled the transcriptional activation domain of some transcription factor genes. Many of the proline residues were adjacent to serines, which are potential sites for phosphorylation. Immediately downstream of the proline-rich region, three clusters of cysteine-rich regions were identified that resembled the zinc-finger motif of many DNA-binding proteins. An α-helical region was found farther downstream of the cysteine-rich motif. This region has the potential to form a coiled-coil structure and is structurally similar to the leucine zipper region of the *fos* oncogene family of proteins. Another

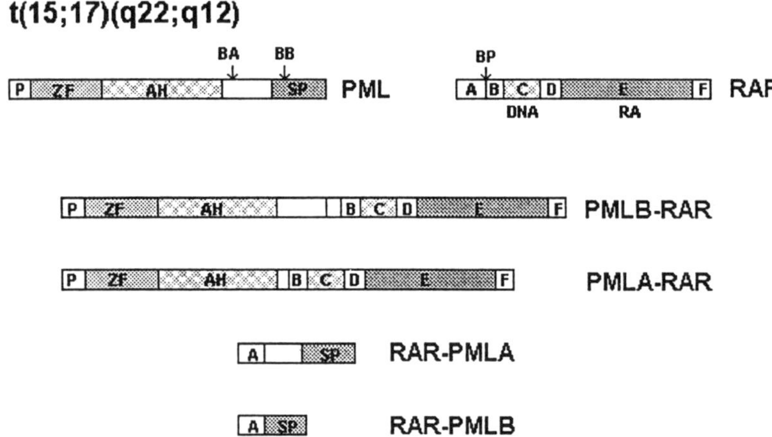

Fig. 8.4: Structural features of the PML, RARα, and the fusion proteins encoded from the t(15;17) translocation breakpoints. Four major forms of fusion proteins, PMLA-RARα, PMLB-RARα, RARα-PMLA, and RARα-PMLB, encoded from the t(15;17) translocation are shown. BA and BB represent the two major breakpoint sites on the *PML* gene. BP represents the breakpoint cluster region on the *RARα* gene. DNA and RA represent the two ligands binding domains of the RARα protein. The P indicates the proline-rich region; ZF, zinc-finger; AH, α-helix or coiled-coil; and SP, the serine/proline-rich region.

proline/serine-rich domain was located at the carboxy-terminal portion of the PML protein. This region can also be a potential phosphorylation site. According to these predicted structural features of the PML protein, it appears that PML is a potential transcription factor, and phosphorylation-dephosphorylation of the protein may be important to the regulation of its biological function.

The biological function of RARα as a transcription factor dependent on retinoic acid (RA) has been well documented. It belongs to a steroid hormone receptor superfamily. Some human myeloid leukemia cells can be induced to differentiate by RA, and this effect appears to be mediated through the action of RARα. Transfection of the *RARα* expression plasmid into the RA-resistant HL60 cells completely restored their ability to respond to differentiation induction by RA. However, this RA-responsiveness can also be restored by transfection with any one of the RAR or retinoic X receptor (RXR) family members. This result suggests that disruption of the *RARα* gene in APL by the t(15;17) translocation may not have any direct effect on leukemogenesis. Reduced expression of the *RARα* gene in APL may be compensated for by other RAR or RXR family members.

The etiologic role of PML in APL is not well documented. However, the importance of PML in leukemogenesis of APL was emphasized by the finding that some of the t(15;17)-negative APL patients showed rearrangement of the *PML* gene, but no abnormality of the *RARα* gene was found. Immunofluorescence staining of the PML protein displayed a nuclear speckle pattern resembling the staining pattern of proteins associated with the spliceosome (e.g., snRNP or SC-35). In addition, the amino acid sequence of PML shows homology with the 52-kDa component of an RO/SS ribonucleoprotein particle, indicating that PML may be associated with the ribonucleoprotein complex. The staining pattern of PML in the APL-derived NB4 cells has been changed, presumably due to the staining of the PML-RARα fusion protein and the normal PML protein being localized in a different environment. Interestingly, after 48 hours of ATRA-induced differentiation of NB4 cells, the bright signal of nuclear speckle pattern reappeared. This result indicates that ATRA-induced maturation of NB4 cells may involve a redistribution of the normal PML protein.

The biological function of PML is beginning to unfold by the finding that it is a growth or transformation suppressor, as is evident by the following findings: PML suppresses anchorage-independent growth of NB4 cells in soft agar assay and tumorigenicity in nude mice, and PML suppresses oncogenic transformation of rat embryo fibroblast by cooperative oncogenes (K.-S. Chang, unpublished results). These findings suggest that disruption of the *PML* gene in APL is one of the critical events in leukemogenesis.

As discussed above, the t(15;17) translocation in APL creates the fusion proteins PML-RARα and RARα-PML (Fig. 8.4). The PML-RARα consists of the major functional domains of both the RARα and the PML proteins and was shown to be able to act as an RA-dependent transcription factor with an altered promoter and cell-type specificity compared with RARα. The fusion transcript *RARα-PML* is not found in some of the APL samples, and it does not seem to carry any important feature of either the PML or the RARα protein. Therefore, it is believed that this fusion protein may not be important for the leukemogenesis of APL. The level of PML-RARα in the APL cells was reported to be much higher than that of RARα and PML, possibly owing to the more stable nature of the fusion protein. PML-RARα is localized mainly in the cytoplasm; however, in the presence of RA it is translocated into the nucleus (12). It was also shown that PML-RARα is able to form a heterodimer efficiently with PML through the coiled-coil region of the PML protein. Thus, it is possible that the combined effects of one *PML* gene being disrupted by the t(15;17) translocation and the heterodimer formation between PML-RARα and PML in the APL cells results in sequestration of its normal function. This result supports the hypothesis that PML-RARα can be a dominant negative inhibitor of PML in the APL cells. It has been found that PML-RARα can form a heterodimer with RXR through the DNA-binding domains of RARα and RXR (13). The heterodimer formation effectively sequestered RXR in the cytoplasm. This observation suggests that PML-RARα can be a dominant negative inhibitor of RXR in the APL cells. This hypothesis was supported by the finding that PML-RARα is able to prevent the binding of the vitamin D_3 receptor (VDR) to its target sequences. This inhibited vitamin D_3-dependent activation of a VDR-responsive reporter gene. Heterodimer formation of RXR with RAR and several other nuclear receptors greatly enhanced their binding to the targeted responsive elements. These nuclear receptors include the thyroid-hormone receptor (TR), VDR, and the peroxisome proliferator activated receptor (PPAR). Therefore, sequestration of RXR function in APL cells should have a significant effect on these nuclear receptors' signaling pathway. In addition, vitamin D_3 has been shown to induce differentiation of human promyelocytes. An inhibition of the vitamin D_3 function by PML-RARα in APL cells may be responsible for the differentiation arrest of the APL cells at the promyelocyte stage of myeloid differentiation.

The molecular mechanism by which APL develops resistance to ATRA is currently unclear. It appears that the fusion protein PML-RARα encoded from the breakpoint junction of the t(15;17) translocation is responsible for the ATRA sensitivity. APL cells that lose their expression of PML-RARα also lose their ATRA inducibility. In addition, transfection of the expression plasmid of PML-RARα into the myeloid cells U937 significantly increased ATRA sensitivity to differentiation induction.

The importance of PML-RARα in APL leukemogenesis is further suggested by the finding that expression of PML-RARα protein in the human myeloid leukemia cell line U937 resulted in a loss of vitamin D_3 and transforming growth factor-β1 inducibility and increased sensitivity to RA-induced differentiation (14). Interestingly, PML-RARα expression inhibits programmed cell death and consequently prolongs cell survival. These changes in the U937 cells as a result of PML-RARα expression represent a recapitulation of the critical feature of the APL phenotype. The final proof that the PML-RARα fusion protein plays a central role in leukemogenesis of APL comes from studies using animal models similar to those used to investigate the bcr-abl fusion protein in the pathogenesis of CML. However, expression of the PML-RARα fusion protein is lethal to the host cells, and it has been difficult to obtain stable transfectants. A modified retroviral vector with PML-RARα driven by an inducible promoter should be useful in this study. Results obtained from an animal model should be available in the near future.

It is well established that carcinogenesis is a multistep event. In t(15;17)-positive APL, the translocation appears to be the only consistent chromosomal abnormality. Mutations of the *ras* and *p53* oncogenes are rare. It is thus possible that multiple events created by the t(15;17) translocation represent multistep events of APL leukemogenesis. The primary event is the t(15;17) translocation resulting in the creation of the PML-RARα fusion protein and disruption of the *PML* gene. Sequestration of PML and RXR by the fusion protein as a result of heterodimer formation could be the secondary event of APL leukemogenesis. From this hypothesis, a model of APL leukemogenesis can be proposed as follows. The fusion protein PML-RARα plays a central role as a dominant negative inhibitor of PML and RXR. Suppression of PML and RXR leads to the inhibition of a growth suppressor function and of differentiation induction at the promyelocyte stage. Leukemogenesis of APL, then, may arise through continuous growth caused by the suppression of PML function and through differentiation arrest at the promyelocyte stage caused by the inhibition of RXR function.

Other Forms of Chromosomal Abnormalities in Acute Promyelocytic Leukemia

About 5% to 10% of the APL patients do not have a t(15;17) translocation. It appears that in some of these patients the *PML* gene, but not the

RARα gene, is rearranged by Southern blot analysis. The presence of a masked translocation that is undetectable by the conventional method of cytogenetic analysis has been detected in CML and t(8;21)-negative AML (see above). Similarly, masked translocations have also been demonstrated to exist in APL, as indicated by the formation of fusion transcripts *PML-RARα* in several t(15;17)-negative APL cases. This may be the result of a small genomic fragment being translocated from chromosome 15 (the *PML* gene) to 17 at the *RARα* locus, or from chromosome 17 (the *RARα* gene) to 15 at the *PML* locus. This type of submicroscopic translocation is expected to provide a precise joining between the second intron of the *RARα* gene and the bcr of the *PML* gene in order to create the identical fusion transcript similar to the t(15;17)-positive APL. Molecular cloning and a detailed verification of this type of translocation breakpoint have not yet been reported.

A rare case of variant translocation t(11;17)(q23;q21) was also found in APL. The translocation breakpoint in this type of APL was recently cloned and characterized (15). The *RARα* and a novel *Kruppel*-like zinc-finger gene designated as *PLZF* (for promyelocytic leukemia zinc-finger) are involved at the breakpoint site. The *PML* gene in this patient was normal. No sequence homology or similarity between PML and PLZF was found except that PLZF also encodes a putative zinc-finger transcription factor with a stretch of nine repetitive zinc-finger motifs (Fig. 8.5). Both fusion transcripts *RARα-PLZF* and *PLZF-RARα* are expressed in this APL patient. The predicted RARα-PLZF fusion protein consists of the A region of RARα and seven of the zinc-finger motifs derived from PLZF. This may represent a new form of DNA-binding protein with an unknown function. The PLZF-RARα fusion protein consists of the regions B to F of RARα including the DNA- and RA-binding domains (the C and E regions, respectively) truncated downstream of the amino-terminal portion of PLZF with two zinc-finger motifs adjacent to the RARα. No other significant feature farther upstream of the PLZF protein was found. A sequence of 123 amino acids with a proline-rich domain was found in the PLZFb isoform, presumably resulting from alternative splicing. Both fusion transcripts PLZFa-RARα and PLZFb-RARα are expressed in the APL patient. *PLZF* gene expression appears to be myeloid-specific and has some sequence homology with the *MZF-1* gene, a zinc-finger protein expressed preferentially in myeloid cells. As t(15;17) and t(11;17) translocation breakpoints share the same target region of the *RARα* gene, RARα appears to be critical for APL pathogenesis. However, it remains to be determined whether sequestration of RXR by PLZF-RARα occurs in this type of APL cell, or whether inactivation of the *PLZF* gene by the translocation plays a role in the leukemogenesis.

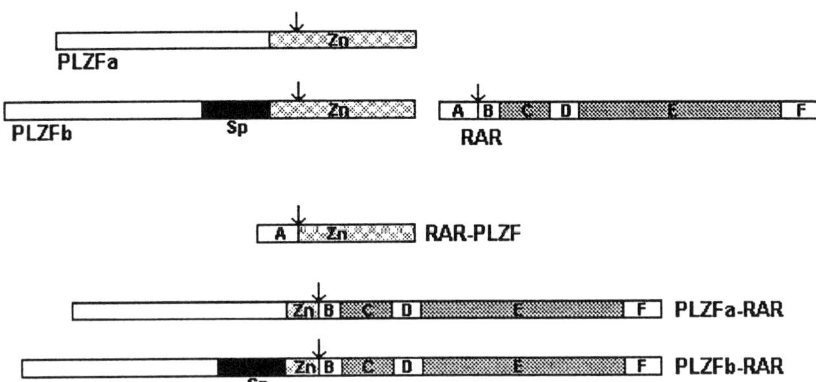

Fig. 8.5: Structural features of PLZF, RARα, and the fusion proteins encoded from the t(11;17) translocation breakpoint. Sp indicates the site of alternative splicing, and Zn the putative zinc-finger motif of the PLZF protein. Two different forms of PLZFa-RARα and PLZFb-RARα are encoded from the breakpoint site as a result of alternative splicing. Inverted arrowheads indicate the sites of the translocation breakpoint.

Acute Myelomonocytic Leukemia with Bone Marrow Eosinophilia

Acute myelomonocytic leukemia has a characteristic dysplastic bone marrow usually associated with the presence of bone marrow eosinophils with abnormal granulation (AML-M4Eo). This distinctive morphological subtype of acute myelomonocytic leukemia (AMML or AML-M4) is predictive of a more favorable course with longer remission and survival (16–18). Inversion (16)(p13;q22) and the related t(16;16)(p13;q22) were originally observed in such leukemic cells following the report of deletion (16)(q22) in similar patients (16,17,19). In several studies, treated patients had the inversion chromosome disappear upon remission (17,18). Therefore, a cytogenetic-clinicopathological association appears to exist between a subset of AMML cases and these chromosomal abnormalities.

It remained for the associated breakpoints to be cloned and studied at the molecular level so that the genetic events associated with these chromosomal aberrations, and their relation to leukemogenesis, could be identified. An early report (20), implying that these chromosome 16 rearrangements split the metallothionien gene cluster at 16q22 and suggesting that event as important in the pathogenesis of leukemia, remains unconfirmed. The long-arm breakpoint of inv(16) had been mapped between two anonymous DNA sequence markers found to be within 450 kb of each other (21). By FISH, the p-arm breakpoint was mapped between anonymous cosmids located in band 16p13.13 separated by an unknown distance (22), and it was suggested that the breakpoint is within a chromosome 16–specific repeat sequence which may play a role in the origin of chromosome 16 rearrangements in leukemia (23,24).

A human × Chinese hamster ovary (CHO) cell somatic cell hybrid cell line selected for retention of the human DNA repair gene, *excision repair cross complementing 4 (ERCC4)*, at 16p13.3-p13.2 (25) and containing only the p-arm of chromosome 16 was used as starting material for making FISH probes by inter-Alu-PCR (26). When they were applied to patients' bone marrow preparations, rapid and unambiguous cytogenetic diagnosis of the inversion was achieved by noting that the fluorescent probe was split between the p- and q-arms on the inv(16) chromosome (27). The hybrids were then used to isolate and identify cosmid clones (obtained from the Los Alamos National Laboratory and usually containing between 30 and 50 kb of human genomic insert) sufficiently close to the inversion breakpoint to identify macrorestriction fragments altered by the event. Unique sequence (that does not contain repetitive sequences) from one such cosmid was used for the isolation of four YACs, which contained large inserts of human DNA flanking the region of the cosmid sequences used to identify them. The human inserts in the YACs ranged in size from 100 to 780 kb. The three largest of these were determined to contain the inv(16) p-arm breakpoint by virtue of the fact that an inter-Alu-PCR FISH probe made from them was split to give signal on the p- and q-arms of the inv(16) chromosome in patient material.

The molecular cloning of the p-arm breakpoint allowed the rapid identification of the genes affected by the inversion as reported by Liu and associates (28). The first step was to narrow the human chromosome 16 genomic DNA containing the p-arm breakpoint. For this, a cosmid (16C3) whose human sequence was contained in all three YACs was isolated. This cosmid itself contained the breakpoint, as indicated by the fact that when it was used as a FISH probe, signal was produced on both the p- and q-arms of the inv(16) chromosome. A unique sequence fragment (16C3e) isolated from this cosmid was used as a probe on Southern blots made from the DNA of AML-M4 cells. If it is near the p-arm breakpoint, this probe should detect new restriction fragments generated by the inversion. It did so, making 16C3e a valuable diagnostic tool for the identification of p-arm breaks in the DNA from putative AML-M4Eo cells. Just as importantly, 16C3e could then also be used to probe a cosmid library made from AML-M4Eo cells. Some cosmids from that library would most likely be from the normal chromosome 16, and others from the inv(16). A cosmid from the inv(16) chromosome should have a segment from the q-arm region rearranged into the 16C3 region as a result of the inversion. Such a cosmid, upon restriction enzyme digestion, would most likely produce a fragment not seen when 16C3 was digested. That fragment would then be the key to the cloning of the q-arm breakpoint. That experiment did identify such a fragment, which was used to screen the Los Alamos National Laboratory's chromosome 16 cosmid library and led to the isolation of genomic clones spanning the q-arm breakpoint.

With cosmids available that spanned either the p-arm or q-arm breakpoints, human expression cDNA libraries could then be screened to isolate the genes at the breakpoints and therefore very likely rearranged by the inversion event. When this was done, the gene at the p-arm breakpoint coded for a smooth muscle myosin heavy chain gene named *MYH11*. The q-arm cosmid identified cDNA clones that showed a high degree of sequence similarity to a newly described mouse gene coding for a DNA-binding transcription factor CBFβ (29). The human gene was named *CBFB*. Mapping of the p- and q-arm cosmids indicated that the inversion breakpoints of each gene fell in introns.

As the breakpoints were located in introns of both the p- and q-arm genes, a fusion transcript could be generated in the AML cells. Such an event would have consequences both for the diagnosis of, and for understanding the basis of, leukemogenesis of this malignancy. To determine whether a fusion transcript was produced in the leukemic cells, PCR primers were designed from the middle of the *CBFB* coding sequence and the 3′ region of the *MYH11* coding sequence (28). Reverse transcription was conducted on the total RNA from AML-M4 cells derived from six patients to make cDNA from the RNA message produced in the cells. If a fusion transcript was produced and converted to cDNA by the reverse transcription process, then PCR of that cDNA (RT-PCR), using the primers described above, should produce a product. Product was generated from all six AML-M4 RNAs, but no product was ever obtained using RNA from non-AML-M4 cells. This approach was extended by Claxton et al. (30) to a total of 36 viably preserved cell RNAs from AML-M4 patients. Included were seven cases with the related translocation t(16;16)(p13;q22). In all cases (including the translocations!) fusion products were obtained composed of upstream *CBFB* sequences and downstream *MYH11* sequences. No such products were obtained from 10 patients with AML not showing chromosome 16 alterations. These data also indicated that the molecular consequences of the t(16;16) were similar to those of inv(16) (see Fig. 8.6). The assay was shown to be very sensitive by detecting chimeric transcript in as little as 20 pg of total cellular RNA—a quantity equivalent to two to five cells. Therefore, the technology used to develop an understanding of the molecular genetic

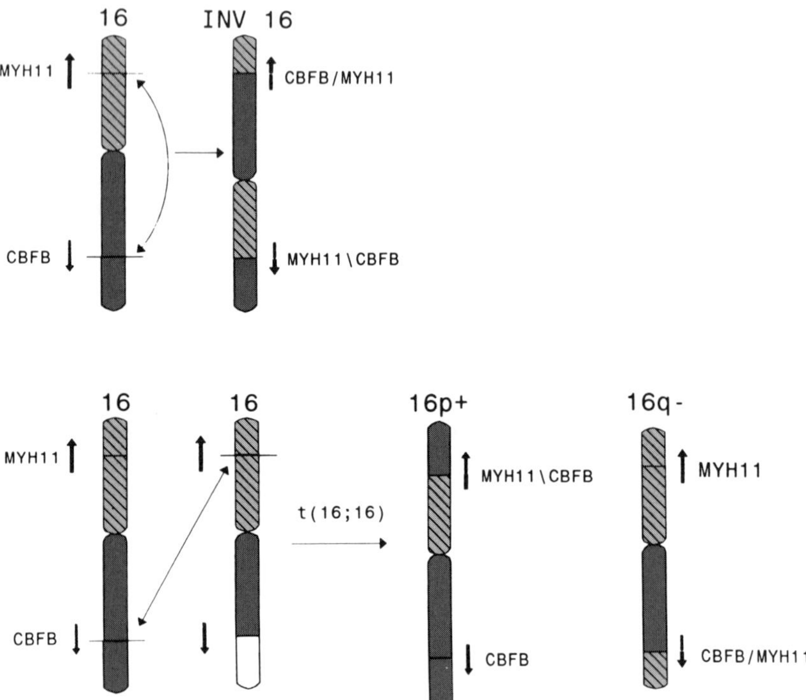

Fig. 8.6: Schematic view of the consequences of inv(16) and t(16;16). The inv(16) fuses most of the *CBFB* coding sequence encoded at 16q22 upstream of the 3′ end of *MYH11* on 16p13 of the same chromosome. In t(16;16), breakpoints on different chromosome 16 homologues translocate the distal p-arm region of the first homologue to the distal q-arm region of the second, while the distal q-arm region of the second homologue is translocated back to the distal p-arm region of the first. This results in a chimeric *CBFB/MYH11* gene similar to that in inv(16) cases. Although the reciprocal fusion gene *MYH11/CBFB* should also be generated, *MYH11* sequences upstream of the breakpoint appear to be deleted in some cases that are clinically indistinguishable from those without the deletion. This suggests that the *CBFB/MYH11* chimera is the operative factor in the disease. (*Reproduced by permission from Claxton DF, Liu P, Hsu HB, et al. Detection of fusion transcripts generated by the inversion 16 chromosome in acute myelogenous leukemia. Blood 1994;83:1750–1756.*)

consequences of the chromosome rearrangements also provided a sensitive method for detecting minimal residual disease in patients being treated for this disease.

Sequencing of the RT-PCR products revealed that they were amplified from in-frame fusion gene transcripts in which the *CBFB* breakpoints were all close to the 3′ end of the coding region, with only the last 17 of the 182 amino acids deleted from the resultant putative chimeric protein. Liu et al. (28) identified three different breakpoints in the *MYH11* coding region, and a fourth was identified in the expanded set of patients seen by Claxton et al. (30). Therefore, four different classes of breakpoints have been identified, A, B, C, and D, relative to the position of the breakpoints at *MYH11* cDNA base pair positions 1921, 1528, 1201, and 994, respectively. The great majority of patients, 29, had type A rearrangements, while four had type C, two had type D, and one had type B. Six of the patients with translocations had type A, and one had type D. There appeared to be no correlation of fusion type with any clinical or cytogenetic parameters.

These data surely suggest the involvement of the CBFB-MYH11 fusion protein in AML-M4Eo, but what of the putative reciprocal products of the inversions and translocations—MYH11-CBFB? There are three reasons at this point to think that such a fusion product does not play a role in leukemogenesis of this disease: RT-PCR products of transcripts in that configuration have not been observed; MYH11 (whose 5′ end would lead the transcription of such a fusion gene) has not been observed to be transcribed in cells of the affected lineage; and inversion AML-M4Eo patients have been identified with deletions proximal to the p-arm breakpoint, suggesting that the 5′ end of the *MYH11* gene is not present even though disease symptoms are present.

The most striking feature of the results thus far has to do with the rearrangement of the *CBFB* gene from the q-arm of chromosome 16. As indicated above, the gene has been studied in rodent systems and shown to encode a polypeptide that interacts with a polypeptide, CBFα, encoded by a different gene, to produce a protein (CBF) that functions as a heterodimeric transcription factor (29). The gene encoding CBFα is the same gene on human chromosome 21 that has been shown to be rearranged by the 8;21 translocation leading to AML-M2 (29) and has come to be named *AML1* (see AML-M2 section, above). CBF binds to the core site of murine leukemia virus and also to the enhancers of the T cell receptor genes. It appears to be a major genetic determinant of the tissue specificity of leukemia induced by the murine leukemia virus (29). These functions are not inconsistent with the hypothesis that alterations in the genes specifying its subunits play a role in leukemogenesis. Liu et al.

(28) have therefore suggested that the separate subunits of the transcription factor can be involved in different leukemias and that CBF is most likely crucial to the control of cell division or differentiation of the myeloid lineage.

So what is the involvement of *MYH11*, the myosin heavy chain gene, in all of this? The transcript fusion points occur between codons for each of the two genes (*CBFB* and *MYH11*) rearranged by the inversions or translocations in AML-M4. Therefore, Claxton et al. (30) have postulated that only *MYH11* exons beginning with a complete codon are suitable partners for fusion to the *CBFB* transcript at the only site at which it has shown to be rearranged. The coding region of *MYH11* may thus have some functional importance for the biological activity of the fusion gene, even if that involves nothing more than stabilizing the resultant chimeric protein (28). The specific role of the chimeric transcription factor in the leukemogenesis process and the roles of the genes discussed here in normal myeloid differentiation are issues yet to be resolved.

Chromosome Translocation t(6;9)(p23;q34) in Acute Myeloid Leukemia

The t(6;9) translocation is found in a specific subtype of AML. Its presence means a poor prognosis for mainly young adult patients. These patients are mainly classified as AML-M2 or M4 and occasionally as M1 or acute undifferentiated leukemia (AUL). This translocation is also found in patients with myelodysplastic syndrome (MDS). Since the chromosome 9 component of this translocation is in proximity to the t(9;22) in CML, the c-*abl* gene was used as a hybridization probe to identify the breakpoint of t(6;9). It was found that a DNA fragment 360 kb downstream of the c-*abl* gene is within the breakpoint cluster region of the translocation.

The t(6;9) breakpoint on chromosome 9 clustered within an 8-kb region of a large gene (>130 kb) designated as *CAN* (31) (Fig. 8.6). The *CAN* gene transcribes a 7-kb normal mRNA that encodes a CAN protein of 220 kDa. Most translocations fuse the *CAN* gene with an unknown gene called *DEK* within a 9-kb region on chromosome 6. A fusion transcript *DEK-CAN* of 5.5 kb was identified that encodes a putative DEK-CAN fusion protein of 165 kDa (Fig. 8.6). The 5′ cDNA probe of the *CAN* cDNA was unable to detect an abnormal mRNA on Northern blot, indicating that the reciprocal translocated chromosome did not transcribe a *CAN-DEK* fusion transcript. The *DEK* gene transcribes an mRNA of 2.7 kb that encodes a predicted DEK protein of 43 kDa. The *CAN* gene is expressed in most tissue except the liver, ovary, and placenta. The *DEK* gene is expressed in all tissues tested. CAN protein appears to be mainly cytoplasmic, but DEK is a nuclear protein. The fusion of DEK and CAN resulted in a nuclear localization. Structural feature analysis of the CAN protein revealed a putative leucine zipper located upstream, and a potential domain of dimerization located downstream, of the bcr. The carboxy-terminal region of CAN has a domain with some sequence homology to the human estrogen receptor. As shown in Figure 8.6, the fusion protein DEK-CAN consists of almost the entire DEK and the dimerization domain of CAN. It was speculated that replacement of the amino-terminus of the CAN sequence with DEK may activate the transformation potential of CAN.

In one case of a t(6;9) translocation in AUL, the *CAN* gene was fused with another gene designated as *SET* (32). The *SET* gene encodes two mRNAs of 2.0 and 2.7 kb as a result of alternative polyadenylation and encodes a protein of 32 kDa. A fusion transcript *SET-CAN* of 5 kb found in this patient encodes a chimeric protein of 155 kDa. Similar to the DEK-CAN fusion protein, SET-CAN consists of almost the entire sequence of SET and the dimerization domain of CAN. Analysis of the SET sequence revealed sequence homology with the yeast nucleosome assembly protein NAP-1. Similar to DEK, SET is highly acidic, which is a common characteristic of nuclear proteins. The highly acidic region of these two proteins may serve as transcription-activation domains. Similar to the *DEK* gene, the *SET* gene is expressed in all tissues tested. These studies suggest that the *CAN* gene may function as an oncogene and can be activated by fusion with the *SET* or the *DEK* genes. The involvement of the *SET* or the *DEK* genes in the translocation may determine the leukemia cell arrest at a specific stage of myeloid differentiation.

To further understand the molecular mechanism of leukemogenesis of the t(6;9)-positive AML, the biological function of CAN, DEK, and the SET proteins or their cellular roles in normal cells will have to be determined. The possibility that the fusion proteins DEK-CAN or SET-CAN are oncogenic should also be critically evaluated by transformation assay or in an animal model system.

Acute Myeloid Leukemia with Chromosome Translocation Involves the Chromosome 11q23

Chromosome translocations within the *ALL-1* (also designated as *MLL* or *HRX*) gene on chromosome 11q23 have been found in a wide variety of acute leukemias (33,34). These chromosome translocations include t(4;11)(q21;q23), t(9;11)(p22;q23), t(11;19)(q23;p11), t(10;11)(p12;q23), t(11;17)(q23;q25), t(1;11)(p32;q23), t(6;11)(q27;q23), t(2;11)(p21;q23), and t(X;11)(q13;q23). The most prevalent translocations involving the *ALL-1* gene are t(4;11), t(9;11), and t(11;19). Both T and B cell lineage and AML have been found to be associated with these types of leukemias. The t(9;11) translocation occurs mainly in patients with AML-M5, but it does occasionally occur in early B-ALL. The t(4;11) and t(11;19) translocations are found mainly in B-ALL or early B-ALL, respectively, but AML and mixed-lineage leukemia with these translocations have also been reported.

The *ALL-1* gene is believed to be involved in the determination of a different lineage in the hematopoietic system. Analysis of the breakpoint sites in various types of translocations involving the *ALL-1* gene demonstrated a highly conserved breakpoint cluster region irrespective of the leukemia phenotype. The translocation breakpoint clustered within an 8-kb region between exons 6 and 9.

ALL-1 is a very large gene that spans over 100 kb of DNA sequence with at least 21 exons. Two mRNA species of 13 and 15 kb are transcribed by the *ALL-1* gene with an open reading frame capable of encoding a protein of more than 3910 amino acids. The predicted size of the ALL-1 protein is about 431 kDa. Significant sequence homology was found between *ALL-1* and the *Drosophila trithorax* gene in three different regions. Since trithorax is involved in the segment determination of *Drosophila*, strong sequence homology between these two genes implies that *ALL-1* may be involved in a similar function, such as lineage determination in the human hematopoietic system. The involvement of this gene in mixed-lineage leukemia supports this notion. Two potential DNA-binding domains were found in the ALL-1 protein, a cysteine-rich zinc-finger motif and an "AT-hook" motif related to the DNA-binding motif in the HMG proteins. These structural features indicated that ALL-1 is most likely involved in transcriptional regulation.

Relatively little information is currently available to facilitate our understanding of lineage determination in the hematopoietic system. Understanding the molecular mechanism of leukemogenesis in acute leukemia that involves the *ALL-1* gene on chromosome 11q23 will provide valuable information about how early multipotential progenitors are committed to a specific lineage. However, the full-length cDNA of the *ALL-1* gene is over 13 kb, and manipulation of such a large cDNA clone may be technically difficult. Nevertheless, this information will undoubtedly emerge in the near future.

MUTATIONS AND CHROMOSOME DELETIONS

Acute Myeloid Leukemia Associated with Chromosome 5q− Abnormality

Carcinogenesis as a result of a loss of gene function has been well documented in solid tumors such as Wilms' tumor and retinoblastoma. In

AML, loss of chromosome 5 and interstitial deletions within the chromosome 5q loci are frequently found and associated with a poor prognosis. Patients with therapy-related leukemia (AML or MDS) are prone to this type of chromosome abnormality. Analysis of a large number of patients with 5q− chromosome abnormalities concluded that the 5q31 loci are consistently involved in the deletion, suggesting that genes within this segment of the chromosome are critical for the pathogenesis of this type of AML (35). As shown in Fig. 8.7, many genes important for myeloid cell growth and regulation have been localized to this region of the chromosome. These genes include the interleukins 3, 4, 5, and 9 (*IL-3*, *IL-4*, *IL-5*, *IL-9*), granulocyte-macrophage colony-stimulating factor (*GM-CSF*), colony-stimulating factor 1 receptor (*CSF1R*), interferon factor 1 (*IRF1*), early growth response gene 1 (*EGR1*), fibroblast growth factor A (*FGFA*), and the glucocorticoid receptor (*GRL*). IL-3, IL-4, IL-5, IL-9, and GM-CSF are normally produced in T cells or the endothelial cells, which do not carry the chromosome 5q31 deletion. Thus, deletions of these genes in the AML cells should not be critical for pathogenesis.

The search for candidate genes, which possibly encode a novel tumor suppressor, being inactivated by the 5q− abnormality has been inconclusive. The interstitial deletion of the 5q31 region spans approximately 17,000 kb of DNA; however, the critical gene responsible for the pathogenesis of AML has not yet been identified. The D5S89 locus (an anonymous site) was consistently found to be lost in patients with 5q− chromosome (35). Several YAC clones spanning the D5S89 locus have been identified, and multiple potential transcription units were found within this genomic DNA fragment. Further characterization and expansion of these clones might eventually lead to the identification of the critical gene responsible for the pathogenesis of this type of AML.

Mutation of Oncogenes and Tumor Suppressor Genes in Acute Myeloid Leukemia

The N-*RAS* Oncogene Mutation Mutations of the *RAS* family proteins (H-*RAS*, K-*RAS*, N-*RAS*) have been found in a wide variety of tumors. The role of these mutations in tumorigenesis has been well documented in animal tumor models. In AML, mutations within the *RAS* family protein have been extensively studied, and 20% to 40% of patients were found to carry N-*RAS* mutations. Mutations in other *RAS*-family members, H-*RAS* or K-*RAS*, are not common. The biological function and mechanism of action of *RAS* will be discussed in other chapters of this book. Here we will discuss only our current understanding of N-*RAS* in the leukemogenesis of AML.

Mutation of the N-*RAS* gene that alters the critical amino acid at codon 13 was initially discovered by transfection of patients' DNA into NIH3T3 cells followed by a tumorigenicity assay in nude mice. Using this method, the striking finding was that DNA from four of five AML patients analyzed produced tumors in nude mice (36). Mutations in the N-*RAS* genes in these patients' DNA were confirmed by hybridization with a specific oligonucleotide. Since then extensive analysis of mutations within the *RAS* genes in AML have been performed. Mutations that altered the amino acids of codons 12, 13, and 61 of the ras protein have been found consistently (37). Mutations in other *RAS* genes rarely occur in AML.

The *RAS* gene encodes a 21-kDa protein that is part of the GTPase family. It normally binds guanine triphosphate (GTP) and catalyzes hydrolysis of the substrate by a complex mechanism. A switch from the active GTP-bound form to the inactive GDP-bound form, which involves several other protein factors, plays an important role in signal transduction. The mutant RAS protein is unable to catalyze GTP hydrolysis and forms a stable GTP-bound complex that is unable to convert to the inactive GDP-bound form. The active GTP-bound form is believed to interact with an unknown protein factor to promote cell growth and proliferation. It thus appears that mutation of the N-*RAS* gene in AML may be involved in the development of leukemia. However, in a considerable number of AML patients, the presence of *RAS* mutations did not correlate with disease progression and relapse. In these patients the N-*RAS* mutations were detected at clinical presentation, but the mutations were lost at relapse. This finding suggests that N-*RAS* mutations are not a prerequisite for leukemogenesis. Because carcinogenesis is a multistep event, mutations of the N-*RAS* gene may represent only one of the events. The relapsed patients that lost the mutation may have acquired other mutations or chromosome abnormalities. Detection of the N-*RAS* mutation in individual colonies cultured in vitro from AML samples indicated that the proportion of mutation-positive colonies varies from 5% to 57% in different patients. A subset of the colonies, in which normal allele was lost, was found in a majority of the cases. In some cases two different mutations were found in different colonies derived from the same patients. As it is well documented that AML arises from clonal expansion of a single initiating cell, the above results suggest that N-*RAS* mutation is a secondary event of leukemogenesis and does not contribute to the initiation of leukemia.

Loss of one allele of the N-*RAS* gene was also found in some AML cases. But the significance of this lost allele in the pathogenesis of this group of AML patients is unknown.

Mutation of Other Oncogenes Several human myeloid leukemia cell lines were found to carry mutations in the *p53* tumor suppressor gene detected by RT-PCR and single-strand conformation polymorphism (SSCP) (38). In a total of nine AML cell lines analyzed, two did not express a detectable level of *p53* mRNA. Southern blot analysis of these cells demonstrated rearrangement and deletions in both alleles of the *p53* gene. Six of the other cell lines were found to express only the mutated *p53* mRNA. In some cases insertion and deletion was found as a result of abnormal splicing, probably as the result of mutations within the intron region. These studies concluded that eight of the nine myeloid leukemia cell lines analyzed failed to express the normal *p53* mRNA, indicating that mutation of this tumor suppressor gene may be involved in the establishment of these cell lines.

Of 10 AML patients with 17p monosomy, four were found to carry *p53* gene mutations that encode abnormal p53 protein, indicating that mutation within this gene may have a role in the pathogenesis of this type of AML. However, *p53* gene mutation rarely occurred in other types of AML patients.

REFERENCES

1. Sawyers CL, Denny CT, Witte ON. Leukemia and the disruption of normal hematopoiesis. Cell 1991;64:337–350.

2. Gao J, Erickson P, Gardiner K, et al. Isolation of a yeast artificial chromosome spanning the 8;21 translocation breakpoint t(8;21)(q22;q22.3) in acute myelogenous leukemia. Proc Natl Acad Sci USA 1991;88:4882–4886.

3. Shimizu K, Ichikawa H, Miyoshi H, et al. Molecular assignment of a translocation breakpoint in acute myeloid leukemia with t(8;21). Genes Chromosom Cancer 1991;3:163–167.

4. Yunis JJ, Tanzer J. Molecular mechanisms of hematologic malignancies. Crit Rev Oncog 1993;4:161–190.

5. Meyers S, Downing JR, Hiebert SW. Identification of a AML-1 and the (8;21) translocation protein (AML1/ETO) as sequence-specific DNA-binding proteins: the runt homology domain is required for DNA binding and protein-protein interactions. Mol Cell Biol 1993;13:6336–6345.

6. Nucifora G, Begy CR, Erickson P, Drabkin HA, Rowley JD. The 3:21 translocation in myelodysplasia results in a fusion transcript between the AML1 gene and the gene for EAP, a highly conserved protein associated with the Epstein-Barr virus small RNA EBER1. Proc Natl Acad Sci USA 1993;90:7784–7788.

7. Huang M, Ye YC, Chen BR, et al. Use of all-trans retinoic acid in the treatment of acute promyelocytic leukemia. Blood 1988;72:567–572.

8. Warrell RP, Frankel SP, Miller WH, et al. Differentiation therapy of acute promyelocytic leukemia with tretinoin (all-trans-retinoic acid). N Engl J Med 1991;324:1385–1393.

9. Warrell RP, de The H, Wang ZY, Degos L. Acute promyelocytic leukemia. N Engl J Med 1993;329:177–189.

10. Grignani F, Fagioli M, Alcalay M, et al. Acute promyelocytic leukemia: from genetics to treatment. Blood 1994;83:10–24.

11. Chang KS. Molecular pathology of acute promyelocytic leukemia. Hematol Pathol 1993;7:209–223.

12. Kastner P, Perez A, Lutz Y, et al. Structure, localization and transcriptional properties of two classes of retinoic acid receptor α fusion proteins in acute promyelocytic leukemia (APL): structural similarities with a new family of oncoproteins. EMBO J 1992;11:629–642.

13. Perez A, Kastner P, Sethi S, Lutz Y, Reibel C, Chambon P. PML RAR homodimers: distinct DNA binding properties and heteromeric interaction with RXR. EMBO J 1993;12:3171–3182.

14. Grignani F, Ferrucci PF, Testa U, et al. The acute promyelocytic leukemia–specific PML-RARα fusion protein inhibits differentiation and promotes survival of myeloid precursor cells. Cell 1993;74:423–431.

15. Chen Z, Brand NJ, Chen A, et al. Fusion between a novel Kruppel-like finger gene and the retinoic acid receptor α locus due to a variant t(11;17) translocation associated with acute promyelocytic leukemia. EMBO J 1993; 12:1161–1167.

16. Arthur DC, Bloomfield CD. Partial deletion of the long arm of chromosome 16 and bone marrow eosinophilia in acute nonlymphocytic leukemia, a new association. Blood 1983;61:994–998.

17. Le Beau MM, Larson RA, Bitter MA, Vardiman JW, Golomb HM, Rowley JD. Association of an inversion of chromosome 16 with abnormal marrow eosinophils in acute myelomonocytic leukemia: a unique cytogenetic-clinical pathological association. N Engl J Med 1983;309:630–636.

18. Larson RA, Williams SF, Le Beau MM, Bitter MA, Vardiman JW, Rowley JD. Acute myelomonocytic leukemia with abnormal eosinophils and inv(16) or t(16;16) has a favorable prognosis. Blood 1986;68:1242–1249.

19. Hogge DE, Misawa S, Parsa NZ, Pollak A, Testa JR. Abnormalities of chromosome 16 in association with acute myelomonocytic leukemia and dysplastic bone marrow eosinophils. J Clin Oncol 1984;2:550.

20. LeBeau MM, Diaz MO, Karin M, Rowley JD. Metallothionein gene cluster is split by chromosome 16 rearrangements in myelomonocytic leukemia. Nature 1985;313:709–711.

21. Callen DF, Kuss B, Willman CL, et al. Precise localization of the long arm breakpoint of the inv(16) of ANNL M4Eo and progress towards cloning this breakpoint [abstract]. Am J Hum Genet 1992;51:A57.

22. Wessels JW, Mollevanger P, Dauwerse JG, Cluitmans FHM, Breuning MH, Beverstock GC. Two distinct loci on the short arm of chromosome 16 are involved in myeloid leukemia. Blood 1991;77:1555–1559.

23. Dauwerse JG, Jumelet EA, Wessels JW, et al. Extensive cross-homology between the long and the short arm of chromosome 16 may explain leukemic inversions and translocations. Blood 1992;79:1299–1304.

24. Stallings RO, Doggett NA, Okumura K, Ward DC. Chromosome 16 specific repetitive DNA sequences that map to chromosomal regions known to undergo breakage rearrangement in leukemia cells. Genomics 1992;13:332–338.

25. Liu P, Siciliano J, Legerski R, et al. Regional mapping of human DNA excision repair gene ERCC4 to chromosome 16p13.13-p13.2. Mutagenesis 1993; 8:199–205.

26. Liu P, Siciliano J, Seong D, et al. Inter-Alu-PCR primers and conditions for isolating human chromosome painting probes from hybrid cells. Cancer Genet Cytogenet 1993;65:93–99.

27. Liu P, Claxton DF, Marlton P, et al. Identification of yeast artificial chromosomes containing the inversion 16 p arm breakpoint associated with acute myelomonocytic leukemia. Blood 1993;82:716–721.

28. Liu P, Tarle SA, Hajra A, et al. Fusion between transcription factor CBFβ/PEBP2β and a myosin heavy chain in acute myeloid leukemia. Science 1993;261:1041–1044.

29. Wang S, Wang Q, Crute BE, Melnikova IN, Keller SR, Speck NA. Cloning and characterization of subunits of the T-cell receptor and murine leukemia virus enhancer core-binding-factor. Mol Cell Biol 1993;13:3324–3339.

30. Claxton DF, Liu P, Hsu HB, et al. Detection of fusion transcripts generated by the inversion 16 chromosome in acute myelogenous leukemia. Blood 1994;83:1750–1756.

31. Linden MV, Fornerod M, Baal SV, et al. The translocation (6;9) associated with a specific subtype of acute myeloid leukemia results in the fusion of two genes, *dek* and *can*, and the expression of a chimeric leukemia-specific dek-can mRNA. Mol Cell Biol 1992;12:1687–1697.

32. Linden MV, Baal SV, Wiegant J, Raap A, Hagemeijer A, Grosveld G. Can, a putative oncogene associated with myeloid leukemogenesis, may be activated by fusion of its 3′ half to different genes: characterization of the *set* gene. Mol Cell Biol 1992;12:3346–3355.

33. Gu Y, Nakamura T, Alder H, et al. The t(4;11) chromosome translocation of human acute leukemias fuses the *ALL*-1 gene, related to drosophila trithorax, to the *AF*-4 gene. Cell 1992;71:701–708.

34. Tkachuk DC, Kohler S, Cleary ML. Involvement of a homolog of drosophila trithorax by 11q23 chromosomal translocations in acute leukemias. Cell 1992;71:691–700.

35. Nagarajan L, Zavaldi J, Claxton D, et al. Consistent loss of the D5589 locus mapping telomeric to the interleukin gene cluster and centromeric to EGR-1 in patients with 5q− chromosome. Blood 1994;83:199–208.

36. Bos JL, Tokboz D, Marshall CJ, et al. Nature 1985;315:726–730.

37. Bos JL. The ras gene family and human carcinogenesis. Mutat Res 1988;195:255–271.

38. Sugimoto K, Toyoshima H, Sakai R, et al. Frequent mutation in the p53 gene in human myeloid luekemia cell lines. Blood 1992;79:2378–2383.

CHAPTER 9

Acute Lymphoblastic Leukemia

Geoffrey R. Kitchingham

Acute lymphoblastic leukemia (ALL) results from the clonal expansion of a transformed hematopoietic progenitor cell. The orderly pattern of events typical of B cell or T cell differentiation is disrupted by changes in the expression or function of a variety of genes. These genes have been identified more rapidly than genes in many other tumor types because of the relative ease with which informative karyotypes can be obtained. The leukemic cells from thousands of patients have been karyotyped, and a catalog of nonrandom chromosomal translocations (translocations for which there are at least five independent reports) has been generated from this immense database. Genes located at the breakpoints of virtually all of the major, nonrandom chromosomal translocations associated with ALL have been identified. Although the function of many of these genes has not been determined experimentally, functions can often be inferred by comparison with homologous genes studied in other organisms. Data from such studies provide a framework for understanding the origin of lymphoid leukemias.

B AND T CELL DIFFERENTIATION AND LEUKEMIAS

Monoclonal antibodies that specifically recognize antigens on the surface of a restricted set of cell types can be used to identify discrete stages of B and T cell differentiation. Groups of monoclonal antibodies that recognize the same antigen or antigens have been identified and assigned to cluster groups of differentiation (CD) by the International Workshops on Human Leukocyte Differentiation Antigens. Using such antibodies, six types of B lineage ALL have been identified from the pattern of expression of HLA-DR, CD19, CD10, CD20, cytoplasmic immunoglobulin (cIg), and surface immunoglobulin. These subtypes of ALL correspond to six stages of normal B cell differentiation, as the immunophenotypes of B lineage ALLs have been found to parallel closely those of normal bone marrow B cells (see Fig. 9.1).

T cell ALLs are similarly categorized according to the stages of thymocyte maturation (Fig. 9.1). Stage I thymocytes express CD7, CD2, CD5, CD38, and CD71. The transition to stage II ("common" or "double-positive") thymocytes involves the loss of CD71 expression and the acquisition of surface expression of CD1, CD4, and CD8. Further differentiation of thymocytes involves loss of CD1, selective expression of either CD4 or CD8, and acquisition of CD3 (stage III). Although CD3 expression on the cell surface is limited to stage III thymocytes, it can be detected in the cytoplasm as early as stage I.

It is assumed, and has been proved in some experimental systems, that cytogenetic abnormalities such as translocations, inversions, insertions, and deletions are involved in tumor development and progression. In this chapter, the major focus is on genes located near the breakpoints of chromosomal translocations, as most of the available data come from studies of these genes, which are clearly implicated in the development of

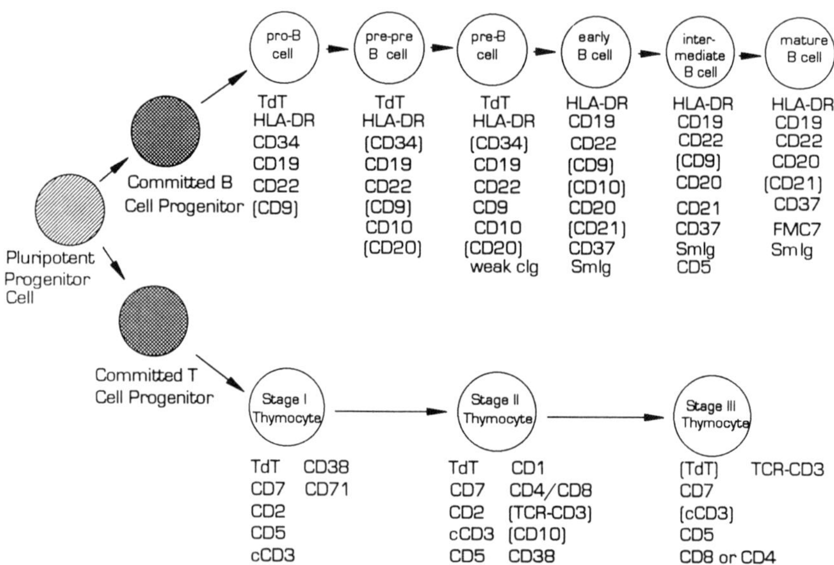

Fig. 9.1: Differentiation scheme for B and T lymphoid cells. The stages of differentiation as defined by the pattern of antigen expression are shown for bone marrow B cells and thymocytes.

leukemia. Little is known about genes affected by nonrandom inversions and deletions. Subcytogenetic changes in the genome, such as small deletions and mutations, are known to play a role in some types of leukemia. Changes in the structure of three genes affected by such changes, TAL-1, p53, and RB, will be discussed. Because analysis of protooncogenes in various forms of leukemia has failed to provide convincing data to correlate their overexpression with leukemia, these studies will not be discussed.

TUMOR SUPPRESSOR GENES

Tumor suppressor genes ("antioncogenes") are genes whose functional loss confers an increased proliferative capacity upon a cell. The list of tumor suppressor genes is growing, and includes the p53, retinoblastoma (RB), deleted in colon cancer (DCC), adenomatous polyposis coli (APC), and mutated in colon cancer (MCC) genes. The importance of these genes in various types of malignancies and their normal cellular functions are covered elsewhere, but a brief discussion of their role in acute leukemias is merited here.

The p53 Gene

The tumor p53 suppressor gene is often mutated in solid tumors, but rarely in lymphoid leukemias, although the incidence of mutation is higher in adult ALL than in childhood ALL. Mutations in p53 may be more frequent in adult ALLs because Burkitt-type ALL (L3) is more prevalent in adults and p53 mutations are associated with this subtype of ALL. Overall, between 1% and 10% of ALLs in adults and children involve mutations in exons 5 through 8 of the p53 gene. Mutated p53 genes are probably involved in the progression, rather than the initiation, of disease, in that they are found more often in patients with relapsed ALL than in those with newly diagnosed leukemias. Most of the base pair changes in p53 genes are transitions, indicating that they probably arise by spontaneous mutation mechanisms rather than exposure to carcinogens.

The RB Gene

The retinoblastoma gene, located on chromosome 13, band q14, was the first human tumor suppressor gene to be identified and cloned. Loss of this region has been reported in cases of both ALL and chronic lymphocytic leukemia. Deletions within the RB gene have been identified in one of five T cell ALL cell lines and in one of 26 cases of primary T cell ALL. Differences in the level of RB expression have also been found in some hematopoietic malignancies, but the significance of these findings is unknown. Mutations of RB are more prevalent in high-grade than in low-grade neoplasms, indicating that they are more likely to be involved in progression than in initiation of lymphomas and leukemias. The studies of RB mutations in lymphoid neoplasms have relied on relatively insensitive techniques that are not capable of identifying small deletions or point mutations. The use of more sensitive techniques, such as the reverse transcriptase–polymerase chain reaction (RT-PCR) and single-strand conformational polymorphism analysis, should help determine the role (if any) of RB mutations in the pathogenesis or progression of acute leukemias.

Expression of the RB gene is normally associated with the late stages of B cell differentiation, and is rarely present in B cells that do not express surface immunoglobulin. The RB protein is hypophosphorylated in resting B and T lymphocytes, but is phosphorylated in response to mitogens as these cells enter S phase, suggesting that RB phosphorylation is required for S phase progression. The effect of mutations associated with lymphoid malignancies on the phosphorylation state of RB and, therefore, on function is unknown. The genes that code for the kinases and phosphatases that regulate RB may be altered in lymphoid neoplasms, thereby disrupting the regulatory role of the RB protein. However, no data are available to support or refute this idea.

Studies thus far have failed to implicate tumor suppressor genes in lymphoid leukemias, with the possible exception of p53 in some of the more mature B cell leukemias. Data regarding changes in the expression or structure of the DCC, MCC, and APC genes in lymphoid neoplasms are not available. If tumor suppressor genes are routinely involved in B and T lineage leukemias, these genes have yet to be discovered. Thus, solid tumors, which are often associated with the deletion or mutation of several tumor suppressor genes, may differ in fundamental aspects of transformation from hematopoietic tumors, in which such genes are currently viewed as having a minimal role in transformation.

GENES NEAR THE BREAKPOINTS OF NONRANDOM CHROMOSOMAL TRANSLOCATIONS

Karyotyping of the majority of de novo acute leukemias has identified numerous nonrandom chromosomal breakpoints. Cytogenetic abnormalities have been identified in 66% to 90% of ALLs. The use of probes for genes known to be near the breakpoints, or of more labor-intensive methods of gene localization, has resulted in the cloning of many of the genes situated near chromosomal breakpoints. The nonrandom chromosomal translocations found in acute lymphoid leukemias are listed in

Table 9.1. These genes can be divided into two broad categories: those whose coding sequences are basically intact, but whose inappropriate expression has been enforced by the presence of an immunoglobulin (Ig) or T cell receptor (TCR) enhancer; and chimeric genes, which encode proteins derived from genes located on both translocated chromosomes. Some of the chromosomal translocations are reciprocal, and both derivative chromosomes may encode chimeric proteins, with either one or both potentially contributing to the transformed state of the lymphoid cells. Translocations in which genes are activated by the Ig and TCR enhancers will be discussed in the next section, and those translocations involving the formation of chimeric genes will be discussed after that.

Genes Activated by Enforced Expression

The Ig and TCR genes contain enhancer elements that promote tissue-specific transcription of those genes. Enhancers can work over long distances and in an orientation-independent manner, so genes brought near them by chromosomal translocations will have their expression activated by these enhancers. The first translocated gene to be cloned, c-*myc*, was found to be translocated into the Ig heavy chain gene locus during the cloning of rearranged heavy chain genes from Burkitt's lymphomas and mouse plasmacytomas. c-*myc* is expressed in normal B and T cells, but its uncontrolled expression alters cell cycle regulation, leading to uncontrolled growth. Other genes activated by Ig or TCR enhancers are not normally expressed in B or T cells, so it is their inappropriate expression that contributes to the altered growth properties of the leukemias. Each gene involved in a chromosomal translocation with an Ig or TCR gene will be considered individually in the following section, emphasizing the potential function of the gene based on its homology with other genes.

Rhombotin Genes: Translocations Containing 11p13 and 11p15 Two adjacent regions of chromosome 11p are involved in chromosomal translocations with members of the TCR gene family. Most often the TCR-δ gene on chromosome 14, band q11, is involved, although translocations to the TCR-β locus on chromosome 7 have also been found. These two loci, 11p13 and 11p15, contain the related genes rhombotin 2 (Ttg-2) and rhombotin 1 (Ttg-1). The genes were so named after rhombotin 1 was found to be expressed mainly in the rhombomeres during fetal development. A third member of the rhombotin gene family has been cloned but has yet to be linked with chromosomal translocations. Chromosome 11p13 breakpoints all map within a 25-kb region upstream of the rhombotin 2 gene, most within a 7-kb region that is hypersensitive to DNase I digestion.

Table 9.1: Nonrandom Chromosomal Rearrangements in Lymphoid Leukemias

Gene Activation Mechanism	*Chromosomal Breakpoints*	*Leukemia Phenotype*	*Protein Product(s)*
Inappropriate expression activated by Ig heavy or light chain gene enhancers	t(8;14)(q24:q32) t(8;22)(q24;q11) t(2;8)(p11;q24)	ALL-L3 (B-ALL); also found in Burkitt's lymphoma	MYC, a nuclear HLH protein
	t(5;14)(q31;q32)	Pre-B ALL	Interleukin-3 growth factor
Inappropriate expression activated by TCR-α, β, or γ gene enhancers	t(11;14)(p15;q11)	T-ALL	Rhombotin-1/Ttg-1, an LIM domain protein, related to transcription factors but lacks DNA-binding domain
	t(11;14)(p13;q11) t(7;11)(q35;p13)	T-ALL	Rhombotin-2/Ttg-2, an LIM domain protein, related to transcription factors but lacks DNA-binding domain
	t(10;14)(q24;q11) t(7;10)(q34;q24)	T-ALL	HOX11/TCL3, a homeobox-containing gene probably involved in transcriptional control
	t(1;14)(p32;q11) t(1;7)(p32;q34)	T-ALL	TAL-1/SCL/TCL5, an HLH protein that binds to E-box transcription factors
	t(7;9)(q34;q34)	T-ALL	TAL-2, an HLH protein
	t(1;7)(p34;q34)	T-ALL	LCK, a tyrosine kinase, member of *src* family
	t(7;9)(q34;q34.3)	T-ALL	TAN-1, human homologue of *Drosophila* Notch gene, an integral membrane protein
	t(7;19)(q35;p13)	T-ALL	LYL-1, an HLH protein
	t(8;14)(q24;q11)	T-ALL	MYC, an HLH protein

Rhombotin 1 is expressed at low levels in mouse thymus, lung, kidney, uterus, and placenta, and RT-PCR can detect expression in human T cell lines and in many other tissues. The rhombotin 1 gene has two distinct promoters and two first exons, generating two messenger RNA species of 1.2 and 1.4 kb. Use of the two promoters varies with developmental stage in the nervous system and the thymus. Although the majority of translocations occur outside the gene in T cell ALLs, a translocation that truncates the first promoter and generates a very high level of expression from the second has been found in one T cell line. The majority of T cell ALLs with translocations containing rhombotin 1 express high levels

Table 9.1: *Continued*

Gene Activation Mechanism	*Chromosomal Breakpoints*	*Leukemia Phenotype*	*Protein Product(s)*
Inappropriate expression controlled by a non-Ig, TCR gene	del(1)(p32)	T-ALL	SIL deregulates TAL-1; fusion mRNA formed, but not fusion protein
Fusion genes	t(1;19)(q23;p13)	Pre-B ALL, less often with early B cell ALL	E2A-PBX: E2A is an HLH protein; PBX is a homeodomain protein
	t(17;19)(q22;p13)	Early B cell ALL associated with disseminated intravascular coagulation	E2A-HLF: HLF is a transcription factor containing a leucine zipper
	t(9;22)(q34;q11)	Early B cell ALL often associated with mixed-lineage phenotype	BCR-ABL: BCR, GTPase-activating protein for $p21^{rac}$ ABL, tyrosine kinase; ABL-BCR fusion product is also formed
	t(4;11)(q21;q23) t(11;19)(q23;p13) t(1;11)(p32;q23)	Early B cell ALL often associated with mixed-lineage phenotype	ALL-1/MLL-1: gene on chromosome 11, serine-proline-rich proteins on chromosomes 4, 9, and 19 form hybrid transcription factors

of mRNA from this gene, which may directly contribute to malignant transformation.

Rhombotin 2 is normally expressed in nerve tissue and in fetal liver, spleen, and kidney, but not in activated T cells and minimally, if at all, in mouse thymus. Although most leukemic T cells with translocations containing rhombotin 2 produce high levels of its mRNA, there have been several cases of such cells with low or undetectable expression. Hence, the importance of rhombotin 2 expression in these T cell leukemias is unclear, although these data could be interpreted to imply a role for this gene in the initiation, rather than progression, of the leukemia (i.e., expression is down-modulated after oncogenic transformation). Conversely, there have been several cases of rhombotin 2 overexpression with no detectable changes in the gene. As there are several methylation-free zones upstream and downstream of the rhombotin 2 gene, implying the presence of other transcribed genes, it is possible that another gene near the breakpoint is the active participant in T cell leukemogenesis. Whether rhombotin 2 is involved at an early or late stage of leukemogenesis, if it is involved at all, will be determined only by analyzing the expression of other genes near the chromosomal breakpoint and by studies in transgenic mice.

Enforced expression of either rhombotin 1 or 2 in the T cell compartment of transgenic mice has no effect on development, and only a small percentage of animals develop leukemias, and then only after a long latency period. Although such studies demonstrate that these genes can produce T cell leukemias that, to a certain extent, recapitulate the original disease, the long latency periods indicate that other genetic changes are required for malignant transformation. The identity of the other gene (or genes) in which changes occur is presently unknown. Approaches that have succeeded in identifying cooperating genes in transgenic mice carrying other chromosomal translocations have been unsuccessful when applied to translocations involving rhombotin 1.

The proteins encoded by the rhombotin genes each contain a region with a duplicated cysteine-rich motif, homologous to the LIM motif of transcription factors. The LIM proteins, named for *lin*-11, *isl*-1, and *mec*-3—the first three proteins to be found with the cysteine-rich domain—also have homeodomains and activation domains typical of some transcription factors, but absent in the rhombotin proteins. The role of the cysteine-rich LIM-homology domains in the rhombotin family may be in protein-protein interactions, and could negatively regulate transcriptional activation by homeobox or LIM-domain protein partners. Inappropriate expression of the rhombotins in T cell ALLs may result in the repression of genes involved in controlling cell division. Alternatively, as the LIM domain is homologous to ferrodoxins, the rhombotins may act via modification of LIM-domain factor redox states, which could have positive or negative effects on gene transcription. Other cellular proteins to which the rhombotins bind, and the effects of this binding on gene expression, will have to be identified to resolve these questions.

HOX11/TCL-3: Translocation t(10;14)(q24;q11) Approximately 7% of T cell ALLs are characterized by the t(10;14)(q24;q11) translocation, in which the HOX11 gene on chromosome 10 is juxtaposed with the TCR-δ-chain gene on chromosome 14. The breakpoints on chromosome 10 are clustered within a 15-kb region that is outside the HOX11 coding sequences. A variant translocation, t(7;10)(q35;q24), involves a breakpoint on chromosome 10 that is 12 to 13 kb upstream of the t(10;14) breakpoint and involves the TCR-β locus on 7q35. In all cases, it is likely that the TCR gene enhancer element drives inappropriate expression of the HOX11 gene.

The HOX11 gene is relatively compact, with three exons and two introns dispersed over a 7-kb DNA region. HOX11 mRNA is detectable as a 2.1-kb transcript in normal human liver tissue, and small quantities of transcripts can be found in normal human T cells. Some leukemic T cell lines that do not harbor the t(10;14) chromosomal translocation also express HOX11 mRNAs. High levels of expression are observed only in acute leukemic cells with the t(10;14) or t(7;10) translocations.

Homeodomain-containing proteins bind to specific DNA sequences to activate transcription, but DNA-binding targets of the HOX11 protein are not known. HOX11 is a nuclear protein of approximately 40 kDa whose expression appears to be cell cycle regulated, with the highest levels of expression at the G_1/S-phase boundary. Cell-cycle-regulated expression probably does not occur in T cell ALLs, in which expression of HOX11 is under the control of the TCR gene enhancers. Taken together, these data suggest that the HOX11 gene product may function as a transcription factor for G_1 cell cycle progression. However, several transcription factors, active only briefly during the cell cycle, are controlled by phosphorylation, and thus far it appears that the HOX11 protein is not phosphorylated and is not a substrate for any protein kinase. Therefore, how the expression and function of HOX11 is normally controlled is not known, nor is it known which other proteins or factors the HOX11 interacts with to initiate gene transcription.

Transgenic mice bearing the HOX11 gene under the control of a T-cell-specific promoter develop T cell lymphomas by 7.5 months of age. The gene has growth-promoting activity for mouse hematopoietic cells in culture, but in transgenic mice the overall level of thymocytes is reduced by 90%. These mice have normal levels of stage I thymocytes, but the numbers of stage II and stage III cells are severely reduced. Thus, while enforced expression of the HOX11 gene in thymocytes does not affect differentiation of stem cells to stage I, it apparently blocks progression to stage II. The fact that T cell ALLs with t(10;14) translocations do not cluster in the stage I thymocyte compartment indicates either that this translocation occurs after differentiation to a more mature phenotype, or that the transgenic mouse system does not accurately reproduce human T cell ALL.

Basic Helix-Loop-Helix Proteins Three members of a well-recognized group of transcription factors, the so-called basic helix-loop-helix (bHLH) proteins, are activated by translocations. These three related bHLH genes, LYL-1, TAL-1/SCL, and TAL-2, are capable of DNA binding, and each can potentially form protein heterodimers.

TAL-1/SCL: *Translocation t(1;14)(p32-34;q11)* In the t(1;14)(p32;q11) translocation, found in about 3% of T cell ALL cases, the TAL-1 gene is translocated from chromosome 1 to the TCR-α/δ-chain gene locus on chromosome 14. Although initially described as disrupting the TAL-1 coding sequences, these translocations involve either 5′ or 3′ untranslated regions and thus do not alter the TAL-1 gene product. A few patients with translocations involving TAL-1 and the TCR-β locus on chromosome 7q35 have been described. In these translocations, the TAL-1 gene presumably comes under the transcriptional control of the TCR gene enhancers.

The TAL-1 locus is quite interesting in that about one-fourth of patients with T cell ALL have a 90-kb deletion of TAL-1 genomic DNA that is not detectable by karyotyping. This deletion brings the TAL-1 coding sequences close to the first exon of the SIL gene, leading to the production of a fusion mRNA in which the 5′ untranslated sequences of TAL-1 are replaced by the 5′ untranslated region of SIL. A chimeric protein is not generated, but the control of TAL-1 transcription is probably mediated by SIL promoter sequences.

The same enzyme complex thought to be responsible for the inappropriate rearrangement of the TAL-1 locus into one of the TCR loci, the VDJ recombinase, probably is also responsible for the site-specific 90-kb deletion. The breakpoints for this deletion are tightly clustered and flanked with sequences resembling the heptamer and nonamer sequences recognized by VDJ recombinase during the normal process of Ig or TCR gene rearrangement. Rearrangement into the TCR locus or the 90-kb deletion results in the expression of TAL-1 in T cells, in which it normally is not found.

TAL-1 mRNA has two alternative first exons, Ia and Ib, resulting in the production of two distinct mRNAs (type A and type B). The t(1;14) translocation and the site-specific 90-kb deletion both result in the loss of exon Ia, which is substituted with the first exon of SIL in the 90-kb deletion. In the mouse, *tal*-1 mRNA is normally expressed to high levels in erythroid and mast cells, and also in some immature myeloid cell lines. Chemically induced differentiation of murine erythroid cell lines leads to an increase in *tal*-1 mRNA. The expression pattern in humans is similar to that in mice in that TAL-1 mRNA is found in erythroid cell lines, but not in B cell or myeloid cell lines. TAL-1 expression cannot be induced by phytohemagglutinin treatment of peripheral blood T cells or in normal thymocytes, but TAL-1 transcripts have been detected in human fetal liver and in regenerating bone marrow of adults.

Expression of the TAL-1 gene parallels that of the GATA-1 transcription factor, indicating possible complementary roles for these two proteins. Both genes are expressed in megakaryocytes and erythroblasts as well as in basophilic granulocytes, their expression being down-modulated during erythroid and megakaryocytic differentiation. One of the two TAL-1 promoters contains consensus binding sites for GATA-1, suggesting that GATA-1 can modulate TAL-1 expression during differentiation.

The TAL-1 gene product has been expressed in bacteria and purified, and its interaction with other HLH-containing proteins studied in vitro. The bacterially expressed protein can interact with the ubiquitous HLH-containing proteins E12 and E47, which are products of the E2A gene on chromosome 19 involved in the t(1;19) chromosomal translocation found in pre-B cells. TAL-1 and either E12 or E47 can form heterodimers that recognize the sequence CANNTG within the Ig heavy chain enhancer (the so-called E-box transcriptional enhancer motif). Although the ability to bind the Ig enhancer motif per se is probably not related to TAL-1's role in leukemogenesis, these data provide insight into the normal functioning of the protein and might help identify genes that are inappropriately expressed in leukemias as a result of TAL-1 activation.

Expression of TAL-1 leads to the production of two phosphoproteins, pp42 and pp22, that represent phosphorylated versions of the full-length protein (residues 1 to 331) and a carboxy-terminal protein (residues 176 to 331). Because the bHLH features of the protein are located in the carboxy-terminus, both proteins contain the bHLH motif, the DNA-binding domain, and the protein dimerization region. The major site of phosphorylation is at serine residue 122, which is present only in the full-length protein. Preliminary experiments in tissue culture cells suggest that the phosphorylation of this amino acid can be induced by epidermal growth factor (EGF), and the timing of this phosphorylation event is such that the ERK/MAP2 protein kinases may be involved. Furthermore, purified ERK1 can phosphorylate Ser-122 in vitro. There are no published functional studies of the TAL-1 protein before and after phosphorylation at Ser-122, so the effect of this modification on protein function is still unknown. Nevertheless, phosphorylation at Ser-122 may be the critical modification controlling the functions of TAL-1 in response to external stimuli such as EGF.

The remote possibility remains that perturbation of SIL gene expression is also involved in these T cell malignancies. The gene is over 70 kb in length, has 18 exons, and shows alternative splicing at the 5′ end. The SIL/TAL-1 rearrangement leads to deletion of the body of the SIL gene on the rearranged allele. Transcription from the SIL gene leads to the production of a 5.5-kb mRNA, which has an open reading frame of 1287 amino acids and a predicted molecular size of 143 kDa. SIL mRNA is expressed at low levels in hematopoietic cell lines and tissues; in mouse embryos, expression is restricted to the hematopoietic compartment. The predicted 143-kDa SIL protein does not show homology to proteins in the databanks, but does contain a motif similar to the active sites of several eukaryotic topoisomerase I proteins. Although it remains possible that SIL functions as a tumor suppressor gene, and that deletion of one copy of SIL contributes to malignant transformation, there are no data to support this possibility. However, until the structure of the remaining SIL allele in these T cell ALLs is examined for mutations, and the T cell ALLs with t(1;14) translocations are analyzed for altered SIL gene expression, this possibility cannot be ruled out.

TAL-2, a gene homologous to TAL-1, is involved in a nonrandom but rare chromosomal translocation, t(7;9)(q34;q32). The TAL-2 gene is located about 30 kb from the breakpoint.

LYL-1: Translocation t(7;19)(q35;p13) The LYL-1 gene on chromosome 19 is related to the TAL-1 gene and is associated with the t(7;19) translocation found in T cell ALL. This translocation in the supT7 T cell line results in the head-to-head orientation of the LYL-1 gene with a TCR-β-chain gene, with loss of the first intron of the LYL-1 gene and possibly of its regulatory sequences as well. Truncated LYL-1 transcripts that contain the entire coding region of LYL-1 accumulate to high levels in ALLs bearing this translocation.

Lyl-1 is expressed in most murine myeloid, erythroid, and B cell lines, but little or no expression occurs in most T cell lines or anywhere outside the hematopoietic system. Like other genes found at chromosomal breakpoints in T cell ALLs, LYL-1 has alternative 5′ exons and alternative promoters. Several transcripts of different sizes have been found. They arise, at least in part, from differential exon usage in the 5′ untranslated region. As with the TAL-1 gene, utilization of the promoters varies in different hematopoietic cell types. The normal role of the LYL-1 gene product in hematopoiesis and its role in T cell differentiation and proliferation have yet to be elucidated.

TAN-1/TCL3: Translocation t(7;9)(q34;q34.3) Another relatively infrequent chromosomal translocation, found in less than 1% of patients with T cell ALL, t(7;9)(q34;q34.3) involves the TCR-β gene and a second gene variously referred to as TCL3 and TAN-1, on chromosome 9 (Fig. 9.2). The breakpoints on chromosome 9 are within the TAN-1 gene, leading to the production of truncated versions of TAN-1 mRNA.

TAN-1 mRNA is expressed in a variety of normal human fetal tissues, and the murine homologue is expressed in many adult tissues, with the highest levels of expression in both species occurring in the lymphoid compartments. The TAN-1 gene encodes a protein highly homologous to

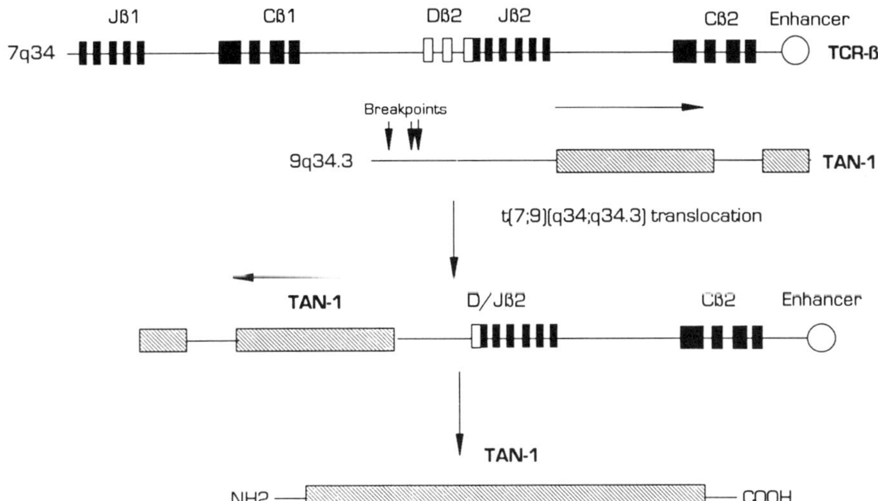

Fig. 9.2: Chromosomal translocation involving the TCR-β-chain gene on chromosome 7 and the TAN-1 gene on chromosome 9. The region of 7q34 containing the TCR-β-chain gene from just upstream of the Jβ1 region to the enhancer is shown on the top line, while the TAN-1 gene from chromosome 9q34.3 is shown on the second line. The TCR-β-chain gene has undergone Dβ2-to-Jβ2 joining. Breakage of chromosome 9 upstream of the TAN-1 gene (indicated by the arrows) with joining to the rearranged Dβ2-Jβ2 leads to a head-to-head orientation of the translocated genes. Transcription of the TAN-1 gene is usually from its own promoter, but the activity of the promoter is now controlled by the TCR-β enhancer.

the *Drosophila* Notch gene product (TAN stands for translocation-activated Notch gene). The *Drosophila* protein is an integral membrane protein that is important for the developmental determination of embryonic cell fates. As would be expected, the TAN-1 protein shows homology to other Notch-related genes such as the *Xenopus* homologue Xotch, and two genes from *Caenorhabditis elegans*, *lin*-12 and *glp*-1.

Clues to the potential function of the TAN-1 protein come from comparison with the related proteins Notch and *lin*-12, both of which contain a motif of highly conserved cysteine residues, as well as motifs similar to ones found in EGF. The binding partner for the *Drosophila* Notch protein is also a transmembrane protein containing extracellular EGF-like repeats. This suggests that interaction of these two proteins may mediate cell-cell adhesion, or receptor signaling through cell-cell interactions.

The t(7;9) chromosomal translocation affects the 5′ end of TAN-1, and could thus have an effect equivalent to removing the extracellular domain of the normal TAN-1 protein. Cell adhesion could thereby be disrupted, contributing to the malignant behavior of the T cells. Alternatively, changes in the external portion of the protein could activate the signal transduction properties of TAN-1 in the absence of binding to its partner, as has been found for several protooncogenes. It is possible that TAN-1 functions normally as a negative growth regulator, and that truncation of its extracellular domain renders it inactive. If this is the case, the truncated version of TAN-1 would be dominant over the wild-type protein, given that the second copy of TAN-1 in T cell ALLs carrying the t(7;9) is unlikely to be mutated. Whatever the mechanism, TAN-1 is the only example in human lymphoid neoplasms of an integral membrane protein that has been altered by chromosomal translocation. Identifying the normal protein's role in signal transduction should provide significant insights into lymphoid neoplasia.

c6.1a/b: Translocation t(X;14)(q28;q11) The TCR-α gene is involved in translocations with an interesting locus on the X chromosome at band q28. The genes involved in this translocation have been cloned from two patients with prolymphocytic T cell leukemia, one of whom also had ataxia telangiectasia. The Xq28 gene is called c6.1a, and the translocations in the two patients occurred in two different introns of this gene. The c6.1a gene is located in a CpG island with another gene, c6.1b, which is transcribed in the opposite direction. Both genes are located within 70 kb of the Factor VIII gene.

It is unlikely that c6.1a is the gene responsible for the leukemia. The t(X;14) translocation leads to the production of a chimeric mRNA that contains 5′ sequences of c6.1a as well as the TCR-α constant region. However, the fusion generated an in-frame c6.1a-Cα mRNA in only one of the two cases studied. The structure of the c6.1b gene was not affected in either case, and was transcriptionally active in both. Active transcription of both genes is probably mediated via the TCR-α enhancer, located downstream of the Cα constant region. Complicating the picture is the fact that, if the DNA sequencing is correct, the c6.1b gene does not encode an open reading frame of any significance, and it bears no significant homology to any other DNA sequenced thus far. The absence of an open reading frame presents a real paradox if c6.1b is to be implicated in leukemogenesis. There are several examples of noncoding RNAs affecting gene expression, a prime example being the X-inactivation-specific transcript XIST, which helps silence transcription from the inactive X chromosome. It is possible that studies of this translocation have helped to identify another noncoding RNA that is involved in gene expression.

LCK: Translocation t(1;7)(p34;q34) A translocation that has been found in one case of T cell leukemia involves activation of the LCK gene by the TCR-β gene enhancer. LCK, a member of the *src* family of cytoplasmic protein kinases, is involved in signal transduction in T cells. How transcription of this gene, under the control of the TCR-β enhancer in T cells, can provide a signal for leukemic transformation is not yet known.

Interleukin-3: Translocation t(5;14)(q31;q32) The interleukin-3 (IL-3) gene is activated in the t(5;14)(q31;q32) translocation that has been found in two cases of pre-B cell acute leukemia. In these two cases, the IL-3 gene and the Ig heavy chain locus were positioned head-to-head, with transcription of the intact IL-3 gene being driven presumably by the heavy chain enhancer located several kilobase pairs upstream. The leukemic lymphoblasts secrete large amounts of IL-3, resulting in eosinophilia.

Chromosomal Translocations Leading to the Production of Fusion Genes

E2A/PBX: Translocation t(1;19)(q23;p13) The t(1;19) chromosomal translocation occurs in approximately 25% of childhood pre-B cell ALLs. Although it is often described as specific to this subtype of leukemia, there have been reports of cIg-negative leukemias with the t(1;19). The translocation involves the E2A gene on chromosome 19, which encodes the transcription factors E12 and E47, and a previously unidentified gene on chromosome 1 called PBX1 (or PRL). In most cases, only the derivative chromosome 19 is retained. The breakpoints on chromosome 19 are clustered in the intron between exons 13 and 14, and those on chromosome 1 are within a single large 50-kb intron (Fig. 9.3).

The t(1;19) produces a chimeric mRNA that has been detected in essentially all cases studied to date. The E2A gene is ubiquitously expressed, while expression of PBX1 is undetectable by Northern analysis in pre-B cells but has been found in a variety of normal fetal and adult cell types by RT-PCR. The chimeric transcript contains coding sequences for the amino-terminal two-thirds of the E2A protein and the carboxy-terminal region of PBX1.

Both of the wild-type proteins, E2A and PBX1, contain features characteristic of transcription factors. The E12 and E47 proteins have been shown to be HLH-containing proteins that bind to an E-box element in the κ light-chain enhancer. The PBX1 protein contains a

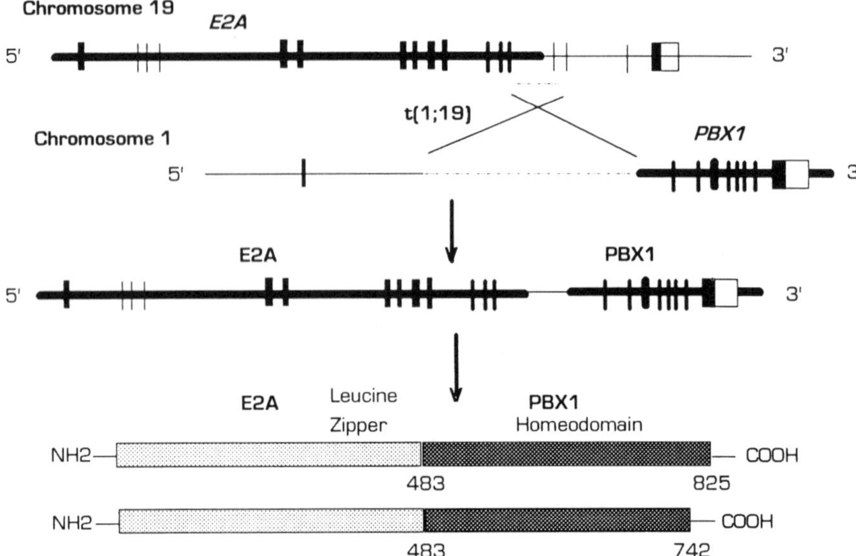

Fig. 9.3: Chromosomal translocation involving the E2A gene on chromosome 19, and the PBX1 gene on chromosome 1. The intron-exon structure of the E2A gene is shown on the top line, with the exons depicted by vertical lines. The PBX1 gene is depicted on the second line, with the dashed line indicating that this region is not drawn to scale. The heavier lines indicate the segments of the two genes that end up on the derivative 19 chromosome. The structure of the chimeric gene is shown on the third line. There is some variability in the location of the breakpoints on both chromosomes 1 and 19, so the thin line between the E2A and PBX1 genes can vary in length considerably between individual translocations. The chimeric gene encodes two chimeric proteins, an 825 amino acid polypeptide containing the leucine zipper of E2A and the homeodomain of PBX1, and a 742 amino acid version that is truncated at the carboxy-terminus.

homeodomain that is distinct from previously described mammalian homeoboxes but bears some homology to the yeast mating factor gene products, MAT a1 and MAT α2. Two E2A-PBX1 fusion proteins of 85 and 77 kDa are produced that contain the activation domain of E2A, but its basic DNA-binding and HLH dimerization domains have been replaced by the putative DNA-binding homeobox domain of PBX1.

Retroviral constructs coding for the chimeric p85 and p77 E2A-PBX1 proteins transform NIH3T3 cells, rendering them tumorigenic in nude mice. Reconstitution of lethally irradiated mice with normal bone marrow transduced with these retrovirus constructs caused acute myeloid leukemias. The tumors arise after a fairly long latency, indicating that other genetic changes are necessary for cell transformation. In light of these findings, it is somewhat puzzling that the translocation is found only in early B cell leukemias, especially given that expression of the chimeric gene is driven by the E2A promoter, which is not tissue specific.

Transgenic mice containing an integrated copy of the E2A-PBX1 fusion gene under the control of an Ig heavy chain promoter and the heavy chain enhancer unexpectedly express the transgene predominantly in the thymus. The μ enhancer is normally active in T cells, but levels of expression from an Ig heavy chain promoter under the control of this enhancer in T cells are usually several orders of magnitude lower than in B cells. The reason for the preferential expression of the transgene in thymocytes is unknown. Despite the unexpected phenotype, the effect of the transgene on lymphoid differentiation does allow for hypotheses to be generated regarding the role of E2A-PBX-1 in lymphoid malignancies. Animals showed a profoundly altered lymphoid differentiation, with the number of thymocytes and bone marrow B cell precursors being reduced to 20% of normal. Analysis of the thymocytes showed that a larger percentage were cycling in the transgenic mice, but there was also an increased percentage of cells undergoing apoptosis, suggesting that the increased cell death reduced the overall number of thymocytes. All of the animals developed T cell malignancies with a stage II phenotype. The apparently opposing roles the chimeric proteins have in T cells, as stimulators of proliferation and of cell death, are similar to those demonstrated for c-MYC. Current thinking is that a second signal is required for proliferation, and those cells not receiving this signal undergo programmed cell death. In the transgenic animals, a second signal for proliferation might be related to the signal for commitment to either the CD4 or CD8 lineage, while in pre-B ALLs it might be related to signaling events involved in Ig heavy chain gene rearrangement.

The mechanism by which the chimeric E2A-PBX1 gene disrupts normal cellular processes and (potentially) induces cell proliferation is not known, but at least two different mechanisms can be envisioned. First, the chimeric proteins may compete with the wild-type PBX1 protein (or proteins) for DNA binding sites and interfere with their regulation of PBX-responsive genes. Alternatively, because the fusion proteins retain the leucine zipper of E2A, which is required for protein dimerization, they might form nonfunctional complexes with wild-type E2A protein, effectively depleting the cell of this protein, which is required for normal differentiation.

The PBX1 gene belongs to a family that includes PBX2 and PBX3, which appear to be ubiquitously expressed. An interesting feature of all three genes is that they encode almost identical homeobox regions, indicating that they may share the same DNA sequence-binding specificity. Although they also share extensive amino acid identity in some of the nonhomeobox regions, the amino- and carboxy-terminal domains have diverged considerably.

E2A-HLF: Translocation t(17;19)(q22;p13) The E2A gene is involved in a second chromosomal translocation, with sequences from chromosome 17. The t(17;19) translocation is relatively uncommon, and patients with acute leukemias containing this translocation have an interesting phenotype that includes disseminated intravascular coagulation—a feature rare in ALLs lacking this translocation.

The t(17;19) produces a chimeric transcription factor consisting of the amino-terminal portion of the E2A protein, fused to the basic DNA-binding and leucine zipper dimerization motifs of a novel hepatic protein, HLF (hepatic leukemia factor). HLF, which is not normally transcribed in lymphoid cells, belongs to the recently described PAR subfamily of basic leucine zipper (bZIP) proteins. This family also includes DBP (albumin D-box protein) and TEF (thyrotroph embryonic factor)/VBP. Wild-type HLF can bind DNA specifically as a homodimer, or as a heterodimer with other PAR factors. Structural alterations of the E2A-HLF fusion protein markedly impair its ability to bind DNA as a homodimer. However, E2A-HLF can bind DNA as a heterodimer with other PAR proteins, suggesting a novel mechanism for leukemogenic conversion of a bZIP transcription factor. The TEF/VBP protein is known to be transcribed in lymphoid tissues of rat and chicken, and could potentially be a dimerization partner in ALL cells carrying the t(17;19).

The bZIP domain of HLF is predicted to mediate binding of the chimeric protein to the promoters of genes normally regulated by wild-type HLF, while the regulatory sequences of the E2A gene, which is expressed in B cell progenitors, may account for expression of the chimeric protein in leukemic cells. The amino-terminal region of E2A contains a transcription activation domain and may thus contribute a transactivating function to the chimeric protein. The data clearly indicate that the amino-terminal region of the E2A protein is involved in transformation of B cells as part of the E2A-PBX1 and E2A-HLF chimeric genes, but further studies are required to define the contributions made by E2A to the potential transcriptional and transforming properties of these chimeras.

BCR-ABL: Translocation t(9;22)(q34;q11) The first chromosomal translocation recognized was the so-called Philadelphia (Ph) chromosome, a shortened chromosome 22 arising from the reciprocal translocation (9;22)(q34;q11). The t(9;22) not only is present in the vast majority of cases of chronic myelogenous leukemia (CML), but also has been found in acute lymphoid and myeloid leukemias in both adults and children. The breakpoints for this translocation differ somewhat in the acute and chronic leukemias, and those features relevant to acute leukemias will be discussed in this chapter. A more complete discussion of the t(9;22) in CML can be found in Chapter 7 by Kuzrock and associates.

The participating gene on chromosome 9 is the human homologue of the transforming gene present in the Abelson murine leukemia virus, v-*abl*. Most of the breakpoints on chromosome 9 occur 5′ to the coding sequences for c-ABL in the very large first intron. On chromosome 22, in some cases of acute leukemia (and in most cases of CML), the translocation breakpoints are clustered within a 5.8-kb region termed the major breakpoint cluster region (BCR). About half of the acute leukemias with t(9;22) have breakpoints in the major cluster region, while the other half have breakpoints in the BCR gene located outside of this region. Breakpoints in the minor cluster regions give rise to different-sized mRNA and protein products and are rare in CML. Some studies indicate that the location of the breakpoint may predict clinical outcome, but these are equally balanced by reports that breakpoint location is nonpredictive. The finding of unique breakpoints in Ph+ ALL giving rise to distinct protein products ended the debate as to whether these leukemias were actually CMLs in which the chronic phase had gone unnoticed.

Approximately 25% of adults, but less than 5% of children, with ALL have the Ph chromosome. The presence of the Ph chromosome indicates a poor prognosis. Ph+ ALLs tend to have mixed-lineage phenotypes; about 50% express myeloid antigens, especially CD13, regardless of the location of the breakpoint in the BCR gene.

The t(9;22) translocation into the major breakpoint cluster region of the BCR gene leads to the production of a chimeric BCR-ABL mRNA of 8.5 kb. Translocation into the minor cluster region, which is within the first intron of the BCR gene, leads to the transcription of a 7-kb fusion mRNA encoding a protein containing coding sequences from only the first exon of the BCR gene. The 8.5-kb mRNA is translated into the CML-type fusion protein, $p210^{pBCR\text{-}ABL}$, while the 7-kb mRNA produces $p185^{pBCR\text{-}ABL}$. The $p185^{pBCR\text{-}ABL}$ product is more actively transforming than the $p210^{pBCR\text{-}ABL}$ product.

The normal cellular ABL protein (145 kDa) has a latent tyrosine kinase activity that can be activated by either deletion of the SH3 domain (src homology domain 3) or substitution of the ABL first exon with the BCR gene in the t(9;22). The BCR gene encodes a member of the guanosine-triphosphatase-activating protein (GAP) family, whose substrate appears to be the *ras*-related guanosine triphosphatase (GTP) binding protein $p21^{rac}$. The peptide encoded by the first exon of BCR can bind to the ABL SH2 requlatory domain and activate the tyrosine kinase activity of ABL. Tyrosine phosphorylation of p160 BCR by $p210^{pBCR\text{-}ABL}$, and the stable physical interaction of these proteins, may perturb normal BCR functions and may be leukemogenic. The major sites of tyrosine phosphorylation in vitro are contained within the first exon of p160 BCR, while BCR-ABL autophosphorylation occurs predominantly at tyrosines within BCR exon 1 sequences. The results raise the possibility that the activated ABL protein kinase of BCR-ABL proteins modulates the putative signal transduction activities of p160 BCR by tyrosine phosphorylation of exon 1 sequences.

In addition to having a tyrosine kinase activity, the wild-type c-ABL protein is also able to bind to F-actin, an activity that is enhanced by the BCR sequences in the fusion protein. Mutations in a conserved domain at the carboxy-terminus of c-ABL disrupt the ability of the protein to bind to actin filaments in vivo and in vitro; this region contains a consensus motif found in several other actin-crosslinking proteins. The ability to bind to F-actin is functionally significant, as BCR-ABL proteins that are unable to associate with F-actin have a lesser ability to transform cells and to abrogate growth factor requirements. Expression of BCR-ABL in transformed cells causes a collapse of the actin filaments, and F-actin becomes associated with condensed structures around the periphery of the nucleus. Thus, two functions, tyrosine kinase and F-actin binding, are important for the pathogenic and physiologic functions of the BCR-ABL and c-ABL proteins.

The t(9;22) translocation has been used as a diagnostic marker for leukemic cells and as a means of monitoring response to chemotherapy. Traditionally cytogenetics has been used to monitor response to chemotherapy, but RT-PCR is quickly becoming the standard technique. Several studies have demonstrated that RT-PCR can identify leukemias with the Ph chromosome when these cells appear to have normal karyotypes. The prognosis for patients with Ph+ ALL is poor, despite the fact that the rates of achieving complete remission are similar to those of patients without the Ph chromosome.

Most research has focused on the BCR-ABL fusion gene and its gene product, but the translocation is usually reciprocal, and the chimeric gene ABL-BCR is also present in the cells of 65% of CML patients and in 88% of Ph+ ALL patients. Three different fusion mRNA transcripts have been found: exon Ib of the ABL gene and BCR exon 2(Ib-e2), ABL exon Ia and BCR exon 2 (Ia-e2), and ABL exon Ib and BCR exon 4. The differentially spliced ABL-BCR mRNAs lead to the production of proteins predicted to consist of the first exon of ABL joined to BCR exons including the GAP domain or to the GTP-exchange factor plus the GAP domain. Some, but not all, of the fusion proteins contain the amino-terminal myristoylation site in the BCR protein, which could redirect them to the inner surface of the plasma membrane. The preliminary data indicating that both partners in the reciprocal translocation may be important in the pathogenesis of Ph+ ALL are sufficient to warrant further investigation into the biological properties of ABL-BCR.

Attempts to study the effect of expression of the BCR-ABL gene encoding the p210 protein under control of the BCR gene promoter in transgenic mice have been unsuccessful, as expression is apparently lethal to the developing embryo. The normal BCR gene is expressed during embryogenesis and, by implication, the chimeric BCR-ABL gene also would be expressed when under the control of the BCR promoter. The developing fetuses exhibited several obvious morphological abnormalities, but there was no evidence of neoplasia. Establishment of a transgenic mouse model for the $p210^{BCR\text{-}ABL}$ type of leukemias will probably require the use of a different promoter-enhancer combination. In contrast, lymphoblastic leukemias develop in mice transgenic for the chimeric BCR-ABL gene that produces the $p190^{BCR\text{-}ABL}$ protein under the control of the metallothionein promoter. The onset is fairly rapid, with overt leukemia or lymphoma detectable on average 70 to 80 days after birth. The leukemias are clonal, and fairly accurately reflect the phenotypes found in human Ph+ ALLs. These mice should provide a system in which to evaluate new treatments.

In CML, the t(9;22) is thought to occur in a pluripotent progenitor cell. Colony-forming assays of cells from patients with Ph+ ALL were carried out to determine whether the Ph chromosome could be detected in cells of other lineages, which would suggest the involvement of a prognitor cell in those leukemias as well. The results indicated that Ph+ ALLs can arise from either a pluripotent precursor or a lymphoid lineage committed cell. Too few cases have been studied in this manner to determine if the differentiation state of the progenitor cell is prognostic, but it might be expected that Ph+ ALLs arising from a pluripotent precursor would be more difficult to cure.

The production of a chimeric mRNA and protein in Ph+ ALL provides two potential targets for directed therapy, the junctional nucleotides in the mRNA by ribozymes and the junctional amino acids by the immune system. Several groups have produced ribozymes directed against BCR-ABL mRNA that can cleave the chimeric mRNA in a cell-free, in vitro system. If the ribozymes are to be effective against leukemic cells, a technique must be developed to deliver the ribozyme into cells in therapeutic quantities. A model in vitro system using a DNA-RNA ribozyme incorporated into liposomes has been used to introduce the ribozyme into a CML cell line, where it inhibited both expression of the BCR-ABL gene product, p210, and cell growth. Although these results are promising, the delivery of ribozymes into circulating leukemic cells in vivo presents a series of challenges. A potentially promising use of ribozymes ex vivo is to purge the marrow of Ph+ patients with ALL or CML who are in remission, for use in autologous transplants. In a preclinical study, a 1:1 mixture of Ph+ leukemia cells and normal bone marrow cells was exposed to a combination of low-dose mafosfamide and BCR-ABL antisense oligonucleotides and assayed for growth ability in clonogenic assays and in immunodeficient mice. BCR-ABL transcripts were not detected in residual colonies, and cytogenetic analysis of individual colonies revealed a normal karyotype. Normal, but not leukemic, hematopoietic colonies of human origin were also detected in marrows of immunodeficient mice one month after injection of the treated cells. These ex vivo results are encouraging. The drug and ribozyme combination appeared highly specific for the leukemic cells and produced minimal toxicity in normal cells. Bone marrow transplant during the first remission is the treatment of choice for Ph+ ALL at many centers, but matched related donors are rare. Effective removal of residual leukemic cells from autologous marrow could potentially improve cure rates.

The second tumor-specific therapeutic target is the fusion peptide unique to the BCR-ABL protein. Antibodies to this fusion peptide have been produced, but little has been done to test the feasibility of priming the immune system to react with this peptide and to destroy the malignant cells. Further, whether this peptide is displayed on leukemic cells in such a way that it could be recognized by cytotoxic T cells is not known. While theoretically presenting a tumor-specific target for both B-and T-cell-mediated immunity, there are no data to indicate that the fusion peptide can be used to target Ph+ leukemias.

MLL/ALL-1/HRX: Translocations Containing 11q23 Perhaps the most promiscuous site in the genome for chromosomal translocations is on chromosome 11, band q23. Translocations have been identified involving this locus and at least 13 other sites in the genome including 1q32, 4q21, 6q27, 7p15, 9p21-24, 15q15, 16p13, and 19p13. The breakpoints on 11q23 in these translocations fall within the MLL gene, which has also been called ALL-1, HRX, and Htrx-1. The 11q23 breakpoint is frequently associated with four subtypes of acute leukemias: infant ALL (70%), infant AML (60%), secondary AML (50%), and monoblastic AML in young children (90%). Overall, translocations involving 11q23 are found in about 6% of the cases of ALL and AML in older children and adults. A chromosomal translocation involving 11q23 and 14q32 that is found in certain types of lymphomas has an 11q23 breakpoint that lies at least 110 kb outside of the MLL gene and probably does not involve the MLL gene.

Translocations into the MLL gene generate chimeric mRNAs and proteins, regardless of the origin of the partner chromosome. Virtually all of the breaks within the MLL gene have been found to occur between exons 5 and 11, an area spanning about 8.3 kb. Because the normal MLL gene spans approximately 100 kb and contains at least 21 exons, the translocations result in fusion proteins containing only the amino-terminal one-fourth to one-half of the MLL protein. The breakpoints on the partner chromosomes also are somewhat clustered, suggesting that specific fusion genes are required for cell transformation to occur. The mechanism responsible for translocations into the 11q23 locus is not known. Sequences around the breakpoints in some 11q23 translocations have been reported to have heptamer-like and nonamer-like regions potentially recognized by the VDJ recombinase, and nongermline sequences indicative of Ig-type and TCR-type rearrangements are found at the junction in some cases. However, since 11q23-containing translocations are also found in myeloid leukemias with no evidence of recombinase activity, this potential mechanism would seem to be limited to lymphoid cells. Chi-like elements, known to be involved in recombination in prokaryotes, have also been observed around the breakpoints. A consensus site for topoisomerase II binding and cleavage has been identified near the breakpoint in a patient with the t(9;11). Treatment with topoisomerase II–binding drugs has been implicated in the pathogenesis of secondary AMLs containing 11q23 translocations, and the presence of these sequences near the breakpoint further implicates these drugs as causative agents for secondary malignancies. Whether one or more of these possible mechanisms is responsible for generating the 11q23-containing translocations will have to await further study.

The pattern of expression of the MLL gene has not been fully explored, but multiple, large mRNA species, from 11 to 15 kb, can be detected at low levels in normal hematopoietic cells. Occasionally, much smaller transcripts, as short as 1.5 kb, can also be detected. Because it is difficult to identify the chimeric mRNAs present in cells with the 11q23 translocation against this background of normal transcripts, much of the most informative data regarding the structure of the chimeric mRNAs has been obtained using RT-PCR. The picture that emerges is one of variability among the structures of the mRNAs owing to differences in the exons involved in the breakpoints in both chromosomes. For example, breakpoints within the MLL gene occur in front of exons 6, 7, and 8 in most 11q23-containing translocations, but those preceding exons 6 and 7 seem to predominate in the t(4;11), and those in front of exon 8 in the t(11;19). Breakpoints in the t(9;11), which is associated with AML-M5 leukemias, also occur between exons 4 and 9 of the MLL gene. Thus far, no clear association between the phenotype of the leukemic cells and the location of the breakpoint has been found. This is best illustrated by studies of t(11;19), where breakpoints within the same introns in both genes were found in both B and T lymphoid, as well as myeloid leukemias. Furthermore, no association between age and a particular junction sequence has been identified, nor do the breakpoints correlate with the differentiation stage of the leukemia, although, in general, leukemias with 11q23-containing translocations tend to be less differentiated than acute leukemias.

The breakpoints on the partner chromosomes are also heterogeneous; thus far, three different AF4 sequences have been found at the junctions in chimeric t(4;11) mRNAs, and two ENL (or LTG19) junctions have been found in cases with the t(11;19). Sequences encoded within the first four to six exons of the MLL gene, which contain a minor groove DNA-binding motif (the so-called AT-hook motif) and the carboxy-terminal regions of the fusion partners, contribute to the function of the oncogenic fusion proteins. For the t(4;11), the region of the AF4 protein from codon 491 (containing nuclear localization and GTP-binding motifs) to the carboxy-terminus is considered to be crucial for oncogenesis.

The normal MLL gene encodes a protein of more than 3910 amino acids and contains three regions of seguence homology to the *Drosophila* trithorax gene, including a cysteine-rich region able to fold into six zinc-finger-like domains, a motif with homology to the zinc-binding domain of a DNA methyltransferase (cytosine-5 methyltransferase), and the AT-hook motif of high-mobility-group proteins. Wild-type AF-4 transcripts, expressed in all cell lines tested as a major and a minor transcript, are predicted to encode a 140-kDa basic protein, rich in prolines, serines, and charged amino acids. The AF-4 protein contains a nuclear localization signal and a consensus GTP-binding site. The region of the AFX1 gene from the t(X;11) translocation that has been sequenced codes for a serine/proline-rich region analogous to that of the ENL gene on chromosome 19. Thus, the genes at three of the breakpoints associated with 11q23 chromosomal translocations code for proteins containing similar motifs, implying that attachment of serine- and proline-rich regions to the minor groove binding domain of MLL is required to generate the oncogenic protein. The translocations also appear to have a deletion requirement of up to four of the six zinc-fingers of MLL, and apparently they must occur in cells at early stages of hematopoietic differentiation. It remains to be determined whether the generation of the chimeric proteins results in gain or loss of function for the MLL protein.

Knowledge of the molecular changes brought about by 11q23-containing translocations has led to the design of PCR primers for leukemic cell analysis. In one study of infant ALL, about half of the patients had cytogenetic abnormalities involving 11q23, but two-thirds were shown by RT-PCR to have rearrangements of 11q23. Infants with no molecular evidence of 11q23 abnormalities had event-free survival rates comparable to those of patients with standard-risk ALLs, while only 15% of those with such abnormalities were in complete continuous remission at 46 months. Patients with 11q23 changes had other adverse prognostic factors, so the relative contribution of this abnormality to the poor treatment outcome is not known. Residual disease studies in patients with 11q23-containing translocations using breakpoint primers should provide significant insight into the dynamics of the leukemic cell population during treatment, possibly aiding in the design of new therapeutic regimens for this poor prognosis subgroup.

The normal function of the human MLL gene is unknown, and it remains to be determined whether the chimeric proteins result in functional loss or gain for the MLL protein. MLL's *Drosophila* homologue, trithorax (*trx*), regulates expression of specific genes expressed during the latter stages of *Drosophila* differentiation. The *trx* gene is one of several positive regulators of two homeobox-containing genes. Genetic data suggest that it probably interacts with brahma, which, as judged by its primary sequence, is probably a member of the DNA helicase family of proteins. The particular group of DNA helicases to which it belongs assist in the activation of transcription by helping to overcome the repressive effects of chromatin. Thus, trx (MLL) and brahma (mammalian homologues have been isolated) could associate in protein complexes that function to alleviate chromatin repression, thereby facilitating transcription. While all of this is speculation based on work in nonvertebrate systems, such studies pave the way for work in human systems to understand the mechanisms by which 11q23-containing translocations disrupt normal cellular growth and differentiation.

MOUSE MODELS OF HUMAN LEUKEMIAS

Virtually every gene implicated in the development of human lymphoid leukemias has been identified by its presence near the breakpoint of a nonrandom chromosomal translocation, and most of those present in nonrandom translocations have been cloned. Most ALLs studied thus far have only had a single genetic change identified, and, given the multistep process of cellular transformation, other genes are clearly involved. New approaches will be needed to identify other genes involved in leukemogenesis. One method widely used in solid tumors, probing for loss of heterozygosity (LOH), has not been widely applied to leukemias as yet, and may identify other genes involved in leukemia. A rather different approach is to identify cooperating genes in mouse models of human leukemias, then use these genes to search human leukemias for alterations in their structure or expression. In this approach, constructs mimicking chromosomal translocations are introduced into transgenic mice, and genes that will cooperate with the transgene to increase the frequency and decrease the latency of leukemias are identified by insertional mutagenesis. Several genes capable of cooperating with genes known to be activated in human leukemias have been identified by this technique, and these will be briefly described.

Transgenic Mice: Enforced Expression of Genes Implicated in Human Leukemias

Transgenic mice carrying genes that mimic chromosomal translocations found in leukemias usually develop hematopoietic abnormalities only after long latent periods, suggesting the requirement for other mutations. One approach to identifying these other genes has been insertional mutagenesis using murine leukemia viruses (MuLV). Although they do not carry oncogenic sequences in their genomes, these viruses can integrate adjacent to cellular genes, thereby altering the regulated expression of those genes. When the target gene is important for cell cycle control, either directly or indirectly, integration can lead to transformation. In the transgenic mouse system, integration of proviruses would serve to identify genes that can cooperate with the transgene. Identification of the genes present at common integration sites indicates their relevance in tumorigenesis.

When the *myc* gene, under the control of the Ig heavy chain gene enhancer (Eμ-*myc*), is introduced into the germline of mice, hematopoietic tumors develop, usually after 6 months of age. Infection of transgenic, neonatal mice with MuLV shortens the latency period. From these mice several common integration sites have been identified. In one study, MuLV integrations upstream of the *pim*-1 and *bmi*-1 genes were each found in about 35% of the mice, the *pal*-1 gene in 28%, and the *bla*-1 gene in 14%. The monoclonal leukemias that developed contained MuLV insertions into more than one site in a small percentage of the animals. MuLV insertions into both the *pim*-1 and *bmi*-1 loci, and both the *pal*-1 and *pim*-1 genes were found in a few animals. Certain combinations were never found, indicating that some of the genes may belong to the same complementation group. A similar approach led to the identification of the *emi*-1 gene, which appears to be a nontranscribed locus. Multiple MuLV integrations were also found in a number of monoclonal tumors in these studies, including combinations of *bmi*-1 plus *pim*-1, and *emi*-1 plus either *pim*-1 or *pim*-2. The *pal*-1, *fat*-1, *tia*-1, and *tic*-1 genes are also activated by MuLV, but at lower frequencies than the previously mentioned genes.

PIM 1 The human PIM-1 gene, which contains six exons, is located on chromosome 6 at band p21, near both the breakpoint for the t(6;9) chromosomal translocation of CML and the site of 6p deletions in T cell lymphomas. While the effects of the 6p breakpoint or the deletion on the expression of the PIM-1 gene have not been analyzed, normal expression of the gene has been studied in some detail. Transcripts of PIM-1 are abundant in a subset of human myeloid leukemias and B cell lymphomas, but are not usually found in T cell or nonhematopoietic cell lines. The PIM-1 protein, a cytoplasmic protein with serine kinase activity, is capable of autophosphorylation; it cannot be assigned to any of the previously identified families of protein kinases.

When the transgenic mouse experiments were repeated with the resident transgene being Eμ-*pim*-1, MuLV insertions near the c-*myc*, n-*myc*, and *bmi*-1 loci led to rapid-onset leukemias, essentially confirming the data from the Eμ-*myc* mice. When mice were made doubly transgenic for Eμ-*pim*-1 and Eμ-*myc*, pre-B-cell leukemias developed in utero at 17 to 19 days of gestation. Thus, changes in other genes probably were not essential components of the leukemic process. However, it is likely that further mutations do occur, as the leukemias are initiated as polyclonal pre-B-cell proliferations that become monoclonal owing to selection for the fastest growing clone. Thus, enforced expression of *myc* and *pim*-1 in early hematopoietic cell progenitors leads to a loss of differentiation potential, and the acquisition of significant proliferation potential.

Both copies of the *pim*-1 gene can be deleted from the genome with little apparent effect on the development or growth of mice. An exception is that bone-marrow-derived mast cells lacking *pim*-1 do not grow as well as mast cells from wild-type mice when IL-3 is present, but have a proliferative potential equal to that of *pim*-1-expressing cells when stimulated by IL-4, IL-9, or the c-*kit* ligand. It is of interest that while *pim*-1 may be involved in IL-3 signal transduction, its expression appears to be regulated by IL-3 (and IL-2).

Thus, in this mouse model system, activation of both *pim*-1 and *myc* leads to the rapid proliferation of B cell progenitors and the development of frank leukemias. Although the pattern of PIM-1 expression in hematopoietic tumors has been examined, a link between its inappropriate expression and human leukemias has yet to be established. Its ability to cooperate with *myc* in the transgenic mouse models indicates that Burkitt's lymphoma would be the malignancy most likely to have activated PIM-1.

BMI-1 Activation of *bmi*-1 by MuLV integrations in Eμ-*myc* transgenic mice is frequent, and results in the development of pre-B-cell lymphomas. The transforming potential of *bmi*-1 appears to be limited to B cells within the lymphoid compartment, as it has never been found to be activated by integration of MuLV in T cell lymphomas in either normal or Eμ-*pim*-1 transgenic mice.

The human BMI-1 protein shares over 98% homology with the mouse protein, and nucleotide sequences in the 5′ and 3′ untranslated regions are over 80% homologous. Like PIM-1, BMI-1 maps to a chromosomal region involved in translocations associated with leukemia, 10p13. A strong association exists between the presence of chromosomal abnormalities involving 10p13 and translocations involving 11q23. In one study, 6 of 56 infants with leukemia had translocations involving 10p13, and half of these also had 11q23 translocations. Like 11q23-containing translocations, 10p13 breakpoints are more frequent in infants, as a study of childhood ALL showed that only approximately 1% showed 10p13 breakpoints. Enticing as these associations are, there are no published data demonstrating an effect of the 10p13 translocations on either the structure or expression of the BMI-1 gene.

The murine *bmi*-1 gene sequence has an open reading frame coding for a protein of 324 amino acids, and a protein of 45 to 47 kDa can be immunoprecipitated from cells with an antibody to *bmi*-1. Bmi-1 mRNA can be detected in most organs, with the highest expression levels being found in the thymus, heart, brain, and testis. The protein has many hallmarks of a transcription factor: 1) a nuclear localization motif and a nuclear location, 2) a novel, putative zinc-finger region similar to that found in other proteins involved in gene regulation, as well as DNA recombination and repair, 3) acidic and basic domains, 4) a putative helix-turn-helix motif, and 5) a carboxy-terminal PEST sequence. None of these motifs would enable the *bmi*-1 to interact directly with *myc*, so these two proteins are currently thought to act at different steps in growth signaling pathways.

There is a gene immediately upstream of the murine *bmi*-1 gene called *bup*. The genes are in the same transcriptional orientation, so they could share transcriptional control mechanisms, and could even possibly form chimeric polypeptides. However, the levels of *bup* mRNA in cells with MuLV inserted within the *bim*-1 locus are not affected in the same manner as *bim*-1 mRNA levels are. There are presently no other data regarding the normal patterns of expression of *bup* in humans or mice, and there are no data to indicate whether or not either of these genes is involved in human leukemias. Nevertheless, their chromosomal location and their potential association with infant leukemias and with the 11q23 translocation make them candidates for transforming genes in ALL.

Insertional Mutagenesis Leading to Leukemia or Lymphoma

Insertional mutagenesis by MuLV in normal mice has been used to identify genes involved in murine leukemias and lymphomas. More than 25 genes have been identified as being consistently activated by insertion of retroviruses into hematopoietic cells, with the activated genes driving the cells to become monoclonal malignancies. Five genes, *ahi*-1, *bla*-1, *bmi*-1, *mlvi*-1, and *pal*-1, are consistently associated with B lineage leukemias and lymphomas, while *dsi*-1, *gfi*-1, *gin*-1, *mlvi*-2, *mlvi*-3, *mlvi*-4, *pim*-1, *pim*-2, *tpl*-1, *tpl*-2, and *vin*-1 are found to be activated in T cell leukemias and lymphomas. Several of the genes in this group also cooperate with *myc* in the transgenic mouse experiments described in the previous section, providing two separate lines of experimentation supporting their potential importance in leukemia. Analysis of their biological functions should provide important insights into how retrovirally activated genes contribute to the leukemic phenotype. Because these genes are not found at chromosomal translocation breakpoints in human leukemias, their involvement is more likely to be as a result of point mutation than by enforced or inappropriate expression.

Are Protein Kinases Involved in Leukemia?

Many protein kinases have been identified, and a number of them are important in normal hematopoiesis. One member of the extended *src* family of tyrosine kinases, *lck*, has been identified at the junction of a chromosomal translocation, but this translocation is rare. The ABL tyrosine kinase is activated in Ph+ ALLs and CML, but no other kinase gene has been associated with the development of leukemia, despite their critical contribution to hematopoiesis. The *blk*, *lyn*, and *hck* kinases are expressed in differentiating B cells, and their expression generally increases with maturity. Similarly, *lck* and *fyn* are expressed in T cells, and critical roles for several of these *src* kinase family members in hematopoiesis have been demonstrated through the use of knock-out mice. One potential explanation for this is that where expression has been examined in tumors, it is generally low. Perhaps these kinases are involved in promoting differentiation and not proliferation. This would imply that an involvement in leukemia would require a loss of activity, perhaps through mutation, rather than enforced expression or gain of function. Given their critical role in hematopoiesis, it is likely that disruption of their activity, or the signaling pathways in which they function, will eventually be found to be important in leukemogenesis.

CONCLUSION

Tumor-specific chromosomal abnormalities are somatically acquired alterations that have helped to identify the locations of genes important in the development of acute leukemias. The vast majority of the breakpoints of nonrandom chromosomal abnormalities have been cloned, and the genes at the breakpoints have been identified. Overall, two general mechanisms are involved in altering the expression and function of breakpoint genes: enforced expression in inappropriate cell types or disregulation of controlled expression, and loss or gain of function owing to the production of chimeric proteins. Several chromosomal abnormalities found in B lineage leukemias are associated with particular phenotypes, but no such association is apparent with T lineage leukemias. The multiplicity of proteins with different potential functions and subcellular locations that are affected by chromosomal translocations points to the involvement of several signal transduction pathways in the development of leukemia.

Leukemias are probably initiated with fewer genetic lesions than most solid tumors. Transgenic mouse studies indicate that frank neoplasms arise with the enforced expression of as few as two protooncogenes. By contrast, changes in two of the six genes involved in colon cancers lead to only benign proliferations. This may reflect differences between the life cycles of epithelial and lymphoid cells. Epithelial cells begin differentiation and lose all proliferation potential. The cell probably accomplishes this through the activation of a number of pathways to inhibit future growth, and each of these pathways must be circumvented for benign and malignant proliferation to occur. In contrast, differentiating B and T cells are destined to proliferate again following antigen stimulation. The block to proliferation during differentiation must therefore be reversed, so it is likely that lymphoid cells have to alter fewer signal pathways to recommence proliferation. Thus, in leukemias with chromosomal translocations, only one other genetic event may be necessary for malignant transformation; other mutations (such as the p53 and RB) may occur during outgrowth, but these mutations are associated with progression rather than initiation.

The transgenic mouse studies implicate other genes that might be involved in human leukemias, but some caution must be used in interpreting these results. Most of the studies involved the identification of genes that would cooperate with c-*myc* in transformation, but except for Burkitt's lymphoma, c-*myc* has not been consistently connected with any other form of leukemia or lymphoma. These studies also demonstrated that only certain combinations of genes would cooperate in transforming lymphoid cells; therefore, whether any of the mouse genes identified so far would cooperate with genes such as rhombotin or TAL-1 is unknown. Many genes potentially involved in murine leukemia and lymphoma have been identified by insertional mutagenesis, but thus far they have not been implicated in human leukemias.

Thus, while the number of genes associated with human leukemias is larger than for any other form of cancer, we do not know which combinations of mutations bring about any single type of leukemia. Because of the mouse studies there is no dearth of candidate genes, and the next few years should yield the identity of genes that cooperate with those involved in cytogenetic abnormalities to disrupt normal hematopoiesis and bring about the uncontrolled growth characteristic of leukemias.

BIBLIOGRAPHY

Askew DS, Bartholomew C, Ihle JN. Insertional mutagenesis and the transformation of hematopoietic stem cells. Hematol Pathol 1993;7:1–22.

Hirsch-Ginsberg C, Huh YO, Kagan J, Liang JC, Stass SA. Advances in the diagnosis of acute leukemia. Hematol Oncol Clin North Am 1993;7.

Pui C-H, Behm FG, Crist WM. Clinical and biologic relevance of immunologic marker studies in childhood acute lymphoblastic leukemia. Blood 1993;82:343–362.

Varmus HE, Lowell CA. Cancer genes and hematopoiesis (perspective). Blood 1994;83:5–9.

CHAPTER 10

Clinical Implications of Molecular Events in Lung Cancer

David S. Schrump
Jack A. Roth

Tobacco consumption is believed to be responsible for more than 90% of lung cancers in men and approximately 80% in women. Presently, the relative risk of lung cancer is 20 times greater for men and 10 times greater for women who are smokers than for nonsmokers, and is proportional to the cumulative amount of tobacco exposure. However, the fact that only 15% of smokers develop lung cancer implies that other genetic or environmental factors contribute to the pathogenesis of this neoplasm.

Epidemiologic studies have demonstrated a familial risk of lung cancer, and suggest that the development of this disease at an early age (before age 50) is related to mendelian inheritance of a rare autosomal codominant allele, although this phenomenon has been observed less frequently in more typical elderly patients. Despite intense efforts to identify polymorphisms involving protooncogenes, tumor suppressor genes, or genes encoding enzymes involved in aberrant metabolism of chemical carcinogens, consistent correlations between particular gene polymorphisms and lung cancer risk have not been demonstrated.

Traditionally lung cancer has been divided into two major groups on the basis of morphological, biochemical, and clinical criteria. Small cell lung cancer, which comprises approximately 25% of all lung carcinomas, is characterized by aggressive proliferation and disseminated disease at the time of presentation. Treatment generally consists of chemotherapy and radiation therapy, although surgery once again is being advocated as part of aggressive multimodality regimens. Although small cell lung cancer is exquisitely sensitive to chemotherapy or radiation therapy, the vast majority of patients with this neoplasm succumb to recurrent metastatic disease within 2 years of diagnosis.

Non-small cell lung cancer, comprising squamous cell, adeno, and large cell carcinomas, constitutes the major portion of the remaining 75% of lung cancers. These tumors tend to be locally invasive; therefore surgery remains an important component of therapy. Generally non-small cell lung cancers tend to be resistant to both chemotherapy and radiation, and because the majority of patients with non-small cell lung cancer have mediastinal metastases at the time of presentation, overall survival for this group is approximately 15%.

The fact that lung cancer mortality has remained relatively unchanged despite aggressive medical efforts underscores the magnitude of the lung cancer problem. Appreciation of the molecular events corresponding to malignant transformation in the tracheobronchial tree may improve the diagnosis and clinical management of patients with advanced disease, as well as providing an experimental foundation on which to base preventive interventions in high-risk patients. This chapter will emphasize aspects of the molecular biology of lung cancer that may be of clinical significance in the near future.

GROWTH FACTORS AND GROWTH FACTOR RECEPTORS

Lung cancers have been associated with abnormal expression of a variety of growth factors and growth factor receptors that may be relevant to the biology and treatment of these neoplasms (see Table 10.1). Growth factors secreted by tumor cells may influence distant cells (endocrine stimulation) or adjacent cells (paracrine stimulation), or they may stimulate the cell from which they have been secreted by autocrine mechanisms. Cells that replicate independently of growth factor support are termed acrine.

Gastrin-Releasing Peptide

Small cell lung cancers secrete a number of neuropeptides and hematopoietic factors of which gastrin-releasing peptide (GRP) appears most signifi-

Table 10.1: Growth Factors, Oncogenes, and Tumor Suppressor Genes Associated with Lung Cancer

Growth Factors	*Oncogenes*	*Tumor Suppressor Genes*
GRP	myc	3p
EGF	ras	PTP-γ
c-erbB2		Rb
PDGF		p53
IGF		

cant. GRP is a 27 amino acid homologue of an amphibian hormone referred to as bombesin, which has been found in neural, bronchial endocrine, and fetal lung tissues. High-affinity GRP receptors are present on small cell, but are absent on non-small cell lung cancer cells. Classic, but not variant, small cell lung cancer lines secrete GRP that is mitogenic for small cell lung cancers as well as normal human bronchial epithelial cells. In vitro proliferation and tumorigenicity of small cell lung cancer can be inhibited by antibombesin monoclonal antibodies or by pharmacological antagonists of bombesin, thereby implicating GRP in autocrine-mediated growth of small cell lung cancer.

Epidermal Growth Factor Receptor

The epidermal growth factor (EGF) receptor is a 170-kDa tyrosine kinase glycoprotein consisting of extracellular, transmembrane, and intracellular domains. A truncated sequence containing of the membrane and cytoplasmic components is the oncogene product carried by the avian erythroblastosis virus. Activation of the EGF receptor by ligand binding results in receptor dimerization, autophosphorylation, and subsequent activation of early response genes leading to cell proliferation. Overexpression of normal EGF receptor is sufficient to transform NIH3T3 cells.

EGF receptor overexpression has been identified in cell lines and specimens derived from primary lung cancers, as well as adjacent normal bronchial epithelia. Although the precise mechanism of EGF receptor overexpression in lung cancer is not clear, gene amplification, overexpression of a normal EGF receptor gene, as well as expression of a mutant EGF receptor protein have been cited in studies involving cell lines or specimens from primary non-small cell lung cancers.

Non-small cell lung cancers containing elevated levels of EGF receptor do not synthesize EGF, but rather produce transforming growth factor-α (TGF-α), which is structurally related to EGF, and can bind to EGF receptor as free ligand or as membrane-bound pro-TGF-α. Autocrine stimulation of EGF receptor by TGF-α may occur at the cell surface as well as intracellularly via reaction between unsecreted ligand and unprocessed EGF receptor. Although high-affinity EGF receptors have recently been identified on small cell lung cancer lines, the number of these receptors per cell is less than recorded for non-small cell lung cancers, and the implication of these observations for autocrine growth stimulation of small cell lung cancers is unclear.

Analysis of a large number of lung cancer and normal lung specimens has revealed that relative to normal tissues, approximately 45% of cancers overexpress EGF receptor, and 60% of tumors overexpress TGF-α; no EGF has been detected in normal or cancerous lung tissues. EGF receptor overexpression has been observed in adenocarcinomas as well as squamous cell cancers, and recent studies have revealed a statistically significant association between EGF receptor overexpression and diminished survival in non-small cell lung cancer patients.

ErbB2/neu

The c-erbB2/neu gene encodes a 185-kDa transmembrane tyrosine kinase receptor molecule structurally related to the EGF receptor which is present on normal ciliated epithelium, mucous cells, and type II pneumocytes of the adult lung. Small polypeptides designated heregulins appear to be the naturally occurring ligands for p185, stimulating proliferation of some, but not all, cells containing this receptor molecule. Overexpression of erbB2 can induce transformation of NIH3T3 cells, which can be suppressed by direct transcriptional repression of the neu gene by the Rb gene product.

Overexpression of erbB2 has been reported in squamous cell carcinomas and adenocarcinomas of the lung, although the precise mechanism responsible for this overexpression is unclear; gene amplification appears to be involved in only a small percentage of cancers. Increased immunoreactivity indicative of p185 overexpression has been detected in approximately 36% of squamous cell cancers, and 38% of adenocarcinomas relative to staining of normal tissues, and multivariate analysis has revealed that p185 expression is associated with reduced survival independent of tumor stage for adenocarcinomas; no such correlation has been documented for squamous cell cancers. The recent observation that topoisomerase II is coamplified with neu in some cancer cells, and the correlation of erbB2 expression with in vitro drug resistance substantiate the biological and clinical relevance of erbB2 in non-small cell lung cancers.

Platelet-Derived and Insulin-like Growth Factors and Their Respective Receptors

The interactions of platelet-derived growth factor (PDGF) receptor and insulin-like growth factor (IGF) receptor with their respective ligands are critical determinants of cell-cycle progression in mammalian cells. PDGF receptor occurs either as a homodimer or as a heterodimer of two structurally related proteins (α and β), whereas the PDGF ligand may exist as a homodimer or heterodimer of A and B chains. The v-sis oncogene product is structurally related to the B chain of PDGF, which can function as an autocrine growth factor in v-sis-transformed cells. Whereas the α receptor can bind all three isoforms of PDGF with high affinity, the β receptor binds PDGF BB isoform with high affinity, and the AB isoform with a lower affinity, but not the AA isoform. PDGF induces receptor dimerization and mitogenic stimulation of a variety of cell types.

Several studies have documented expression of PDGF, PDGF receptor, or both in lung cancer cell lines, and significantly elevated levels of PDGF have been detected in metastatic pleural effusions associated with adenocarcinomas, but not squamous cell lung cancers. Immunohistochemistry and in situ hybridization analysis of c-sis PDGF and PDGF-β receptor expression in lung cancer specimens has revealed that epithelial cells do not express PDGF ligand or receptor transcripts, whereas carcinoma cells express PDGF as well as PDGF receptors. C-sis/PDGF and PDGF-β receptor expression can be induced in epithelial tissues by trauma; although definitive data are lacking, expression of the c-sis protooncogene and PDGF-β receptor conceivably could be induced in bronchial epithelia by noxious stimuli, and deregulation of expression of these genes may be significant with regard to autocrine stimulation of preneoplastic and cancerous lung tissues.

Insulin-like growth factors I and II are structurally similar to proinsulin, and function in growth and metabolism of a variety of tissues. The IGF-I receptor is a tyrosine kinase similar to the insulin receptor, and mitogenic activities of IGF-I, IGF-II, and insulin in lung cancers appear to occur via these receptors. Activation of PDGF receptor induces obligatory quantitative and qualitative alterations in IGF-I receptors, thereby enabling them to transmit mitogenic signals following ligand interaction. Activation of the IGF-I receptor by its ligand appears to be a crucial initiator of cell cycle progression in quiescent cells, and constitutive expression of IGF-I receptor and IGF-I ligand abrogates exogenous growth factor requirements in tissue culture. IGF ligands and receptors have been identified in lung cancer lines, and enhanced expression of these molecules in lung cancers relative to normal lung tissues has been observed. Furthermore, IGF-I, IGF-II, or insulin-stimulated mitogenesis in lung cancer lines can be competitively inhibited by monoclonal antibodies directed against IGF-I receptors or ligands, thereby establishing the role of these molecules in autocrine-mediated growth of lung cancer cells.

DOMINANT ONCOGENES

Ras

The H-ras, K-ras, and N-ras genes are members of an evolutionarily conserved gene superfamily encoding 21-kDa proteins that are localized to the inner plasma membrane via farnesylation, and play a pivotal role in signal transduction from cell-surface receptors to early response genes involved in mitogen-induced proliferation. Ras proteins exhibit guanosine triphosphate (GTP) binding and intrinsic GTPase activities. Inactive ras binds guanosine diphosphate (GDP); activation of ras by the GRB2/SOS complex involves GTP-GDP exchange and conformational alterations, with subsequent activation of raf and other downstream effector molecules. Intrinsic GTPase activity then hydrolyzes GTP to GDP, thereby restoring ras to its inactive conformation. Ras activation is normally tightly regulated by positive and negative control mechanisms. Mutations involving ras stabilize the activated conformation, thereby resulting in unabated growth stimulation.

Ras mutations are among the most common oncogene defects recorded in human cancers, and individual ras genes appear to be preferentially activated in tumors of different histologic types. Essentially all relevant mutations occur in codons 12, 13, and 61 of these genes. As a dominant oncogene, mutation in one ras allele is sufficient to induce transformation of mammalian cells either alone or in concert with other activated oncogenes; v-H-ras can transform normal human bronchial epithelial cells in culture.

The frequency of ras mutations in premalignant lung lesions has not been extensively evaluated. Furthermore, to date no ras mutations have been identified in small cell lung cancer. Interestingly, in small cell lung cancer lines containing amplified C-myc or N-myc (but not L-myc), v-H-ras causes enhanced soft agar growth, diminution of neuropeptide synthesis, and concomitant synthesis of EGF receptor, TGF-α, and PDGF, consistent with the induction of a non-small cell lung cancer phenotype. These results suggest that ras and myc may be involved in normal differentiation of lung epithelium.

Whereas H-, K-, and N-ras activations have been observed only sporadically in squamous, adenosquamous, or large cell cancers, K-ras mutations are relatively common in adenocarcinomas, particularly in patients with tobacco exposure. In a comprehensive study, Rodenhuis and Slebos analyzed 280 lung cancer specimens for ras activation using polymerase chain reaction (PCR) and oligonucleotide hybridization techniques. Of particular interest was the observation that 41 (30%) of 141 adenocarcinomas from patients who smoked had K-ras mutations, the majority of which were G-to-T transversions involving codon 12. In contrast, only 2 (5%) of 40 adenocarcinomas from nonsmokers had K-ras mutations. Other investigations have identified K-ras mutations in 21% to 57% of adenocarcinomas, confirming statistically significant associations between pulmonary adenocarcinomas and K-ras mutations, and between smoking and K-ras mutations. Although no correlation between K-ras mutations and clinical stage has been observed, patients with tumors containing K-ras mutations have a significantly lower survival rate than patients with adenocarcinomas of similar stage containing wild-type K-ras alleles.

In contrast to primary lung cancers, ras mutations appear to be relatively common in non-small cell lung cancer lines. K-ras mutations have been identified in 28%, 33%, and 69% of lines derived from adenocarcinomas, squamous cell cancers, or large cell cancers of the lung, respectively, and appear to coincide with mutations occurring in the primary tumors from which these lines have been derived. Interestingly, the majority of K-ras mutations involving codons 12 or 13 are G-to-T transversions, whereas mutations involving codon 61 are predominantly A-to-T transversions. Such G-to-T transversions in ras are observed in benzo(*a*)pyrene-induced tumors in mice, whereas A-to-T transversions can be experimentally induced by exposure to ethyl carbamate. Although the frequency of K-ras mutations in adenocarcinoma lines appears to correlate with observations regarding the frequency of these mutations in primary lung cancers, the increased frequency of mutations in squamous and large cell lung cancer lines relative to primary cancers may be the result of in vitro selection during establishment of these lines. Nevertheless, the presence of K-ras mutations in established cell lines connotes a statistically significant reduction in survival in the patients from whom these lines were established, regardless of treatment rendered.

Immunohistochemical analysis of lung cancers utilizing monoclonal reagents reactive with H-ras or K-ras, or both, has revealed increased p21 immunoreactivity in approximately 50% of squamous cell cancers, and 75% of adenocarcinomas, correlating in some instances with adverse patient survival regardless of tumor stage or resectability. Although provocative, these immunohistochemistry results are difficult to reconcile in light of previously discussed biochemical data concerning the frequency of ras activations in lung cancer, as well as observations that amplification of ras genes appears to be extremely uncommon in this disease. Whereas the immunostaining techniques may be detecting those tumors having increased activity of ras via constitutive cell surface mitogenic stimulation, other studies have revealed no correlation between ras expression and proliferation or metastatic potential of malignant cells; thus, the biochemical basis and clinical implications of this immunoreactivity await confirmation.

Ras proteins are synthesized as cytosolic precursor molecules that become localized to the inner plasma membrane following post-translational modification. An obligatory modification involves the addition of a farnesyl group to the carboxy-terminal region of the ras molecule in a reaction that is catalyzed by farnesyl protein transferase (FPTase). This step is critical for membrane localization and is essential for normal as well as mutant ras function. Pharmacologic inhibition of FTPase can suppress the growth and malignant phenotype of ras- and src-transformed cells but has no effect on cells transformed by v-raf or v-mos (two downstream effector molecules in the ras pathway). These exciting data suggest that ras can be selectively targeted utilizing pharmacologic reagents which are structurally related to medicines commonly used in clinical practice, and further investigation will in all likelihood yield agents that will be applicable to the clinical management of lung cancer as well as other cancers containing ras mutations.

Myc

The myc gene family consists of three closely related genes (C, L, and N) that are differentially expressed during mammalian development and human carcinogenesis. These genes encode DNA transcription factors containing helix-loop-helix and leucine zipper motifs. Heterodimerization with another helix-loop-helix, leucine zipper structure designated Max enhances DNA binding of myc and is essential with respect to the transformation activities of myc proteins.

Although the normal physiologic roles of the various myc genes have yet to be defined, C-myc is an early response gene which is critical with respect to stimulating movement of quiescent cells into and through the G_1 phase of the cell cycle; activation of myc appears to be sufficient to initiate DNA synthesis. Reduction of C-myc correlates with terminal differentiation in a variety of cell lines, and enforced expression of myc can prevent differentiation in these tissues. Interestingly, C-myc can induce apoptosis in several cell systems, which can be prevented by concomitant expression of bcl-2; the precise mechanism of myc-induced apoptosis remains obscure.

The myc genes consist of three exons, with the major portion of the open reading frame existing in exons two and three, whereas exon one is primarily a noncoding, regulatory sequence. The two major forms of C-myc share the same open reading frame in exons two and three; the larger peptide (67 kDa) results from initiation near the 3′ end of exon 1, whereas a smaller (64-kDa) polypeptide is produced by initiation of transcription within the 5′ region of exon 2. The N-myc polypeptides are produced by initiation from two separate sites within exon 2 of the N-myc gene. In contrast, L-myc protein diversity appears to arise primarily from differential phosphorylation of a single polypeptide, although a minor L-myc protein derived by alternative splicing has been reported.

Myc expression is predominantly deregulated in lung cancer by gene amplification; however, more subtle disturbances in transcriptional control may also be relevant. Nuclear run-off analysis of small cell lung cancer lines has revealed that myc expression can be regulated at the level of initiation (N-myc), attenuation of transcription (L-myc), or a combination of initiation and attenuation of transcription (C-myc). As such, messenger RNA levels may not necessarily correlate with gene amplification in small cell lung cancers.

Although no interchromosomal recombinations involving myc genes have been observed in lung cancers, L-myc has been associated with complex intrachromosomal rearrangements with the creation of a fusion gene involving L-myc and the RLF gene located approximately 800 kb proximal to L-myc. It has been theorized that these rearrangements may deregulate L-myc and may actually precede DNA amplification in those tumors overexpressing this gene.

Aberrant myc expression has been documented primarily in specimens or cell lines derived from small cell lung cancers. Initial studies utilizing Southern analysis of EcoRI digests detected amplification of C-myc in 7 of 13 small cell lung cancer lines, 5 of which were small cell lung cancer variant cell lines. Amplification of C-myc correlated with the presence of double minutes or homogeneously staining regions, and was associated with correspondingly increased levels of C-myc transcripts in these variant cell lines. In contrast, only two of eight classic small cell lung cancer lines and one of five non-small cell lung cancer lines exhibited either amplification of C-myc DNA or increased levels of C-myc mRNA.

Relative to classic small cell lung cancer lines, variant lines have diminished levels of dopa-decarboxylase and bombesin, yet retain elevated levels of CPK-BB and neuron-specific enolase, and exhibit reduced doubling times and increased resistance to radiation. Transfection of C-myc DNA into classic small cell lines has resulted in cells having a large cell phenotype with reduced doubling times yet no significant alteration of dopa-decarboxylase or bombesin levels.

Additional EcoRI bands hybridizing with C-myc probes during initial studies were identified as N-myc previously detected in neuroblastomas, as well as L-myc, which appeared to be a novel myc gene restricted to small cell lung cancers. Amplification of N-myc has been reported in approximately 20% of small cell lung cancer lines, and L-myc or overexpression has been detected with similar frequency in a subset of small cell lung cancer lines lacking either C-myc or N-myc amplification.

In general, different myc genes are not coexpressed within a given tumor, and rather than correlating with histologic subtype per se, myc gene amplification or overexpression appears to be more associated with exposure to chemotherapy. Comprehensive analysis has revealed myc amplification in approximately 10% of cell lines or specimens obtained from patients before chemotherapy compared with 33% of cell lines or tumors obtained from patients after treatment. Although discordant myc expression in cell lines relative to cancer specimens has occasionally been observed, myc abnormalities in cell lines tend to correlate precisely with those observed in primary tumors. In addition, myc amplification tends to occur more frequently in those patients receiving cyclophosphamide-based regimens as opposed to etoposide-cisplatin therapy. Whereas those patients with c-myc amplification have statistically significant reduction in survival relative to those patients without C-myc amplification, no such prognostic value has been ascribed to L- or N-myc abnormalities in small cell lung cancer.

A paucity of data exists concerning expression of myc genes in non-small cell lung cancers. Although analysis has been limited, amplification of L-myc has not been detected in non-small cell lung cancer, and N-myc has been only sporadically observed in adenocarcinoma cell lines. Elevated C-myc expression was observed in four of five large cell carcinoma lines, and 8 of 106 non-small cell lung cancer specimens (5 adenocarcinomas and 3 squamous cell cancers). The relation between C-myc overexpression and chemotherapy exposure and the biological significance of this overexpression in non-small cell lung cancer are unclear at present.

Although the precise mechanisms remain obscure, available data suggest that myc amplification occurs relatively late in the course of carcinogenesis and enhances tumor progression and metastasis. Previously reported associations between L-myc polymorphisms and metastatic potential of primary lung cancer have not been confirmed. However, the observation that the nm23-H2 tumor suppressor gene product can regulate C-myc transcription, and previously described interactions between myc and Rb proteins suggest multiple mechanisms by which mutations involving tumor suppressor genes may deregulate myc gene expression, thereby conferring upon the malignant cell a more aggressive, metastatic phenotype.

TUMOR SUPPRESSOR GENES

3p

Cytogenetic analysis has confirmed that lung cancers are associated with multiple genetic alterations. The majority of these neoplasms are aneuploid with complex karyotypes; however, nonrandom chromosomal abnormalities have been observed in primary lung cancer specimens as well as irradiated bronchial epithelial cells, suggesting that these events are causally related to lung carcinogenesis. In addition to trisomy seven, commonly detected abnormalities include 1p, 3p, 5q, 7q, 9p, 11p, 13q, and 17p.

Deletions involving 3p have been detected in nearly 100% of small cell lung cancers and more than 50% of non-small cell lung carcinomas, as well as a variety of other neoplasms, in particular renal cell cancers. Whereas the region of deletion in non-small cell lung cancer appears to be 3p21, deletions in small cell lung cancer have been observed in 3p14-cen, 3p21.3, and 3p25. Although a number of genes including c-erbA-β (thyroid hormone receptor), the RARβ retinoic acid receptor, as well as an unknown gene designated DNF15S2, map to these regions, homozygous mutations of these genes in small cell lung cancers have not been observed. The phosphotyrosine phosphatase-γ gene (PTP-γ) also has been localized to 3p21 using somatic cell hybrids and Southern analysis, and allelic loss of PTP-γ has been detected in 5 of 10 non-small cell lung cancer specimens; however, definitive proof that PTP-γ is the gene targeted in 3p deletions is lacking at present.

Killary and colleagues generated interspecific microcell hybrids using the A9 murine fibrosarcoma line and portions of 3p and assayed the tumorigenicity of these hybrids in nude mice. Hybrid clones containing a 2-megabase fragment comprising a region of 3p21-22 were observed to have dramatically reduced tumorigenicities. Additional studies have revealed a homozygous deletion of 3p in a small cell lung cancer line corresponding to the 3p21-22 fragment that demonstrated tumor suppressor activity in somatic cell hybrids. Other investigations involving analysis of small cell lung cancer lines using cosmid markers located at 3p21.3-p22 have identified homozygous deletions estimated to be less than 1 megabase in length. It is anticipated that these efforts will ultimately result in the isolation of a tumor suppressor gene which thus far has eluded identification.

The 3p14.2 region is the most sensitive common fragile site in the human genome. Recently this breakpoint has been cloned, and a novel gene designated HRCA1 (hereditary renal cancer associated 1) mapping immediately proximal to the breakpoint has been described. The characterization of this putative tumor suppressor gene and delineation of its significance in lung cancers await further investigation.

Rb

Isolation of the Rb gene in 1986 enabled the elucidation of the molecular mechanisms predisposing to childhood retinoblastoma as originally theorized by Knudson and substantiated by cytogenetic analysis of Rb tumors.

The Rb gene is located on 13q14 and consists of 27 exons spanning 180 kb which encode a 105-kDa nuclear phosphoprotein that is intimately involved in regulation of the G_1/S cell cycle checkpoint. The Rb protein is differentially phosphorylated during cell cycle progression and has complex interactions with a variety of regulatory proteins including cyclin-dependent kinases, transcription factors, and cellular as well as viral oncoproteins. The association of Rb with the E2F transcription factor inhibits transcription of genes required for S phase, and viral proteins bind to Rb, thereby dissociating Rb from E2F, thus permitting transcription in virally infected cells. Microinjection of unphosphorylated (active) Rb protein induces a reversible G_1 arrest that can be abolished with concomitant injections of simian virus 40 T antigen, and wild-type Rb can diminish the proliferation and tumorigenicity of cancer cells bearing these mutations.

In contrast to persons surviving the inheritable form of retinoblastoma, who have an increased risk of osteosarcomas, relatives who carry Rb mutations have a 15-fold increase in lung cancer risk compared with normal individuals. Initial Southern analysis of Rb gene expression in small cell lung cancer specimens and cell lines revealed mutations in approximately 12% of small cell lung cancer specimens, and 18% of small cell lung cancer lines. Gross structural abnormalities detected by Southern analysis correlated with genetic aberrations involving 13q14 in the small cell lung cancer lines examined. Northern analysis of these cell lines revealed that only 23% had normal levels of Rb mRNA; 58% of the cell lines had no detectable Rb transcripts, whereas the remainder had trace levels of Rb mRNA. In contrast, evaluation of non-small cell lung cancer lines demonstrated no Rb gene abnormalities by Southern analysis, whereas 20% had either diminished or absent Rb transcripts.

Other studies revealed that despite apparently normal Rb transcripts as assessed by Northern analysis, 55% of small cell lung cancer lines had no detectable Rb protein and another 33% had Rb proteins with altered migration patterns. Subsequent investigation of two small cell lung cancer lines containing aberrant Rb proteins disclosed point mutations creating or disrupting existing splice donor sites with deletions involving exons encoding regions involved in binding the E1A and SV40 oncoproteins. Furthermore, point mutations resulting in an unphosphorylated protein product incapable of binding SV40T and E1A proteins have been reported, thereby demonstrating that subtle mutations involving this gene can have pronounced effects on the biological activity of the Rb protein.

Available data confirm that the majority of small cell lung cancers harbor abnormal Rb transcripts and lack expression of Rb protein. Furthermore, whereas Northern analysis has detected altered Rb mRNA in

10% of non-small cell lung cancers, immunohistochemistry has demonstrated absent Rb protein expression in approximately 32% of these carcinomas, involving 19% of adenocarcinomas, 38% of squamous cell carcinomas, and 60% of large cell cancers. Although Rb expression relative to clinical stage has been observed in some (but not all) studies, no significant correlation between Rb inactivation and clinical outcome has been reported; thus the implications of Rb mutations with respect to the clinical course of lung cancer await further investigation.

p53

The p53 gene encodes a nuclear phosphoprotein that was originally identified as a 53-kDa polypeptide that coprecipitated with large-T antigen from SV40-infected cells. The p53 protein regulates cell cycle progression in normal and malignant cells via multiple complex mechanisms including sequence-specific DNA binding, transcriptional activation and repression activities, and protein interactions, some of which affect expression of other oncoproteins. Cells containing p53 mutations are incapable of arresting at the G_1/S cell cycle transition point following ultraviolet irradiation, yet regain the ability to delay this transition after transfection of wild-type p53. Whereas p53 mutations by themselves appear inadequate with regard to cell transformation in vitro, they induce a state of genomic instability predisposing to DNA amplification and malignant transformation. Wild-type p53 can restore genomic stability and reduce the growth and tumorigenicity of cells containing these mutations.

The p53 gene contains 10 exons spanning an open reading frame of approximately 20 kb, of which the first 8 to 10 kb contain a noncoding exon and an adjacent intron. As with other neoplasms, the majority of mutations in lung cancers occur in exons 5 to 8, although mutations outside these areas have been noted with relative frequency. Initial comprehensive analysis of p53 mutations in human cancers demonstrated that 57% of all p53 mutations in non-small cell lung cancers involved G:C-to-T:A transversions, all of which occurred on the nontranscribed strand; benzo(*a*)pyrene contained in cigarette smoke is known to induce such mutations. In comparison, G:C-to-A:T transitions were the most common types of p53 mutations observed in small cell lung cancers, comprising 46% of all such mutations detected in these tumors. Transitions at CpC dinucleotides, believed to be indicative of spontaneous mutations, were observed in 10% of non-small cell lung cancer specimens and 31% of small cell lung cancer specimens.

Subsequent analysis of non-small cell lung cancer specimens from Japanese patients revealed that only 25% of all base substitutions were G:C-to-T:A transversions, whereas 52% involved transitions on nontranscribed strands. The discrepancies between the types of mutations identified in this study and those recorded previously may be related to ill-defined environmental or genetic factors.

Polymerase chain reaction–single-strand conformational polymorphism (PCR-SSCP) and direct sequencing analyses have revealed that approximately 50% of all non-small cell and 70% of small cell lung cancers examined thus far contain p53 mutations. Mutations of p53 have been reported in approximately 75% of non-small cell lung cancer lines, tending to occur outside evolutionarily conserved regions in exons 5 through 8 (the typical locations of p53 mutations in other malignancies). A similar and possibly more pronounced phenomenon has been observed in small cell lung cancer lines, virtually all of which contain p53 mutations.

Mutations tend to stabilize and prolong the half-life of the p53 protein. As such, p53 mutations can be evidenced in tissue sections by quantitatively enhanced reactivity with monoclonal antibodies recognizing both wild-type and mutant proteins, or may be detected by monoclonal antibodies recognizing epitopes contained only on mutant proteins. As such, detection of p53 by these monoclonal antibodies may be directly related to the types of p53 mutations, and may also correlate with the biological activities of the mutant p53 proteins. Enhanced immunoreactivity consistent with elevated p53 protein levels might reflect mutations in which wild-type p53 is inactivated by a dominant negative mechanism, whereas low-level immunoreactivity may be compatible with p53 mutations occurring in regions responsible for either transactivation or nuclear localization signaling in which inactivation of p53 occurs by a recessive mechanism.

Aberrant expression of p53 protein as detected by immunohistochemistry techniques has been recorded in approximately 70% of small cell lung cancer specimens, but does not appear to be associated with disease extent or patient survival. In contrast, abnormal p53 expression occurs in approximately 50% of non-small cell lung cancer specimens and appears to correlate with nodal metastases and reduced patient survival. Analysis of non-small cell lung cancers using PCR-SSCP and immunohistochemical techniques revealed that zero of 13 tumors negative for p53 mutations by both methods were metastatic, whereas 17 of 17 metastatic tumors had p53 mutations. A large retrospective study of stage I and II non-small cell lung cancer cases utilizing immunohistochemistry techniques identified abnormal p53 expression in 40% of adenocarcinomas and squamous cell cancers. Expression of p53 correlated significantly with reduced patient survival irrespective of pathological stage. Furthermore, seven patients with stage II disease had negligible p53 immunoreactivity in their primary tumors but had abnormal p53 expression in metastatic lymph nodes; these patients had a mean survival time of 11 months compared with 34 months for patients whose disease stage was the same but whose tumors and lymph nodes had normal p53 expression. These data have been independently confirmed by other investigators who have observed that p53 mutations in primary non-small cell lung cancers as determined by biochemical analysis correlate significantly with reduced patient survival independent of pathological stage.

Several studies involving small numbers of samples have detected p53 mutations in preneoplastic lung tissues, suggesting that these mutations occur relatively early in the course of bronchial neoplasia. Observations reported above suggest that p53 mutations may also occur late in the course of this disease. In all likelihood multiple genetic events may result in neoplastic transformation and metastasis. Mutations of p53 may not necessarily commit premalignant cells to become invasive cancers, but may facilitate the acquisition of other destabilizing genetic events that ultimately culminate in malignant transformation. Similarly, p53 mutations occurring in the context of an established malignancy may act as progression factors, thereby enhancing metastatic potential. In this regard, the issue should not necessarily be the timing of p53 mutations, but rather the biological and clinical implications of such events in lung cancer.

CONCLUSION

Despite an impressive increase in knowledge concerning molecular mechanisms involved in lung cancer, appreciation of the complexities of these genetic events is fragmentary, and the conquest of lung cancer remains a formidable task. Nevertheless, current approaches to lung cancer management can be refined on the basis of available information. For instance, the identification of particular molecular defects such as C-myc amplification in small cell lung cancer, or ras, EGF receptor, or p185 overexpression in non-small cell lung cancers may identify patients at high risk of failure with conventional treatment who should be considered for aggressive experimental protocols. Furthermore, the specific targeting of growth factor receptors such as GRP in small cell lung cancer, or EGF or p185 in non-small cell lung cancer may enhance the specificity of cytostatic or cytocidal agents in the management of these neoplasms. Without question much work is still needed to define the relevance of individual genetic aberrations in premalignant lesions so as to identify individuals at high risk of developing invasive cancers who might benefit from preventive interventions.

Although cancer is a multistep process , correction of one or perhaps two of these alterations may be sufficient to inhibit the relentless proliferation of malignant cells. Dominant oncogenes can be targeted using antisense techniques, whereas tumor suppressor gene defects can be treated by restoration of wild-type gene expression in tumor cells. Recently, the H460 non-small cell lung cancer line containing a K-ras codon 61 mutation was transfected with a 2-kb genomic section of K-ras protooncogene in antisense orientation using the APR1-neo plasmid vector. A β-actin promoter contained in the plasmid vector constitutively drove the antisense K-ras construct. Following G418 selection Southern analysis revealed stable integration of plasmid DNA, and Northern analysis demonstrated the presence of antisense mRNA. Western blot analysis detected only a slight reduction in the total expression of ras proteins, but a 95% diminution of K-ras protein expression in stable clones relative to control cells. Antisense K-ras significantly inhibited the proliferation and tumorigenicity of H460 lung cancer cells.

The above antisense K-ras fragment was inserted into the LNSX retroviral vector. Utilizing the H460 cell line, a 57% transduction efficiency after one infection cycle at 10 colony-forming units (CFU) per cell, which increased to 95% following five exposures to retrovirus, was observed. Expression of antisense K-ras mRNA was confirmed by Northern analysis, and K-ras p21 was specifically inhibited as analyzed by Western blot techniques. A sevenfold reduction of growth in vitro was observed for the H460 cells transduced with LNSX antisense K-ras compared with control cells. Tumorigenicity experiments in nude mice revealed that in contrast to 100% of control mice that had evidence of orthotopic tumor growth, only 20% of mice inoculated with H460 cells containing LNSX antisense K-ras had tumors, and these were much smaller than the tumors in the control animals. Subsequently, the efficacy of intratracheal instillation of LNSX antisense K-ras with respect to inhibition of growth of H460 cells was analyzed using an orthotopic nude mouse model. Three days after being inoculated intratracheally with 10^5 H460 cells, nude mice received three daily tracheal instillations of LNSX retrovirus, LNSX sense K-ras, LNSX antisense K-ras, or medium alone. At autopsy 30 days later, 90% of control mice had tumors, whereas only 13% of mice treated with LNSX antisense K-ras had tumors, and that these tumors were much smaller than those observed in control animals. These provocative findings suggest that antisense K-ras delivered via retroviral vectors can induce tumor regression in vivo, and may be relevant to the treatment of established pulmonary cancers as well as the prevention of lung neoplasms in high-risk individuals.

More recently viral vectors have been utilized to deliver wild-type p53 to lung cancer cells harboring p53 mutations. A LNSX retroviral vector containing a 6.6-kb fragment comprising p53 complementary

DNA and the β-actin promoter has been utilized to transduce H322a, H358, and H460 lung cancer cell lines containing point-mutated, null, or wild-type p53 alleles, respectively. Transduction efficiency exceeded 90% following five exposures to virus at a multiplicity of infection (MOI) of 40 CFU per cell. Wild-type p53 transcripts could be detected within 18 hours of infection in H358 cells, and p53 mRNA and protein levels remained stable during a 6-month study period. Wild-type p53 inhibited proliferation of H322 and H358 cells by more than 90% in comparison with 15% for H460 cells relative to appropriate controls, and significantly diminished the tumorigenicity of lung cancer cells in orthotopic nude mouse models.

In other experiments, a helper-independent replication-defective adenovirus containing the cytomegalovirus E1 promoter, p53 cDNA, and the SV40 polyadenylation signal (Adp53) has been used to deliver wild-type p53 to lung cancer cells. Transduction efficiency in H322a, H358, and H460 cells was approximately 100% following a single exposure to the adenoviral vector at MOI = 30 plaque-forming units (PFU) per cell. Western blot analysis revealed high-level p53 protein expression, which was maximal 3 days after infection and rapidly tapered off over the ensuing 12 days. Wild-type p53 inhibited the proliferation of H322 and H358 cells by approximately 75%, in comparison with 28% for H460 cells relative to control cells. Despite the transient nature of p53 expression, Adp53 significantly inhibited the tumorigenicity of lung cancer cells in orthotopic mouse models.

These exciting findings from our laboratory extend observations by others that specific targeting of critical genes such as Rb, p53, or ras may profoundly influence the proliferation and tumorigenicity of lung cancer cells containing multiple genetic abnormalities. Furthermore, the observations that oncogene and tumor suppressor genes such as myc and p53 are implicated in apoptosis as well as sensitivity to chemotherapeutic agents in cancer cells suggest that novel strategies involving gene therapy alone or in combination with conventional cytotoxic regimens may offer new hope to those patients with advanced lung cancer; more importantly, the targeting of genetic defects responsible for malignant transformation in preneoplastic lesions may prevent cancer in high-risk individuals. Provided that the specificity of delivery and expression of gene constructs can be ensured, the precise targeting of molecular defects in established lung cancers and premalignant lesions via gene therapy will no longer remain a laboratory phenomenon, but will become a clinical reality.

BIBLIOGRAPHY

Amati B, Brooks MW, Levy N, Littlewood TD, Evan GI, Land H. Oncogenic activity of the C-myc protein requires dimerization with max. Cell 1993;72:233–245.

Antoniades HN, Galanopoulos T, Neville-Golden J, O'Hara CJ. Malignant epithelial cells in primary human lung carcinomas coexpress in vivo platelet-derived growth factor (PDGF) and PDGF receptor mRNAs and their protein products. Proc Natl Acad Sci USA 1992;89:3942–3946.

Barbacid M. ras genes. Annu Rev Biochem 1987;56:799–827.

Bargmann CI, Hung M-C, Weinberg RA. The neu oncogene encodes an epidermal growth factor receptor–related protein. Nature 1986;319:226–230.

Baserga R, Rubin R. Cell cycle and growth control. Crit Rev Eukaryot Gene Expr 1993;3(1):47–61.

Bennett WP, Colby TV, Travis WD, et al. p53 protein accumulates frequently in early bronchial neoplasia. Cancer Res 1993;53:4817–4822.

Boldog FL, Gemmill RM, Wilke CM, et al. Positional cloning of the hereditary renal carcinoma 3;8 chromosome translocation breakpoint. Proc Natl Acad Sci USA 1993;90:8509–8513.

Brachmann R, Lindquist PB, Nagashima M, et al. Transmembrane TGR-alpha precursors activate EGF/TGF-alpha receptors. Cell 1989;56:691–700.

Brennan J, O'Connor T, Makuch RW, et al. Myc family DNA amplification in 107 tumors and tumor cell lines from patients with small cell lung cancer treated with different combination chemotherapy regimens. Cancer Res 1991;51:1708–1712.

Cai DW, Mukhopaphyay T, Liu Y, Fujiwara T, Roth JA. Stable expression of the wild-type p53 gene in human lung cancer cells after retrovirus-mediated gene transfer. Hum Gene Ther 1993;4:617–624.

Chesa PG, Rettig WJ, Melamed MR, Old LJ, Niman HL. Expression of $p21^{ras}$ in normal and malignant human tissues: lack of association with proliferation and malignancy. Proc Natl Acad Sci USA 1987;84:3234–3238.

Chiba I, Takahashi T, Nau MM, et al. Mutations in the p53 gene are frequent in primary, resected non-small cell lung cancer. Oncogene 1990;5:1603–1610.

Cuttitta F, Carney DN, Mulshine J, et al. Bombesin-like peptides can function as autocrine growth factors in human small-cell lung cancer. Nature 1993;316:823–825.

Falco JP, Baylin SB, Lupu R, et al. V-ras induces non-small cell phenotype, with associated growth factors and receptors, in a small cell lung cancer cell line. J Clin Invest 1990;85:1740–1745.

Farmer GJ, Bargpmetto J, Zhu H, Friedman P, Prywes R, Prives C. Wild-type p53 activates transcription in vitro. Nature 1992;358:83–86.

Friend SH, Bernards R, Rogelj S, et al. A human DNA segment with properties of the gene that predisposes to retinoblastoma and osteosarcoma. Nature 1986;323:643–646.

Fujiwara T, Grimm EA, Mukhopaphyay T, Cai DW, Owen-Schaub LB, Roth JA. A retroviral wild-type p53 expression vector penetrates human lung cancer spheroids and inhibits growth by inducing apoptosis. Cancer Res 1993;53:4129–4133.

Gazdar AF, Carney DN, Nau MM, Minna JD. Characterization of variant subclasses of cell lines derived from small-cell lung cancer having distinctive biochemical, morphological and growth properties. Cancer Res 1985;45:2924–2930.

Gibbs JB. Ras C-terminal processing enzymes—new drug targets? Cell 1991;65:1–4.

Harbour JW, Lai S-L, Whang-Peng J, Gazdar AF, Minna JD, Kaye FJ. Abnormalities in structure and expression of the human retinoblastoma gene in sclc. Science 1988;241:353–357.

Hateboer G, Timmers HTM, Rustgi AK, Billaud M, van't veer LJ, Bernards R. TATA-binding protein and the retinoblastoma gene product bind to overlapping epitopes on c-myc and adenovirus E1A protein. Proc Natl Acad Sci USA 1993;90:8489–8493.

Heldin CH, Westermark B. Platelet-derived growth factor: mechanism of action and possible in vivo function. Cell Regul 1990;1:555–566.

Hibi K, Takahashi T, Yamakawa K, et al. Three distinct regions involved in 3p deletion in human lung cancer. Oncogene 1992;7:445–449.

Hollingsworth RE Jr, Chen P-L, Lee W-H. Integration of cell cycle control with transcriptional regulation by the retinoblastoma protein. Curr Opin Cell Biol 1993;5:194–200.

Hollstein M, Sidransky D, Vogelstein B, Harris CC. p53 mutations in human cancers. Science 1991;253:49–53.

Holmes WE, Sliwkowski MX, Akita RW, et al. Identification of heregulin, a specific activator of p185 erbB-2. Science 1992;256:1205–1210.

Hunter T. The epidermal growth factor gene and its product. Nature 1984;311:414–416.

Husgafvel-Pursiainen K, Hackman P, Ridanpaa M, et al. K-ras mutations in human adenocarcinoma of the lung: association with smoking and occupational exposure to asbestos. Int J Cancer 1993;53:250–256.

James GL, Goldstein JL, Brown MS, et al. Benzodiazepine peptidomimetics: potent inhibitors of ras farnesylation in animal cells. Science 1993;260:1937–1942.

Kern JA, Schwartz DA, Nordberg JE, et al. p185 neu expression in human lung adenocarcinomas predicts shortened survival. Cancer Res 1990;50:5184–5191.

Killary AM, Wolf ME, Giambernardi TA, Naylor SL. Definition of a tumor suppressor locus within human chromosome 3p21–p22. Proc Natl Acad Sci USA 1992;89:10877-10881.

Kishimoto Y, Murakami Y, Shiraishi M, Hayashi K, Sekiya T. Aberrations of the p53 tumor suppressor gene in human non-small cell carcinomas of the lung. Cancer Res 1992;52:4799–4804.

Kohl NE, Mosser SD, deSolms SJ, et al. Selective inhibition of ras-dependent transformation by a farnesyl transferase inhibitor. Science 1993;260:1934–1937.

Krystal G, Birrer M, Way J, et al. Multiple mechanisms for transcriptional regulation of the myc gene family in small-cell lung cancer. Mol Cell Biol 1988;8:3373–3381.

Kuerbitz SJ, Plunkett BS, Walsh WV, Kastan MD. Wild-type p53 is a cell cycle checkpoint determinant following irradiation. Proc Natl Acad Sci USA 1992;89:7491–7495.

LaForgia S, Morse B, Levy J, et al. Receptor protein-tyrosine phosphatase gamma is a candidate tumor suppressor gene at human chromosome region 3p21. Proc Natl Acad Sci USA 1991;88:5036–5040.

Levine AJ. The p53 tumour suppressor gene and product. Cancer Surv 1992;12:59–78.

Lowe SW, Ruley HE, Jacks T, Housman DE. p53-dependent apoptosis modulates the cytotoxicity of anticancer agents. Cell 1993;74:957–967.

Marchetti A, Buttitta F, Merlo G, et al. p53 alterations in non-small cell lung cancers correlate with metastatic involvement of hilar and mediastinal lymph nodes. Cancer Res 1993;53:2846–2851.

Marcu KB, Bossone SA, Patel AJ. Myc function and regulation. Annu Rev Biochem 1992;61:809–860.

McCormick F. How receptors turn ras on. Nature 1993;363:15–16.

Mitsudomi T, Steinberg SM, Oie HK, Mulshine JL, Phelps R, Viallet J. Ras gene mutations in non-small cell lung cancers are associated with shortened survival irrespective of treatment intent. Cancer Res 1991;51:4999–5002.

Miura I, Siegfried JM, Resau J, Keller SM, Zhou J, Testa JR. Chromosome alterations in 21 non-small cell lung carcinomas. Genes Chromosom Cancer 1990;2:328–338.

Momand J, Zambetti GP, Olson DC, George D, Levine AJ. The mdm-2 oncogene product forms a complex with the p53 protein and inhibits p53-mediated transactivation. Cell 1992;69:1237–1245.

Mukhopadhyay T, Tainsky M, Cavender A, Roth JA. Specific inhibition of K-ras expression and tumorigenicity of lung cancer cells by antisense RNA. Cancer Res 1991;51:1744–1748.

Nakanishi Y, Mulshine JL, Kasprzyk PG, et al. Insulin-like growth factor-1 can mediate autocrine proliferation of human small cell lung cancer cell lines in vitro. J Clin Invest 1988;82:354–359.

Ooi WL, Elston RC, Chen VW, Bailey-Wilson JE, Rothschild H. Increased familial risk for lung cancer. J Natl Cancer Inst 1986;76:217–222.

Ookawa K, Shiseki M, Takahashi R, Yoshida Y, Terada M, Yokota J. Reconstitution of the rb gene suppresses the growth of small-cell lung carcinoma cells carrying multiple genetic alterations. Oncogene 1993;8:2175–2181.

Pezzella F, Turley H, Kuzu I, et al. bcl-2 protein in non-small-cell lung carcinoma. N Engl J Med 1993;329:690–694.

Postel EH, Berberich SJ, Flint SJ, Ferrone CA. Human c-myc transcription factor PuF identified as nm23-H2 nucleoside diphosphate kinase, a candidate suppressor of tumor metastasis. Science 1993;261:478–480.

Quinlan D, Davidson AG, Summers CL, Warden HE, Doshi HM. Accumulation of p53 protein correlates with a poor prognosis in human lung cancer. Cancer Res 1992;52:4828–4831.

Reissmann PT, Koga H, Takahashi R, et al, and Lung Cancer Study Group. Inactivation of the retinoblastoma susceptibility gene in non-small-cell lung cancer. Oncogene 1993;8:1913–1919.

Rodenhuis S, Slebos RJ. Clinical significance of ras oncogene activation in human lung cancer. Cancer Res 1992;52:2665–2669.

Rusch V, Baselga J, Cordon-Cardo C, et al. Differential expression of the epidermal growth factor receptor and its ligands in primary non-small cell lung cancers and adjacent benign lung. Cancer Res 1993;53:2379–2385.

Sellers TA, Bailey-Wilson JE, Elston RC, et al. Evidence for mendelian inheritance in the pathogenesis of lung cancer. J Natl Cancer Inst 1990;82:1272–1279.

Sekido Y, Takahaski T, Makela TP, et al. Complex intrachromosomal rearrangements in the process of amplification of the I-myc gene in small-cell lung cancer. Mol Cell Biol 1992;12:1747–1754.

Seto E, Usheva A, Zambetti GP, et al. Wild-type p53 binds to the TATA-binding protein and represses transcription. Proc Natl Acad Sci USA 1992; 89:12028–12032.

Shopland DR, Eyre HJ, Pechacek TF. Smoking attributable cancer mortality in 1991: is lung cancer now the leading cause of death among smokers in the United States? J Natl Cancer Inst 1991;83:1142–1148.

Smith K, Houlbrook S, Greenall M, Carmichael J, Harris AL. Topoisomerase II alpha coamplification with erbB-2 in human primary breast cancer and breast cancer cell lines: relationship to m-AMSA and mitoxantrone sensitivity. Oncogene 1993;8:933–938.

Suzuki Y, Orita M, Shiraishi M, Hayashi K, Sekiya T. Detection of ras gene mutations in human lung cancers by single-strand conformation polymorphism analysis of polymerase chain reaction products. Oncogene 1990;5:1037–1043.

Takahashi T, Carbone D, Takashi T, et al. Wild-type but not mutant p53 suppresses the growth of human lung cancer cells bearing multiple genetic lesions. Cancer Res 1992;52:2340–2343.

Takahashi T, Suzuki H, Hida TY, Sekido YY, Ariyoshi Y, Ueda R. The p53 gene is very frequently mutated in small-cell lung cancer with a distinct nucleotide substitution pattern. Oncogene 1991;6:1775–1778.

Tefre T, Borresen A-L, Aamdal S, Brogger A. Studies of the L-myc DNA polymorphism and relation to metastasis in Norwegian lung cancer patients. Br J Cancer 1990;61:809–812.

Tsai C-M, Chang K-T, Perng R-P, et al. Correlation of intrinsic chemoresistance of non-small cell lung cancer cell lines with HER-2/neu gene expression but not with ras gene mutations. J Natl Cancer Inst 1993;85:897–901.

Veale D, Kerr N, Gibson GJ, Harris AL. Characterization of epidermal growth factor receptor in primary human non-small cell lung cancer. Cancer Res 1989;4:1313–1317.

Veale D, Kerr N, Gibson GJ, Kelly PJ, Harris AJ. The relationship of quantitative epidermal growth factor receptor expression in non-small cell lung cancer to long term survival. Br J Cancer 1993;68:162–165.

Wagner AJ, Small MB, Hay N. Myc-mediated apoptosis is blocked by ectopic expression of bcl-2. Mol Cell Biol 1993;13:2432–2440.

Weinberg RA. The retinoblastoma gene and gene product. Cancer Surv 1992;12:43–57.

Weinert T, Lydall D. Cell cycle checkpoints, genetic instability and cancer. Semin Cancer Biol 1993;4:129–140.

Wharton J, Polak JM, Bloom SR, et al. Bombesin-like immunoreactivity in the lung. Nature 1978;273:769–770.

Willey JC, Hei TK, Piao CQ, et al. Radiation-induced deletion of chromosomal regions containing tumor suppressor genes in human bronchial epithelial cells. Carcinogenesis 1993;14:1181–1188.

Woll PJ, Rozengurt E. D-Arg, D-Phe, D-Trp, Leu—substance P, a potent bombesin antagonist in murine Swiss 3T3 cells, inhibits the growth of human small cell lung cancer cells in vitro. Proc Natl Acad Sci USA 1988;85:1859–1863.

Yamakawa K, Takahashi T, Horio Y, et al. Frequent homozygous deletions in lung cancer cell lines detected by a DNA marker located at 3p21.3-p22. Oncogene 1993;8:327–330.

Yin Y, Tainsky MA, Bischoff FZ, Strong LC, Wahl GM. Wild-type p53 restores cell cycle control and inhibits gene amplification in cells with mutant p53 alleles. Cell 1992;70:937–948.

Zhang W-W, Fang X, Mazur W, French BA, Georges RN, Roth JA. High-efficiency gene transfer and high-level expression of wild-type p53 in human lung cancer cells mediated by recombinant adenovirus. Cancer Gene Ther 1993;1:5–13.

Zhang Y, Mukhopadhyay T, Donehower LA, Georges RN, Roth JA. Retroviral vector-mediated transduction of K-ras antisense RNA into human lung cancer cells inhibits expression of the malignant phenotype. Hum Gene Ther 1993;4:451–460.

CHAPTER 11

Molecular Basis of Colon Cancer

Marsha L. Frazier

Colorectal carcinoma (CRC) arises as a result of a series of genetic alterations that are paralleled by progression of the normal colonic mucosa through the various histologically identifiable stages of carcinogenesis (Fig. 11.1). It appears that the order of the genetic events is not as important as the number of alterations that have accumulated (1). However, as depicted in Figure 11.1, certain types of events tend to occur earlier, and others are more likely to occur later. Also, the early events in CRC vary among the different types. This will be discussed below in greater detail. Many of the alterations are mutations in oncogenes and tumor suppressor genes. Molecular studies of CRC have been facilitated by the ease with which one can obtain tissue at the various stages of colon carcinogenesis, which are illustrated in Figure 11.1. The earliest histologic transition is from normal mucosa to hyperplasia. Early adenomas can then arise from these areas of hyperproliferation and are likely to be classified histologically as tubular adenomas. These adenomas progress into intermediate adenomas that are likely to have developed a villous component and are thus classified as tubulovillous. As an adenoma progresses to a late adenoma, it is likely to be composed predominantly of villous elements with severe dysplasia, and carcinoma in situ usually arises from this villous component of the adenoma. Eventually, if not removed, malignant cells erode through the basement membrane, then through the stalk of the polyp, and then invade the colorectal lymphatics and nerves as well as breaking through the serosa into the peritoneal cavity or perirectal tissues.

CRC can be classified into three major types on the basis of family history and clinical features: sporadic (individuals with no family history of CRC), adenomatous polyposis coli (APC), and hereditary nonpolyposis colon cancer (HNPCC). APC includes both familial adenomatous

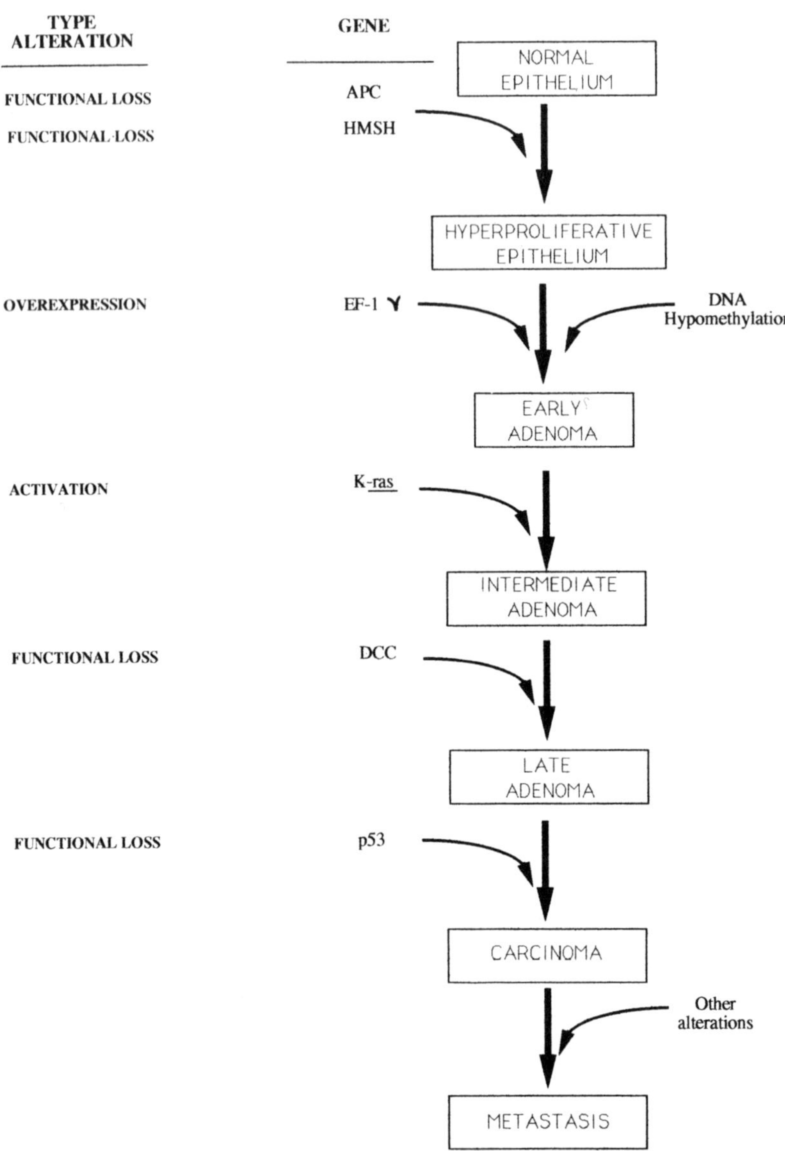

Fig. 11.1: Genetic model for colorectal tumorigenesis. Mutations in the APC gene and hMSH gene are inherited in hereditary colorectal carcinoma. Loss of function of either of these genes leads to tumorigenesis, but the underlying mechanisms involved in the pathways are likely to vary as discussed in the section on sporadic CRC, with tumors arising from loss of APC function resembling those

polyposis (FAP) and Gardner's syndrome (GS) and is inherited in an autosomal dominant fashion (2). Several reviews have addressed the diagnosis, screening, and treatment of the various types of CRC (2–6). The major focus of this chapter will be on the molecular events associated with the pathogenesis of the disease.

ADENOMATOUS POLYPOSIS COLI

APC accounts for approximately 1% of the colorectal cancer in the Western world (7,8). The gene that is mutated in APC has been identified (9–12). It spans over 300 kilobase pairs (kb) of DNA, has 8532 base pairs (bp) of coding sequence contained in 15 exons, and encodes a protein that was previously unidentified. Alternate splicing of the APC transcript gives rise to two messenger RNAs. The more abundant transcript contains a 300-bp exon that is not present in the other mRNA and encodes a 2843 amino acid peptide with a mass of 311.8 kDa. Figure 11.2 is a map of the coding portion of this more abundant form. It shows the positions of the intron-exon boundaries and placement of the amino acid sequence within the exons. Most of the mutations that have been detected in the APC gene are small deletions or insertions of DNA that result in frameshifts in the coding portion of the gene and so give rise to truncated proteins (9–15). As a result of the frameshifts, stop codons downstream of the mutations are now in-frame and so give rise to truncated proteins. In a smaller number of cases, splice junction mutations result in frameshifts in the RNA coding sequence, once again resulting in a downstream stop codon and a truncated protein. Only in a few cases were large deletions of the gene detected, such as those found in the patient studied by Groden et al. (9) and Joslyn et al. (10) for the identification of the APC gene. Nagase and Nakamura (16) have summarized the majority of mutations that have been published thus far. Most of these mutations occur in the 5′ half of the coding region, with about two-thirds falling in a region they call the mutation cluster region (MCR). The MCR spans amino acid 1285 to 1465 of the coding region.

During the studies that identified the APC gene, the mutated in colon cancer (MCC) gene was also identified because it too is located at

Fig. 11.1 Continued: found in patients with adenomatous polyposis coli, and tumors arising from loss of function of hMSH resembling more those found in patients with hereditary nonpolyposis coli. *(Adapted from Fearon ER, Vogelstein B. A genetic model for colorectal tumorigenesis. Cell 1990;61:759–767.)*

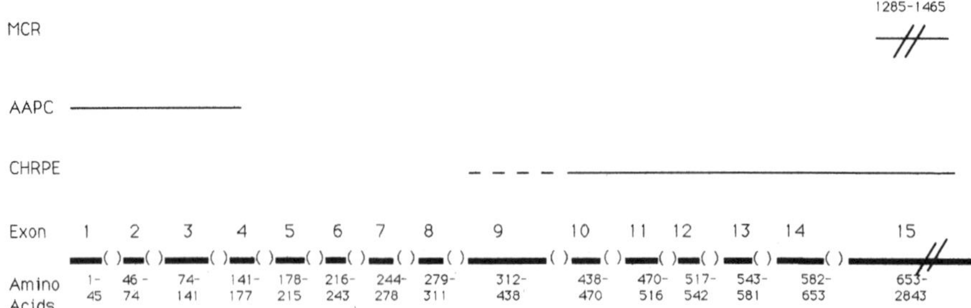

Fig. 11.2: Map of the amino acid coding portion of the more abundant form of the APC mRNA. The 15 exons are shown with their amino acid sequences below. The parentheses mark the positions of the introns. Above this, the locations of the mutations in the patients with congenital hypertrophy of the retinal pigment epithelium are indicated by a solid line. The most 3′ mutation observed was at amino acid 1387. The dashed line indicates the transitional region in exon 9. Above this, the solid line indicates the region containing mutations associated with the attenuated form of APC. The mutation cluster region (MCR), where about two-thirds of the mutations in the APC gene are located, is indicated by a solid line, with the amino acids that this region spans above the line (9–13).

chromosome 5q21. This finding, combined with the observation that somatic mutations in MCC occur in colorectal tumors, made MCC a candidate gene for APC (11). However, Nishisho et al. (12) failed to detect germline mutations in their analysis of 90 FAP families, thus ruling out this gene as a major cause of FAP. Furthermore, in two unrelated APC patients with deletions of DNA at 5q21, the MCC gene was located outside of the deletion region.

APC Mutations Associated with Specific Clinical Features

Comparison of the clinical features of APC disease and the locations of mutations within the APC gene suggests that the location of the APC mutations within the gene influences some of these features. One example of this is the attenuated form of familial adenomatous polyposis coli (AAPC), which was previously shown to be linked to the APC locus (17,18). Patients with AAPC still have a high risk of developing CRC, but they develop fewer polyps (usually < 100, sometimes as few as one or two) than APC patients do. In contrast, APC patients develop hundreds to thousands of adenomatous polyps. Although the average age of onset for FAP is 39 years (2), for AAPC it is 54. Spirio et al. (19) identified APC gene mutations in each of seven AAPC families. Four of the families shared a common mutation, and the other three families had different mutations. As was commonly the case with classic APC, the mutations were all frameshift or nonsense mutations that produced an early stop codon and so gave rise to truncated proteins. Spirio et al. (19) observed that all four mutations were located in the 5′ portion of the coding sequence and were upstream of the mutations identified in classic APC families. The most 3′ mutation was located at codon 156 in the AAPC patients, while the most 5′ of the mutations in the classic APC patients was a 4-bp deletion involving codons 169 and 170. Spirio et al. (19) therefore hypothesized that these mutations define a "functional" boundary within the APC gene that determines the phenotypic differences between APC and AAPC families.

Another example in which the location of an APC mutation may influence a specific clinical feature is a common clinical manifestation of the APC disease that is characterized by the presence of multiple patches of congenital hypertrophy of the retinal pigment epithelium (CHRPE) (20). CHRPE occurs in approximately 70% of patients with APC. Olschwang et al. (14) demonstrated a correlation between CHRPE and the position of the mutation: CHRPE lesions are present when the mutation occurs after exon 9 and absent if the mutation occurs before exon 9. Exon 9 is a transitional exon, with CHRPE lesions being present if a mutation occurs in the 3′ half of the exon, and absent if the mutation occurs in the 5′ half of the exon. The most 3′ mutation observed was at amino acid 1387 in exon 15.

Gardner's syndrome is similar to FAP except that GS patients also have benign extracolonic tumor growths such as osteomas, epidermoid cysts, desmoid tumors, and dental abnormalities (2). It is thought to be a variant of FAP, as the inheritance of both is linked to the same region of 5q21 (21). The results of Nishisho et al. (12) extended these observations by demonstrating that a patient with GS symptoms and a patient with no evidence of extracolonic disease had identical mutations (a C-to-G transition at codon 302). The patient with no extracolonic symptoms was 46 years old and therefore old enough to be expected to display such manifestations if they were going to occur. Thus, the specific codon 302 mutation does not appear to completely specify the extracolonic manifestations of FAP and GS; the exact phenotype is probably the result of other genetic or environmental factors.

Powell et al. (22) observed germline variants of unknown consequence in the APC gene. Three of these resulted in amino acid changes. One variant resulted in a change from an aspartate to a valine and was found in nine patients. In two patients, a second variant was seen: a change from a proline to a serine. A third variant (glycine to valine) was seen in a single patient. These variants occurred at the same frequencies in 100 healthy subjects as in this patient population. Further studies will be necessary to determine if individuals carrying these variants are at increased risk of developing CRC.

Function of the APC Protein

Although the precise function of the APC protein is unknown, recent evidence indicates that it may play a role in intercellular interactions. A series of immunoprecipitation studies demonstrated that α- and β-catenin are associated with the APC protein (23,24). These catenins are known to bind to E-cadherin, and this binding is essential for the proper functioning of E-cadherin. The cadherins are cell-surface molecules that mediate calcium-dependent intercellular interactions and are important for morphogenesis. Although both APC and cadherins associate with catenins, their interactions are not identical. Cadherins appear to associate with similar amounts of α- and β-catenins, whereas APC associates preferentially with β-catenin. It is possible that APC does not interact directly with α-catenin but indirectly by binding to β-catenin. Because both E-cadherin and APC bind to catenins, Su et al. (24) point out the possibility that E-cadherins and APC exist in one complex. However, they were not able to detect any kind of association between the two using either immunoprecipitation or immunoblotting following immunoprecipitation. Rubinfeld et al. (23) noted that the sequence Ser-Leu-Ser-Ser-Leu is in the β-catenin-binding domain of E-cadherin and is also a part of a 20 amino acid repeat found in APC. There are seven imperfect copies of the 20 amino acid repeat. Rubinfeld et al. (23) speculated that β-catenin forms independent complexes with E-cadherin and APC, and that APC and E-cadherin may share a β-catenin-binding motif. Because most truncated APC proteins are missing at least 5 of these repeats, they speculate that the truncated APC has reduced affinity or no affinity for β-catenin. The studies of Su et al. (24) and Rubinfeld et al. (23) suggest that APC is involved in cell adhesion and that adenomas may arise as a result of a loss of intracellular interactions that are dependent upon an interaction between APC and β-catenin. They hypothesize that APC may modulate the interaction between cadherins and catenins, thereby affecting the pathway through which intercellular interactions control cell growth and differentiation.

HEREDITARY NONPOLYPOSIS COLON CANCER

HNPCC (Lynch syndrome I and Lynch syndrome II) is an autosomal dominant disorder (5) that accounts for 4% to 13% of all cases of CRC. Patients with CRC are defined as affected if they have at least two relatives in two generations with CRC and one of the relatives was diagnosed before age 50 (25). This classification scheme is a very conservative one and most likely excludes some families that have the disorder. Unlike patients with APC, those with HNPCC do not have diffuse polyposis. In Lynch syndrome I, colorectal carcinoma is the only tumor to develop (5), but in Lynch syndrome II, cancers are observed at other sites. CRC and endometrial carcinomas are the two most common cancers in HNPCC family members (26–28). Other organs affected are the stomach, pancreas, biliary tract, and urinary tract (26,27).

Because HNPCC produces more than one type of cancer and because there is no way of distinguishing sporadic tumors from those arising as a result of mutation at the HNPCC locus, it was difficult to perform linkage analysis of HNPCC families. Furthermore, adenomas in the general population that do not arise as a result of HNPCC are quite common, making individuals with adenomas in HNPCC families more difficult to classify. Another difficulty in classifying individuals in HNPCC families as affected or unaffected is that the age at which HNPCC tumors appear is quite variable, with the mean age being 45. Despite this problem, two HNPCC loci have been mapped by linkage analysis to chromosomes 2p and 3p.

HNPCC Locus on Chromosome 2p

The first HNPCC locus to be identified on chromosome 2p is the best characterized to date (29). Peltomaki et al. performed a systematic search of the genome using 345 microsatellite markers in two large kindreds (29). Microsatellites are short tandem repeats of DNA that occur randomly throughout the genome. These repeats are highly polymorphic owing to variation in the number of the repeat units. They are flanked by unique DNA sequences, which makes it possible to amplify the sequences by the polymerase chain reaction (PCR) using genomic DNA as a template. These microsatellite polymorphisms have been found to be stable enough in normal persons to be used in linkage analysis (30,31). Figure 11.3 shows an example of a microsatellite polymorphism on chromosome 12p (32). Because it is difficult to classify individuals in HNPCC families as affected or unaffected, Peltomaki et al. classified those with colorectal or endometrial carcinomas as affected, and those with a colorectal adenoma or with a single carcinoma of the ovary, stomach, hepatobiliary system, small intestine, kidney, or ureter as of unknown status for a high-stringency analysis and as affected for a low-stringency analysis (29). They determined that one of the microsatellite markers, D2S123, located at 2p16–15, was linked to cancer in these families.

Aaltonen et al. (33) further examined the association of the anonymous D2S123 microsatellite marker with HNPCC in another 14 smaller families and determined that there was heterogeneity among the families regarding linkage to the 2p16–15 locus. They were able to exclude linkage

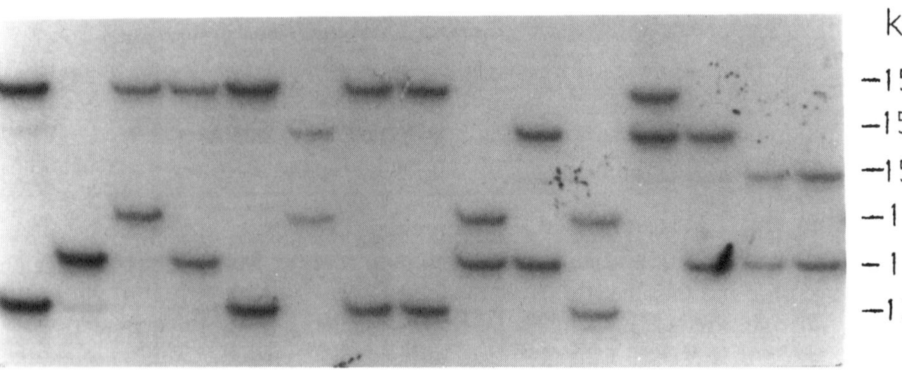

Fig. 11.3: Microsatellite repeat polymorphism located on chromosome 12 in DNA isolated from the peripheral blood leukocytes of normal subjects. Lengths of PCR are indicated in base pairs.

in three of the families. The remaining 11 families had various positive and negative lod (log of the odds) scores. They then examined tumor DNA and normal DNA to determine if loss of heterozygosity (LOH; deletion of one of two alleles of a gene) had occurred at the D2S123 locus. The locus was not deleted in any of the HNPCC tumors they examined, and only one deletion was observed in 46 sporadic tumors. Absence of allelic loss of the D2S123 locus suggested that the gene linked to this locus is not a tumor suppressor gene.

Replication Errors in HNPCC Tumors

While looking for LOH, Aaltonen et al. (33) observed a phenomenon that they call the replication error phenotype (RER+). Tumors that were RER+ displayed shifts in the electrophoretic mobility of microsatellite repeat fragments, suggesting that replication errors had occurred in these sequences during tumor development. Some examples of the RER+ phenotype are shown in Figure 11.4. In the study of Aaltonen et al., DNA from 11 of 14 HNPCC tumors examined showed shifts in at least two of seven microsatellite markers from four different chromosomes (33). Although they only found evidence for chromosome 2 linkage in about half of the HNPCC families, 10 of 13 families studied had RER+ tumors. Of these 10, one showed no evidence of linkage, and in seven linkage analysis was equivocal or impossible. These findings provided evidence that more than one genetic locus can cause this phenotype.

The gene for the HNPCC locus on chromosome 2p has been identified (34,35). It is called hMSH2 and is the human homologue of the bacterial mutS and *Saccharomyces cerevisiae* MSH proteins. In *Escherichia*

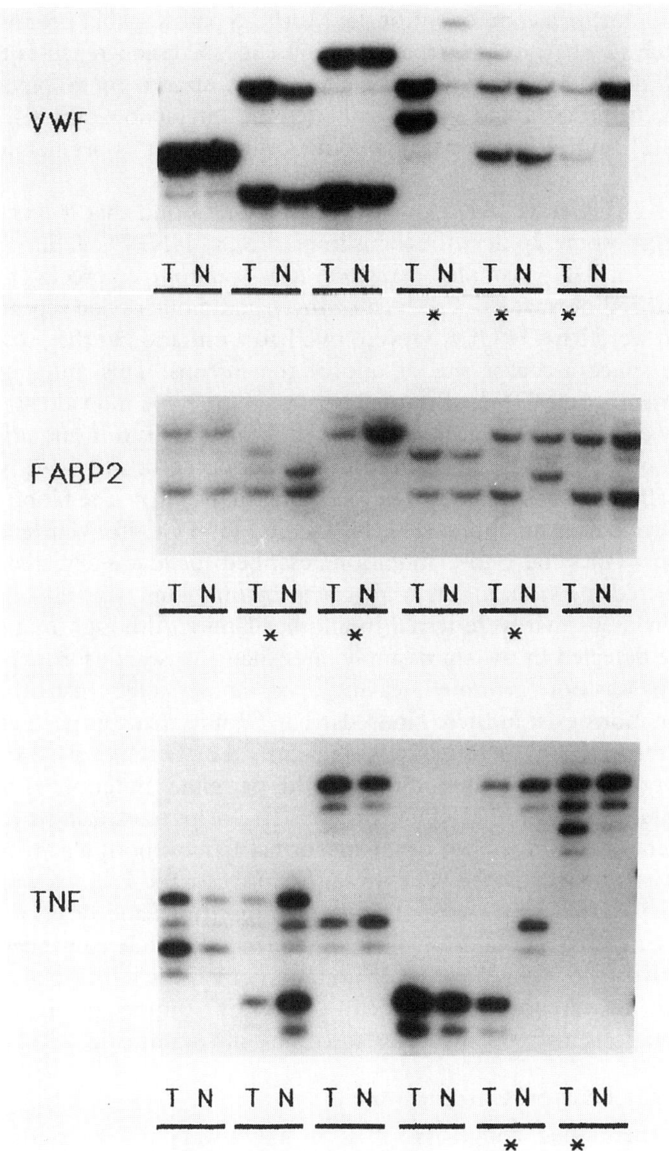

Fig. 11.4: Microsatellite repeat polymorphisms in tumor (T) and normal distant (N) mucosa of the same patient. Asterisks indicate specimens with the RER+ phenotype.

coli, mutS is a component of the MutHLS pathway that promotes a long-patch (2-kb) excision repair reaction (36). Excision repair is targeted to the newly replicated unmethylated DNA strand. mutS binds the mismatched DNA, and mutL facilitates the interaction between mutS and mutH. mutH functions by binding GATC sites that are hemimethylated and makes incisions on the unmethylated strand.

Fishel et al. (34) cloned this gene and found that it was located on chromosome 2p near the locus implicated in HNPCC. They analyzed 26 pairs of DNA samples extracted from random colorectal tumors and matched normal tissue. Seven of these had dinucleotide repeat instability and were RER+. Of the seven, two had a mutation at the −6 position of the splice acceptor site of one of the introns. This mutation was not present in the DNA of normal tissues of the same individuals. Fishel and coworkers note that similar types of T-to-C transition mutations within the polypyrimidine tracts in mRNA splice acceptor sites have been shown to affect mRNA splicing. They examined DNA from the blood of affected individuals from nine small HNPCC and HNPCC-like kindreds for mutations. The same T-to-C mutation described above was detected in all three affected individuals that were tested in one family and in both affected individuals that were tested in another family. Although mutations were not detected in the third family, an exhaustive search for hMSH2 mutations was not performed, leaving open the possibility that other hMSH2 mutations exist in these kindreds. The finding that this gene corresponds to the HNPCC locus on 2p was reaffirmed by Leach et al. (35). They also demonstrated that in addition to the germline mutation, other somatic mutations occur in HNPCC tumors. This helps to explain why the RER+ phenomenon does not occur in normal tissues: both alleles of the gene have to be inactivated. The mutations in yeast and bacteria with defective mutS-related genes are not limited to insertions and deletions at simple repeated sequences, though these sequences provide convenient tools for analysis (36). This means that mutations are likely to occur at several loci, and, as will be discussed below, HNPCC tumors do appear to have mutations in genes that are also commonly mutated in CRC.

HNPCC Locus on Chromosome 3

As mentioned above, HNPCC is not always linked to the locus on chromosome 2, which has led to a search for other genetic loci linked to HNPCC in some families. Lindblom et al. (37) demonstrated linkage of markers on the short arm of chromosome 3 (3p21) to the disease locus. They examined three families and observed significant linkage in one family. In a second family the lod score was below the commonly accepted level of significance but was suggestive of linkage, and in the third family there was evidence against linkage between the markers on chromosome 3 and the disease locus. The disease locus in this third family was also not linked to the HNPCC locus on chromosome 2 or to other candidate loci such as APC and the deleted in colon cancer (DCC) gene, which will be discussed below. However, because a tumor from this family has the RER+ phenotype, one might speculate that the locus encodes another protein involved in a repair pathway. It is worth mentioning that Strand et al. (38) showed that mutations in any one of the yeast genes PMS1, MLH1, and MSH2 can lead to mutations in simple repeats in the yeast genome. These genes are all components of the yeast counterpart of the MutHLS pathway in *E. coli*. PMS1 and MLH1 are homologues of mutL, and MSH2 is a homologue of mutS. The human counterparts to these genes are therefore candidates for other HNPCC genes.

SPORADIC COLORECTAL CARCINOMA: COMPARISON WITH APC AND HNPCC

The major difference between hereditary CRC (APC and HNPCC) and sporadic CRC is that in hereditary CRC the first genetic alteration has already occurred. Otherwise, tumor progression in each case is the result of a series of genetic alterations, and the order of the farther genetic alterations varies. There are some differences in the progression of these three major tumor types. For example, although all forms of CRC are thought to arise from adenomas, the rate of conversion from adenoma to carcinoma is likely to vary, depending on the molecular origins of this adenoma. In HNPCC the incidence of adenomas is not particularly high, but the rate of conversion of adenoma to carcinoma appears to be significantly faster than in APC or sporadic CRC (6,39). In contrast, patients with APC mutations have a dramatically increased rate of adenoma formation, but the rate of conversion of adenoma to carcinoma is normal (40). Other differences will be mentioned below as each of the genetic alterations that occur in sporadic CRC is discussed.

Most patients with CRC are classified as having the sporadic form of the disease; however, Burt et al. pointed out that many of these cases cluster in families and that the first-degree relatives of those with malignant lesions of the large bowel have a twofold to threefold greater risk of developing CRC (41). Whether such observations can be explained by genetic factors, environmental factors, or both remains to be determined.

APC Mutations in Sporadic CRC

The discovery of the APC gene on chromosome 5q21 was important in understanding the molecular basis of sporadic CRC as well as APC. It was already known that 5q21 alleles are often lost from sporadic CRC tumors (42,43). Powell et al. (22) examined 41 sporadic colorectal tumors and determined that 60% (15/25) of the adenocarcinomas and 63% (10/16) of the adenomas had mutations at this locus. Because a similar percentage of adenomas and carcinomas was found to contain APC gene mutations, their evidence suggested that mutations in the APC gene may play an important role in the development of common colorectal tumors early in colorectal carcinogenesis. It is of interest that three of the tumors had more than one APC mutation and five tumors had lost the normal copy of the APC gene and had only the mutated copy. To further substantiate the role of APC, Powell and coworkers examined DNA from five very small adenomas (<1 cm in diameter) that contained APC mutations. Only one of them had a *ras* mutation. *ras* mutations are thought to occur early in the process of tumorigenesis and will be discussed in detail below. Su et al. (24) estimate that mutations in the APC tumor suppressor gene lead to benign colorectal tumors in about one-third of the people in the Western world. This figure was based on the studies by Powell et al. (22) of the mutation rate of APC in adenomas, and on a study by Ransohoff and Lang (44) reporting the incidence of adenomatous polyps in the general population. If not removed, 15% to 20% of these develop into carcinoma.

ras Gene Mutations

ras mutations occur in 40% to 50% of colorectal tumors (45,46). There are three *ras* genes, K-*ras*, N-*ras*, and H-*ras*, and mutations in any of them at codons 12, 13, and 61 produce transforming activity. Bos et al. (45) examined colorectal carcinomas for mutations in codon 12 and 61 of the H-*ras*, K-*ras*, and N-*ras* genes and codon 13 of the N-*ras* and K-*ras* genes, and they found 11 mutations in 27 tumors. Nine tumors had mutations at codon 12 of the K-*ras* gene, one had a mutation in codon 61 of the K-*ras* gene, and one had a mutation in codon 12 of the N-*ras* gene. There were no mutations in the H-*ras* gene.

In an expanded study, Vogelstein et al. (47) examined DNA from 92 carcinomas and 80 adenomas at codons 12, 13, and 61 in the N-*ras* and K-*ras* genes. They detected mutations in 47% of the carcinomas, with 88% of these occurring in the K-*ras* gene, mostly in codon 12. For each gene, mutations were observed at each codon; however, the spectrum of mutations for CRC was different from that found in other tumors, with G-to-A transitions being the most common, while in pancreas and lung tumors G-to-T transversions are more common (48). The explanation for these differences is unknown, but they could represent differences in the mechanisms by which the *ras* mutations are introduced into the different tissues or differences in the ability of different *ras* mutations to contribute to the process of tumorigenesis in the different tissues.

A similar spectrum of *ras* mutations was observed in adenomas. In the non-APC adenomas examined in this study, the frequency of *ras* mutations increased with the size of the adenoma. Only 12% of adenomas under 1 cm in diameter (early adenomas) had a mutated *ras* gene, while 63% of adenomas at least 1 cm in diameter (intermediate adenomas), and 57% of adenomas that had given rise to invasive carcinoma had mutated *ras* genes. These studies suggest that *ras* gene mutations tend to occur most commonly during the transition from early to intermediate adenoma.

p53 Gene Mutations

Vogelstein et al. (47) observed that allelic deletions occurred on chromosome 17p in 75% of CRCs that they examined. In the same study the adenomatous portions of only 24% of large adenomas that had given rise to carcinomas had deletions in this region, and for smaller adenomas these losses were even less frequent (6%). From this Vogelstein and colleagues concluded that 17p deletions tend to occur during the transition from late adenoma to carcinoma. To determine if the allelic deletions occurred in any specific region within chromosome 17p, Baker et al. (49) compared DNA isolated from 58 CRC specimens with DNA from normal adjacent mucosa. Using a panel of 20 DNA probes that detect restriction fragment length polymorphisms on chromosome 17p, they determined that the most common region of deletion was between bands 17p12 and 17p13.3. Because *p53* had been localized to this region of chromosome 17p and had previously been implicated in the process of neoplastic transformation, Baker and colleagues wanted to determine if it was a target gene of the deletion. They were not able to detect gross alterations in the structure or expression of *p53* in tumors but did detect point mutations, including one in a complementary DNA (cDNA) isolated from a nude mouse xenograft (C×3) of a primary colorectal tumor. Nucleotide sequence analysis of the cDNA clone identified a T-to-C transition within codon 143 (GTG to GCG), resulting in a change in the amino acid from valine to alanine. A second tumor had a mutation that resulted from a G-to-A transition at codon 175 that produced a substitution of a histidine for an arginine. The mutation was not detected in DNA from normal cells of this individual.

Both of these tumors had lost one copy of chromosome 17p, suggesting in each case that a mutation had occurred in one copy of the *p53* gene, and the normal copy had been lost by deletion. In an expanded study, Baker et al. (50) observed that *p53* gene mutations were rare in adenomas in general and in adenomas and carcinomas containing two copies of chromosome 17 (17% of 30 tumors). Tumors that had lost a copy of chromosome 17 usually had a mutation in the remaining *p53* allele (86% of 28 tumors). The combination of *p53* mutation and loss of the normal *p53* allele accompanies the transition from benign to malignant growth in colorectal carcinogenesis and therefore is likely to play a causal role in this transition.

Unlike *ras* mutations, which occur within three codons in the *ras* genes, the *p53* mutations were distributed throughout the *p53* coding region, with four hot spots in the coding region for amino acids 132–142, 171–179, 239–248, and 272–286 (51), which are the most evolutionarily conserved regions of the gene, suggesting that these regions are important functional domains.

The gene *p53* is a tumor suppressor and functions as a transcription factor (52). As a transcription factor it stimulates the production of a protein called Cdk-interacting protein 1 (Cip-1), also called wild-type p53-activated fragment 1 (WAF1). Cip-1 inhibits the cyclin-dependent kinases, which are important in driving the cell cycle and are active in the presence of cyclin. When *p53* is mutated in cancer, it is unable to effectively stimulate the transcription of the Cip-1 gene. The end result is uncontrolled cell growth, as is seen in CRC.

Comparison of *ras*, *p53* and APC Mutations in HNPCC and Sporadic CRC

To determine if the mutations that commonly occur in sporadic CRC are also observed in the tumors of HNPCC patients, Aaltonen et al. (33) examined 18 patients from 15 different HNPCC families. It had been determined that one family probably had linkage of HNPCC with the chromosome 2 marker, one was unlinked, and the rest were either uninformative or not studied for linkage. The results of their studies are summarized in Table 11.1. Examination of the K-*ras* gene revealed that 61% of the tumors had K-*ras* mutations, a higher percentage than is found among sporadic cases (40%). K-*ras* mutations tend to occur early in tumorigenesis, and *p53* mutations tend to occur later, generally during the transition from the benign to the malignant state. Mutations in *p53* were

Table 11.1: Comparison of K-*ras*, *p53*, and APC Mutations Between HNPCC and Sporadic Colorectal Carcinoma

Tumor Type	*K-ras*	*p53*	*APC*
HNPCC	11/18 (61%)	7/11 (64%)	8/14 (57%)
Sporadic	37/92 (40%)	81/132 (61%)	20/41 (49%)

found to occur in 64% of HNPCC cancers and in 61% of sporadic CRCs. Mutations in the APC gene are thought to initiate a high proportion of sporadic cancers, while families with APC have germline mutations in the APC gene. Aaltonen et al. (33) hypothesized that the HNPCC gene might substitute for the APC gene in families with HNPCC, but this is not true, because 57% of the HNPCC tumors that they examined had APC mutations as compared with 49% of sporadic colon tumors.

Replication Errors in Sporadic CRC

During the studies of RER in HNPCC tumors described above, Aaltonen et al. (33) also examined 46 sporadic tumors for the RER phenotype, using these same seven microsatellite markers. Of these, six tumors displayed band shifts for two or more of the markers. Like the HNPCC tumors, which were most often on the right side of the colon and diploid, all six of the RER+ sporadic tumors were on the right side while only 17 of the 40 RER− tumors were. Chromosome imbalances were also more common in the RER− tumors than in the RER+ tumors.

Thibodeau et al. (53) examined 90 colorectal carcinomas resected from 87 patients at the Mayo Clinic between 1987 and 1988. Using four microsatellite repeat loci on human chromosomes 5q, 15q, 17p, and 18q, they observed a higher rate of RER+ in sporadic tumors; 25 (28%) of the 90 tumors had changes in the number of repeats. They divided the replication errors in RER+ tumors into two categories: large changes in fragment size were classified as type I mutations, and minor alterations, such as a single 2-bp mutation, were classified as type II mutations. Of the 25 tumors with microsatellite alterations, 8 had type I mutations, 12 type II, and 5 both type I and type II. Each of the 13 tumors with type I mutations had several mutations. Of the 12 with only type II mutations, 10 showed mutations at only one site. Thiboudeau and coworkers found that RER+ tumors occurred most frequently in the proximal colon. Using Kaplan-Meier survival curves, they also observed that these RER+-related mutations correlated positively with overall survival.

Ionov et al. (54) have also observed consistent alterations in simple repeated sequences in a subset of sporadic CRC, using a DNA fingerprinting technique in which DNA is subjected to PCR using primers that have been chosen arbitrarily without any previous knowledge of the sequence that is amplified. The bands from DNA isolated from tumors were different from those from normal tissue. Reduction in band size was observed in tumor tissue with 6 of 10 primers. Nucleotide sequence analysis demonstrated that the altered bands resulted from changes in the poly(A) tracts of Alu repeats, which are repetitive DNA sequences about 300 bp long. Deletions were also detected in A:T base pairs in DNA sequences without Alu repeats. From the frequency with which these mutations were seen, Ionov and colleagues calculated that these deletions occur every 1×10^4 to 3×10^4 bp, meaning that there would be 1×10^5 to 3×10^5 deletions per genome.

DCC Gene Deletions

The DCC gene is a tumor suppressor located on the long arm of chromosome 18. Allelic deletions of this gene have been observed in 71% of CRCs examined, reduced expression in 88%, and somatic mutations in 13% (55). Introduction of a normal chromosome 18 into a human CRC cell line suppressed its tumorigenicity, further supporting the role of chromosome 18 in colorectal carcinogenesis (56).

MCC Gene Mutations

The MCC gene was assayed for mutations in 17 MCC exons in 90 CRCs using RNase protection assays, and six point mutations were identified (11,12). The frequency of mutation in the coding regions of the MCC gene in sporadic CRC, based on an estimate of 35% efficiency, is approximately 19%. While the function of the MCC gene is unknown, it is implicated in tumorigenesis because it is mutated.

Deletion of Other Genes

Vogelstein et al. (47) examined polymorphic DNA markers from every chromosomal arm in 56 paired CRC and adjacent normal colonic mucosa specimens. They found that one of the alleles from each polymorphic marker was lost in at least some tumors, and some tumors had lost more than half of their parental alleles. In some cases, new DNA fragments were generated. These new DNA fragments were detected with probes that detected variable numbers of tandem repeats (VNTRs). Whether or not these new DNA fragments are related to the RER+ phenomenon described above is not known. A greater number of the allelic deletions correlated with a worse prognosis when tumors of similar size and stage were compared. A number of studies indicate that LOH correlates with tumor location. Tumors without LOH tend to occur in the proximal colon (57–59). Colorectal tumors from HNPCC patients and sporadic RER+ colorectal tumors also tend to occur in the proximal colon. In addition, LOH is relatively less common in both HNPCC and sporadic RER+ tumors than in sporadic RER− tumors (33,53,60).

The general conclusion one might draw from all of these findings is that patients with HNPCC and patients with sporadic RER+ tumors tend to have less LOH and a better prognosis and that their tumors are likely to occur in the proximal colon, while sporadic colorectal tumors that are

RER− have more LOH, tend to occur in the distal colon, and have a worse prognosis.

ALTERATIONS IN THE LEVELS OF GENE EXPRESSION IN COLORECTAL CARCINOMA

In CRC, alterations in gene expression can occur for a variety of reasons. Understanding how mutations in oncogenes and tumor suppressor genes influence expression of other genes is fundamental to understanding tumorigenesis. As discussed above, the level of Cip-1 is reduced in tumors in which *p53* is inactivated as a result of mutations or deletions. Other examples of genes whose expression is altered in CRC are discussed below.

Overexpression of Cell-Cycle-Dependent Genes

Calabretta et al. (61) hypothesized that *p53* and other cell-cycle-dependent genes were expressed at higher levels in cancer because of the increase in the number of cycling cells in tumors relative to normal adjacent tissue. They reported that p53 protein and RNA levels in cell culture are highest when most cells are in late G_1 or S phase and that increases in p53 parallel increases in expression of other cell-cycle genes. However, contradictory to this hypothesis was their finding that when they examined human fibroblast cell lines and their simian virus 40–transformed counterparts, the levels of p53 mRNA were found to be significantly higher in the SV40-transformed cells. This increase was not accompanied by a proportional increase in histone H3 mRNA, a cell-cycle-dependent mRNA that is normally expressed preferentially in S phase. Therefore, the overexpression in this case is not totally explained by differences in the numbers of cycling cells in the SV40-transformed fibroblasts relative to the SV40-non-transformed fibroblasts, so there must be other mechanisms by which *p53* is overexpressed in malignancies. It is also worth mentioning that because *p53* is frequently mutated in CRC, another explanation for its overexpression at the protein level may be an increase in its stability as a result of the mutation. Expression of the RNA for c-*myc* (another cell-cycle-dependent gene) was found to be increased in the normal and adjacent neoplastic cells of six different colonic neoplasms (62). This increase was accompanied by an increase in the expression of other cell-cycle-dependent genes, but because the ratio of c-*myc* to histone H3 (another cell-cycle-dependent gene) was higher in the adenocarcinoma of one patient relative to the adenoma, Calabretta and colleagues concluded that in the adenocarcinoma, there was an increase in the amount of c-*myc* in addition to the overexpression due to the increase in the number of cycling cells.

Overexpression of Elongation Factor-1γ and Other Translation Factors in Colorectal Adenoma and Carcinoma

Elongation factor-1 (EF-1) is a guanine-binding protein and plays an important role in translation by mediating the transport of amino-acyl transfer RNA to 80S ribosomes. EF-1 contains four major subunits, EF-1α, EF-1β, EF-1γ, and EF-1δ. It was previously demonstrated that human EF-1γ-hybridizing RNA was overexpressed in 25 of 29 colorectal carcinomas and 13 of 25 colorectal adenomas (63,64), but overexpression was not due to the greater number of cycling cells in tumor tissue. The EF-1γ subunit is believed to play a stimulatory role in the functioning of EF-1. In addition, EF-1γ is phosphorylated by a cell-division-controlled protein kinase, $p34^{cdc2}$ (65). The functional role of this modification is not yet known. Because EF-1γ was found to be overexpressed in 6 out of 10 small adenomas ($\leq$1 cm), the overexpression is believed to occur early in tumorigenesis.

There is evidence that overexpression of other translational factors may play a role in the process of tumorigenesis. For example, the overexpression of EF-1α in Balb/c3T3 cells increased their susceptibility to chemically and physically induced transformation (66). It is important to note that EF-1γ is a stimulator of EF-1α activity, suggesting that overexpression of EF-1γ and EF-1α would produce the same effect. Overexpression of eIF-4E, a translational initiation factor, in NIH3T3 cells and Rat 2 fibroblasts resulted in tumorigenic transformation, as determined by three criteria: formation of tranformation foci on a monolayer of cells, anchorage-independent growth in nude mice, and tumor formation in nude mice (67). Studies of the mechanisms by which overexpression of translational factors can effect these changes will be important in understanding tumorigenesis.

Overexpression of Transcription Factor and Transcription Factor–like Genes

The c-*myc* gene, a DNA-binding transcriptional activator (68), is commonly overexpressed in CRC (69,70). The c-*fos* gene is overexpressed in the aberrant crypts induced in rats as a result of treatment with azoxymethane. Aberrant crypts are the earliest detectable preneoplastic lesions in CRC (71). c-*fos* forms a heterodimer with c-*jun* that binds to the AP-1 binding site in the promoter region of some genes. Because some of these genes are involved in cell proliferation, and, in fact, c-*fos* is required for cell proliferation, the overexpression of c-*fos* in aberrant crypt foci may contribute to the increased rate of proliferation in these cells or may be a consequence of the increased cell proliferation.

The DRA (down-regulated in adenoma) mRNA is expressed in normal colonic tissue and at significantly lower levels in adenomas and adenocarcinomas of the colon (72). The DRA protein sequence that has been deduced from the nucleotide sequence of the cDNA has features of a transcription factor or a protein that interacts with transcription factors. It is located on chromosome 7, and as a potential transcription factor, it is also a candidate tumor suppressor.

CONCLUSION

Over the past decade, a great deal of emphasis has been placed on identifying genes that are mutated in colon cancer and determining their relative importance in initiating the early phases of tumorigenesis. Future studies must deal with the mechanisms by which this expanding body of genes contributes to the overall process of carcinogenesis. Studies of alterations in gene expression will become essential in elucidating these mechanisms.

REFERENCES

1. Fearon ER, Vogelstein B. A genetic model for colorectal tumorigenesis. Cell 1990;61:759–767.

2. Bussey HJR, Veale AMO, Morson BC. Genetic and gastrointestinal polyposis. Gastroenterology 1978;74:1325–1330.

3. Bussey HJR. Familial polyposis coli: family studies, histopathology, differential diagnosis and results of treatment. Baltimore: Johns Hopkins University Press, 1975.

4. Boland CR, Itzkowitz SH, Kim YS. Colonic polyps and the gastrointestinal polyposis syndrome. In: Gleisinger MH, Fordtran JS, eds, Gastrointestinal disease, 4th ed. Philadelphia: WB Saunders, 1989:1483–1518.

5. Lynch HT, Watson P, Kriegler M, et al. Differential diagnosis of hereditary nonpolyposis colorectal cancer (Lynch syndrome I and Lynch syndrome II). Dis Colon Rectum 1988;31:372–377.

6. Lynch HT, Smyrk TC, Watson P, et al. Genetics, natural history, tumor spectrum, pathology of hereditary nonpolyposis colorectal cancer: an updated review. Gastroenterology 1993;104:1535–1549.

7. Mulvihill JJ. The frequency of hereditary large bowel cancer. In: Ingall JRF, Mastromarino AJ, eds. Prevention of hereditary large bowel cancer. New York: Alan R. Liss, 1983:61–75.

8. Jarvinen HJ. Epidemiology of familial adenomatous polyposis in Finland: impact of family screening on the colorectal cancer rate and survival. Gut 1992;33:357–360.

9. Groden J, Thliveris A, Samowitz W, et al. Identification and characterization of the familial adenomatous polyposis coli gene. Cell 1991;66:589–600.

10. Joslyn G, Carlson M, Thliveris A, et al. Identification of deletion mutations and three new genes at the familial polyposis locus. Cell 1991;66:601–613.

11. Kinzler KW, Nilbert MC, Vogelstein B, et al. Identification of a gene located at chromosome 5q21 that is mutated in colorectal cancer. Science 1991;251:1366–1368.

12. Nishisho I, Nakamura Y, Miyoshi Y, et al. Mutations of chromosome 5q21 genes in FAP and colorectal cancer patients. Science 1991;253:665–669.

13. Miyoshi Y, Ando H, Nagase H, et al. Germ-line mutations of the APC gene in 53 familial adenomatous polyposis patients. Proc Natl Acad Sci USA 1992;89:4452–4456.

14. Olschwang S, Tiret A, Laurent-Puig P, Muleris M, Parc R, Thomas G. Restriction of ocular fundus lesions to a specific subgroup of APC mutations in adenomatous polyposis coli patients. Cell 1993;75:959–968.

15. Varesco L, Gismondi V, James R, et al. Identification of APC gene mutations in Italian adenomatous polyposis coli patients by PCR-SSCP analysis. Am J Hum Genet 1993;52:280–285.

16. Nagase H, Nakamura Y. Mutations of the APC (adenomatous polyposis coli) gene. Hum Mut 1993;2:425–434.

17. Leppert M, Burt R, Hughes J, et al. Genetic analysis of an inherited predisposition to colon cancer in a family with a variable number of adenomatous polyps. N Engl J Med 1990;322:904–908.

18. Spirio L, Otterud B, Stauffer D, et al. Linkage of a variant or attenuated form of adenomatous polyposis coli to the adenomatous polyposis coli (APC) locus. Am J Hum Genet 1992;51:92–100.

19. Spirio L, Olschwang S, Groden J, et al. Alleles of the APC gene: an attenuated form of familial polyposis. Cell 1993;75:951–957.

20. Blair NP, Trempe CL. Hypertrophy of the retinal pigment epithelium associated with Gardner's syndrome. Am J Ophthalmol 1980;90:661–667.

21. Nakamura Y, Lathrop M, Leppert M, et al. Localization of the genetic defect in familial adenomatous polyposis within a small region of chromosome 5. Am J Hum Genet 1988;43:638–644.

22. Powell SM, Zilz N, Beazer-Barclay Y, et al. APC mutations occur early during colorectal tumorigenesis. Nature 1992;359:235–237.

23. Rubinfeld B, Souza B, Albert I, et al. Association of the APC gene product with β-catenin. Science 1993;262:1731–1734.

24. Su L, Vogelstein B, Kinzler KW. Association of the APC tumor suppressor protein with catenins. Science 1993;262:1734–1737.

25. Vasen HFA, Mecklin JP, Khan PM, Lynch HT. The international collaborative group on hereditary non-polyposis colorectal cancer (ICG-HNPCC). Dis Colon Rectum 1991;34:424–425.

26. Lynch HT, Lanspa S, Smyrik T, Boman B, Watson P, Lynch J. Hereditary nonpolyposis colorectal cancer (Lynch syndromes I and II). Genetics, pathology, natural history and cancer control, part I. Cancer Genet Cytogenet 1991;53:143–160.

27. Mecklin JP, Jarvinen HJ. Tumor spectrum in cancer family syndrome (hereditary nonpolyposis colorectal cancer). Cancer 1991;68:1109–1112.

28. Hakala T, Mecklin JP, Forss M, Jarvinen H, Lehtovirta P. Endometrial carcinoma in the cancer family syndrome. Cancer 1991;68:1656–1659.

29. Peltomaki P, Aaltonen LA, Sistonen P, et al. Genetic mapping of a locus predisposing to human colorectal cancer. Science 1993;260:810–812.

30. Hearne CM, Ghosh S, Todd JA. Microsatellites for linkage analysis of genetic traits. Trends Genet 1992;8:288–294.

31. Weber JL. Informativeness of human (dC-dA)n·(dG-dT)n polymorphisms. Genomics 1990;7:524–530.

32. van Amstel HKP, Reitsma PH. Tetranucleotide repeat polymorphism in the VWF gene. Nucleic Acids Res 1990;18:4957.

33. Aaltonen LA, Peltomaki P, Leach FS, et al. A clue to the pathogenesis of familial colorectal cancer. Science 1993;260:812–816.

34. Fishel R, Lescoe MK, Rao MRS, et al. The human mutator gene homolog MSHA and its association with hereditary nonpolyposis colon cancer. Cell 1993;75:1027–1038.

35. Leach FS, Nicolaides NC, Papadopoulos N, et al. Mutations of a *mutS* homolog in hereditary nonpolyposis colorectal cancer. Cell 1993;75:1215–1225.

36. Modrich P. Mechanism and biological effects of mismatch repair. Annu Rev Genet 1991;25:229–255.

37. Lindblom A, Tannergard P, Werelius B, Nordenskjold M. Genetic mapping of a second locus predisposing to hereditary non-polyposis colon cancer. Nature Genet 1993;5:279–282.

38. Strand M, Prolla TA, Liskay RM, Petes TD. Destabilization of tracts of simple repetitive DNA in yeast by mutations affecting DNA mismatch repair. Nature 1993;365:274–276.

39. Jass JR, Stewart SM. Evolution of hereditary nonpolyposis colorectal cancer. Gut 1992;33:783–786.

40. Bussey HJR. Historical developments in familial adenomatous polyposis. In: Herrera L, ed. Familial adenomatous polyposis. New York: Alan R. Liss, 1990:1–7.

41. Burt RW, Bishop DT, Cannon-Albright L, et al. Population genetics of colonic cancer. Cancer (Suppl) 1992;70:1719–1722.

42. Solomon E, Voss R, Hall V. Chromosome 5 allele loss in human colorectal carcinomas. Nature 1987;328:616–619.

43. Sasaki M, Okamoto M, Sato C, et al. Loss of constitutional heterozygosity in colorectal tumors from patients with familial polyposis coli and those with nonpolyposis colorectal carcinoma. Cancer Res 1989;49:4402–4406.

44. Ransohoff DF, Lang CA. Screening for colorectal cancer. N Engl J Med 1991;325:37–41.

45. Bos JL, Fearon ER, Hamilton SR, et al. Prevalence of ras gene mutations in human colorectal cancers. Nature 1987;327:293–297.

46. Forrester K, Almoguera C, Han K, Grizzle WE, Perucho M. Detection of high incidence of K-ras oncogenes during human colon tumorigenesis. Nature 1987;327:298–303.

47. Vogelstein B, Fearon ER, Hamilton SR, et al. Allelotype of colorectal carcinomas. Science 1989;244:207–211.

48. Smit VTHBM, Boot AJM, Smits AMM, Fleuren GJ, Cornelisse CJ, Bos JL. KRAS codon 12 mutations occur very frequently in pancreatic adenocarcinomas. Nucleic Acids Res 1988;16:7773–7782.

49. Baker SJ, Fearon ER, Nigro JM, et al. Chromosome 17 deletions and p53 gene mutations in colorectal carcinomas. Science 1989;244:217–221.

50. Baker SJ, Preisinger AC, Jessup JM, et al. p53 gene mutations occur in combination with 17p allelic deletions as late events in colorectal tumorigenesis. Cancer Res 1990;50:7717–7722.

51. Nigro JM, Baker SJ, Preisinger AC, et al. Mutations in the p53 gene occur in diverse human tumor types. Nature 1989;342:705–708.

52. Raycroft L, Wu H, Lozano G. Transcriptional activiation by wild-type but not transforming mutants of the p53 anti-oncogene. Science 1990;249: 1049–1051.

53. Thibodeau SN, Bren G, Schaid D. Microsatellite instability in cancer of the proximal colon. Science 1993;260:816–819.

54. Ionov Y, Peinado MA, Malkhosyn S, Shibata D, Perucho M. Ubiquitous somatic mutations in simple repeated sequences reveal a new mechanism for colonic carcinogenesis. Nature 1993;363:558–561.

55. Fearon ER, Cho KR, Nigro JM, et al. Identification of a chromosome 18q gene that is altered in colorectal cancers. Science 1990;247:49–56.

56. Tanaka K, Oshimura M, Kikachi R, et al. Suppression of tumorigenicity in human colon carcinoma cells by introduction of normal chromosome 5 or 18. Nature 1991;349:340–342.

57. Delattre O, Law DJ, Remvikos Y, et al. Multiple genetic alterations in distal and proximal colorectal cancer. Lancet 1989;2:353–355.

58. Kern SE, Fearon ER, Tersmette KWF, et al. Allelic loss in colorectal carcinoma. JAMA 1989;261:3099–3103.

59. Offerhaus GJA, De Feyter EP, Cornelisse CJ, et al. The relationship of DNA aneuploidy to molecular genetic alterations in colorectal carcinoma. Gastroenterology 1992;102:1612–1619.

60. Kouri M, Laasonen A, Mecklin JP, Jarvinen H, Franssila K, Pyrhonen S. Diploid predominance in hereditary nonpolyposis colorectal carcinoma evaluated by flow cytometry. Cancer 1990;65:1825–1829.

61. Calabretta B, Kaczmarek L, Selleri L, et al. Growth-dependent expression of human M_r 53,000 tumor antigen messenger RNA in normal and neoplastic cells. Cancer Res 1986;46:5738–5742.

62. Calabretta B, Kaczmarek L, Ming PML, Au F, Ming SC. Expression of c-*myc* and other cell cycle-dependent genes in human colon neoplasia. Cancer Res 1985;45:6000–6004.

63. Chi K, Jones DV, Frazier ML. Expression of elongation factor 1 γ-related sequence in adenocarcinoma of the colon. Gastroenterology 1992;103:98–102.

64. Ender B, Lynch P, Kim YH, Inamdar NV, Cleary KR, Frazier ML. Overexpression of an elongation factor-1γ hybridizing RNA in colorectal adenomas. Mol Carcinog 1993;7:18–20.

65. Belle R, Derancourt J, Poulhe R, Capony J-P, Ozon R, Mulner-Lorillon O. Purified complex from *Xenopus* oocytes contains a p47 protein, an in vivo

substrate of MPF, and a p30 protein respectively homologous to elongation factors EF-1 gamma and EF-1 beta. FEBS Lett 1989;255:101–104.

66. Tatsuka M, Mitsui H, Wada M, Nagata A, Nojima H, Okayama H. Elongation factor-1α gene determines susceptibility to transformation. Nature 1992;359:333–336.

67. Lazaris-Karatzas A, Montine KS, Sonenberg N. Malignant transformation by a eukaryotic initiation factor subunit that binds to mRNA 5′ cap. Nature 1990;345:544–547.

68. Marcu KB, Bossone SA, Patel AJ. *myc* function and regulation. Annu Rev Biochem 1992;61:809–860.

69. Sikora K, Chan S, Evan G, et al. *c-myc* oncogene expression in colorectal cancer. Cancer 1987;59:1289–1295.

70. Erisman MD, Rothberg PG, Diehl RE, Morse CC, Spandorfer JM, Astrin SM. Deregulation of c-*myc* gene expression in human colon carcinoma is not accompanied by amplification or rearrangement of the gene. Mol Cell Biol 1985;5:1969–1976.

71. Stopera SA, Davie JR, Bird RP. Colonic aberrant crypt foci are associated with increased expression of c-fos: the possible role of modified c-fos expression in preneoplastic lesions in colon cancer. Carcinogenesis 1992;13:573–578.

72. Schweinfest CW, Henderson KW, Suster S, Kondoh N, Papas TS. Identification of a colon mucosa gene that is down-regulated in colon adenomas and adenocarcinomas. Proc Natl Acad Sci USA 1993;90:4166–4170.

CHAPTER 12

Molecular Basis of Prostate Cancer

Leland W. K. Chung
William B. Isaacs
Jan Trapman

The prostate is a walnut-sized gland located at the base of the bladder neck. It has been described as one of the largest organs in the body whose function is unknown. The prostate secretion appearing in the seminal plasma facilitates, but is not required for, sperm maturation, sperm penetration, and subsequent implantation of the fertilized ovum. Prostate secretion is likely to be a "dead-end" product affecting neither prostate growth and the hypothalamic-pituitary-gonad axis, nor secondary sex characteristics and sexual functions.

The prostate gland attracts the attention of laboratory and clinical investigators for two key reasons. First, prostate diseases, such as benign prostate hyperplasia (BPH), prostate cancer, and prostatitis, affect a large percentage of aging males (Table 12.1). For example, 400,000 surgical procedures for BPH are performed per year, making it one of the most common types of surgery performed in men. BPH affects more than 80% of the male population over the age of 80 and accounts for 1.7 million physician visits per year. In turn, prostate cancer is the most common cancer diagnosed (200,000 cases in 1994) and the second leading cause of cancer deaths (38,000 a year) in U.S. men (1). The economic impact of the prostate diseases is estimated to be over 3 billion dollars per year. Second, the prostate gland is easily accessible in experimental animals and highly responsive to sex steroids and thus presents an attractive opportunity to elucidate the molecular basis of androgen action and growth control mechanisms (2). These studies have significant clinical implications because androgen receptor dysfunction may be linked to the clinical progression of prostate cancer (e.g., from androgen-dependent to androgen-independent status) and to several genetic diseases involving muta-

Table 12.1: Impact of Prostate Diseases in the United States

Prostate Disease	*Physician Visits*	*Annual Hospitalization Rates*	*Mortality*	*Cost in Billions of Dollars*
Benign prostate hyperplasia	1,709,053	482,349	2,339	1.82
Prostate cancer	887,341	346,201	36,204	0.97
Prostatitis	1,850,593	108,024	672	0.29
Total	4,446,987	836,573	39,215	3.08

Adapted from Coffey DS. 1992.

tions of the androgen receptor gene and defects of androgen-metabolizing enzymes.

In the past decade, significant progress has been made in the understanding of the molecular mechanisms regulating prostate growth. In this chapter, progress in the following areas of research is highlighted: the androgen receptor in prostate cancer (discussed by J. Trapman); genetic alterations in prostate cancer (discussed by W. B. Isaacs); and stromal-epithelial interaction in prostate cancer growth, differentiation, and metastasis (discussed by L. W. K. Chung).

THE ANDROGEN RECEPTOR IN PROSTATE CANCER

The Androgen Receptor During Development

Sexual differentiation starts relatively late after gestation (day 12 to 14 in the mouse, week 10 to 12 in the human) at the indifferent stage of development (3). At that time the embryonic anlagen of either male or female genital reproductive tissues can be recognized. These structures are the indifferent gonads, the wolffian ducts, the müllerian ducts, and the urogenital sinus.

A cascade of events leading to testicular differentiation and testosterone production is probably triggered by testis determining factor, SRY, which is encoded by the *SRY* gene (on the Y chromosome) (4). As the male reproductive tract develops, anti-müllerian hormone causes the müllerian ducts to regress, and the wolffian ducts differentiate into the epididymis, ductus deferens, and seminal vesicles. The urogenital sinus develops into the bulbourethral gland and the prostate.

Both wolffian duct and urogenital sinus differentiation depend on androgens (e.g., testosterone) (3,5). In some androgen target tissues, the enzyme 5α-reductase converts testosterone into the more potent 5α-dihydrotestosterone. Persons who are 46,XY lacking type 2 5α-reductase develop normal internal male reproductive tissues, but no prostate, and their external genitalia resemble those of females (6); thus testosterone and 5α-dihydrotestosterone have different roles in the development of the male reproductive tract.

Like SRY during early male differentiation (testis development), the androgen receptor is a key molecule during subsequent steps of the development of male reproductive tissues, including the prostate. In target tissues, androgens effect a cellular response by binding with the intracellular androgen receptors. At the beginning of sexual development, androgen receptors are expressed in the mesenchymal cells of the embryonic anlagen of the reproductive tissues. During early development of the various reproductive tissues, androgen receptors are still undetectable in the epithelial cells (7–9). However, it is well established that epithelial cell differentiation is driven by androgens. This means that processes regulated by the activated androgen receptors in the mesenchymal cells are responsible for epithelial cell differentiation. Importantly, reconstitution experiments with mesenchymal cells from normal mice and epithelial cells from Tfm mice, which lack active androgen receptors (10,11), proved that the mesenchyme drives androgen-regulated differentiation of the prostate epithelium (12).

Later steps of prostate development might depend on the androgen receptors present in the epithelial cells, eventually resulting in the production of prostate-specific proteins (13). The molecular mechanism underlying the androgen-regulated mesenchymal cell–epithelial cell interaction in the developing prostate is not well understood (see section below). Probably, androgen-regulated growth factors and extracellular matrix components produced by the mesenchymal cells act on the epithelial cells, stimulating their growth and differentiation in a paracrine mechanism.

In the normal adult prostate, high levels of androgen receptor expression can be found in the secretory epithelial cells. In contrast, the stromal cells show a more variable pattern of androgen receptor expression; the basal epithelial cell layer is almost devoid of androgen receptor expression (14–17). The androgen receptor is essential for maintenance of prostate structure and function. Orchiectomy or application of antiandrogens leads to a rapid loss of the secretory epithelial cells in particular (2,3).

Structure and Properties of the Androgen Receptor

The androgen receptor belongs to the superfamily of ligand-responsive transcription factors (steroid–thyroid hormone–retinoic acid receptor family) (18–20). The androgen receptor exerts its action by binding to a specific sequence called the androgen response element (ARE) in the promoter region of target genes. The receptor interacts with other general and specific transcription factors, forming a stable transcription preinitiation complex, which allows efficient transcription of the target gene.

The androgen receptor is encoded by a messenger RNA species of approximately 11 kb. The open reading frame is 2.7 kb. The mRNA encodes a protein of 910 amino acids with a calculated molecular mass of 98.5 kDa (21–24). Three major functional domains can be distinguished: the amino-terminal regulatory domain, the DNA-binding domain, and the carboxy-terminal ligand-binding domain.

The androgen receptor gene is located on the X chromosome at Xq11-q12 and has a length of more than 90 kb (24–26). The open reading frame is separated over eight exons. The sequence encoding the amino-terminal domain is found in the first, large exon (22); the DNA-binding domain is encoded by exons 2 and 3; the information for the steroid-binding domain is distributed over exons 4 to 8.

In the amino-terminal domain, the human androgen receptor contains several stretches of homopolymeric amino acids. Two of these are very long: a glutamine stretch (Q stretch) starting at amino acid residue 58 and a glycine stretch (G stretch) starting at position 448. The Q stretch is encoded by a (CAG) (CAAn) repeat, and the G stretch by a GGT/C repeat. Both repeats are polymorphic. In DNA from whites, the (CAG) (CAAn) repeat can be composed of 19 to 29 repeat units (27). Five different allelic forms of the GGT/C repeat have been observed, varying in length from 16 to 24 repeat units (27). The polymorphisms offer the possibility of linkage studies in families with suspected defects in androgen receptor function. However, the variability in the lengths of the two repeats has resulted in confusion concerning the exact size of the androgen receptor and the numbering of the amino acids. In this review, the amino acid numbers correspond with an androgen receptor composed of 910 amino acid residues (20 Gln; 16 Gly).

The amino-terminal domain of the androgen receptor, which has a length of over 500 amino acids, is involved in the transactivating function of the receptor (28–30). Transactivating domains are thought to play a role in protein-protein interactions with other transcription factors and coactivators. However, the exact nature and boundaries of the putative transactivating domain (or domains) of the androgen receptor are not well defined.

The steroid-binding domain of the androgen receptor is confined to a carboxy-terminal fragment of approximately 250 amino acids. The amino acid sequence of this domain is completely conserved between the human, rat, and mouse androgen receptors, indicating that its structure is very important in androgen receptor function (21,31). The receptor's structural homology with other members of the steroid receptor family (e.g., glucocorticoid receptor and progesterone receptor) makes it likely that the steroid-binding domain contains sequences important for receptor homodimerization and transactivation (20,24).

A nucleoplasmin-like bipartite nuclear localization signal is found between the steroid-binding and DNA-binding domains (residues 608–624: RKcyeagmtlgaTKlKK) (32). In the absence of its ligand, the androgen receptor is distributed over the nucleus and the cytoplasm; the addition of androgen normally results in a rapid migration of the receptor to the nucleus. However, mutations of the basic amino acids in the 608–624 region cause almost complete cytoplasmic localization of the receptor and inhibition of its transactivating function.

The DNA-binding domain of the androgen receptor reveals approximately 80% homology with those of the glucocorticoid and progesterone receptors (24). From the structure of the DNA-binding domain of the glucocorticoid receptor, it can be predicted that the DNA-binding domain of the androgen receptor is determined by two zinc atoms (resulting in two zinc-fingers) and two perpendicularly situated α-helices (33). Specific amino acid residues in the amino-terminal α-helix (Lys-571, Val-572, and Arg-576) would directly contact the nitrogenous bases of the DNA; other amino acids might interact with the phosphate-sugar backbone of the DNA.

Gly-568, Ser-569, and Val-572 are important for specific recognition of the hormone-responsive element (33). These three amino acids are conserved between the androgen receptor, progesterone receptor, and glucocorticoid receptor. Therefore, it is not surprising that the three receptors recognize the same imperfect palindromic sequence (GGAACAnnnTGTTCT), which contains a 3-bp spacer, the hormone-responsive element (34–36). However, in regulatory regions of natural hormone receptor target genes, many steroid receptor-binding sites diverge considerably from this consensus sequence.

Androgen Receptor Target Genes

Several androgen-regulated genes have been described. From only a few of these it has been established that androgen regulation is the result of interaction of the activated androgen receptor with an ARE in the promoter region of the gene. The best-documented examples are the rat C3 (prostate-binding protein) gene (34,38), the rat probasin gene (39), and the mouse genes Slp (for sex-limited protein) (40–42) and GUS (for β-glucuronidase) (43). Directly androgen-regulated human genes are the PSA (prostate-specific antigen) gene (44,45) and the hGK-1 (human

Table 12.2: Consensus Sequence–like Androgen Response Elements in Androgen Receptor Target Genes

Gene	*ARE*	*Reference*
C3	AGTACG tga TGTTCT	37,38
Probasin	ATAGCA tct TGTTCT	39
Slp	AGAACA gcc TGTTTC	40–42
GUS	AGTACT tgt TGTTCT	43
PSA	AGCACT gca TGTTCT	44,45
hGK-1	AGCACT gca TGTTCC	44,48
"Consensus"	GGAACA nnn TGTTCT	
	A T A C	36

glandular kallikrein-1) gene (44–48, and K. Cleutjens, unpublished observation). Both these genes, which are structurally closely related, are almost exclusively expressed in prostate epithelial cells. Table 12.2 shows sequences in the promoter regions of these target genes that have been proved to be involved in androgen-regulated expression. These sequences all closely resemble the ARE consensus sequence. However, for the majority of the genes listed in Table 12.2, evidence suggests that, in addition to the consensus sequence–like ARE, other sequences are needed for optimal androgen-regulated expression. These sequences might be weak androgen receptor-binding sites or completely different elements that bind different (androgen-regulated) transcription factors. In this way a synergistic cooperation between the various regulatory elements exists.

None of the directly androgen-related genes identified so far has an apparent function in cell growth or differentiation. Identification of such genes in prostate epithelial and mesenchymal cells, as well as other androgen target tissues, is an important goal for the future.

Androgen Receptor Mutations in Androgen Insensitivity and Kennedy's Disease

Mutations in the androgen receptor are involved in several pathological processes. Best documented so far are mutations involved in androgen insensitivity syndrome (49–51). Chromosome X–linked androgen insensitivity is associated with abnormalities in male sexual morphogenesis. These can range from 46,XY individuals with severe or mild defects in the development of the male phenotype (partial androgen insensitivity) to those with an apparently female phenotype (complete androgen insensitivity). There is evidence that androgen insensitivity leads to increased risk of male breast cancer (52,53).

More than 80 different mutations in the androgen receptor have been documented; there are no apparent mutation hot spots (49,51,54). Mutations can range from complete or partial deletions of the gene (which are a rare phenomenon) to point mutations, or small substitutions, insertions, or deletions, in the open reading frame. Most point mutations are missense mutations that result in inhibition of steroid or DNA binding of the androgen receptor. Only a few mutations have been detected in the long amino-terminal region of the receptor; these include nonsense mutations and frameshifts.

In addition to causing androgen insensitivity, androgen receptor mutations cause Kennedy's disease, a rare adult-onset motor neuron disorder that results in atrophy of the spinal bulbar muscles (55). In persons with Kennedy's disease, the length of the Q stretch in the androgen receptor is expanded to 40 to 60 residues. Depending on the cells and promoters tested, the expanded Q stretch has been shown to cause a slight decrease or no change in androgen receptor activity (54,56). However, it is unclear which genes are the affected androgen receptor target genes in motor neuronal cells. It cannot be excluded that the expanded Q stretch results in a gain of function in the target cells.

Expression of the Androgen Receptor in Prostate Cancer

The growth of the majority of prostate tumors, like the development of the prostate, is androgen dependent. Therefore, endocrine therapy by orchiectomy or with antiandrogens is able to inhibit the growth of the tumor. However, this effect is limited, and eventually, prostate tumors become androgen independent and continue to grow.

As indicated above, in the adult prostate the androgen receptor is expressed at a high level in the secretory epithelial cells and at a variable level in the stromal cell compartment. In general, prostate tumors are adenocarcinomas, derived from precursors of luminal epithelial cells; it is assumed that the androgen receptors in these cells are responsible for the androgen-dependent growth of the tumor. However, a role of androgen-driven processes in the stromal cells in the growth of the tumor cells cannot be ruled out, although this role has not been studied in detail. Interestingly, the in vitro growing LNCaP prostate tumor cell is androgen sensitive, even when stromal cells are not present. In model systems, stromal cells derived from the urogenital sinus or adult prostate are able to support growth of human prostate tumor cells in vivo in male nude mice regardless of whether the tumor cells express the androgen receptor (57,58). This finding indicates a sensitivity of the tumor cells to androgen-regulated stimuli in a paracrine fashion, as is observed in normal prostate development. Possibly, as observed during normal development, the androgen receptor in both stromal and tumorigenic epithelial cells is important for the growth of the tumor.

Although advanced-stage prostate tumors no longer respond to endocrine therapy, androgen receptor expression can still be detected in the cells of most tumors; however, the expression levels are more variable (J. A. Ruizeve de Winter, unpublished data). In only 10% of advanced-stage tumors or less, all tumor cells are androgen receptor negative, which explains their unresponsiveness. Interestingly, there is a partial correlation between p53 mutation and lack of androgen receptor expression in these types of tumors (J. Trapman, unpublished observation).

Even during endocrine therapy, the androgen receptor is mainly found in the nucleus. This cellular localization suggests that the androgen receptor is still active as a transactivator. If so, several mechanisms can be proposed. First, it may be that low residual androgen concentrations are able to activate the receptor. Second, it is possible that the receptor is activated via a different pathway, possibly by growth factors or other stimuli, as is proposed for the estrogen and progesterone receptors (59,60). Third, it could be that the receptor has mutated into a constitutively active, hormone-independent receptor or a receptor with modified ligand specificity. However, the possibility that the receptor (activated or not) had no function at all in the growth of endocrine-unresponsive tumors cannot be excluded.

It has long been speculated that mutations in the androgen receptor can contribute to the growth (androgen-dependent or independent) of prostate cancer. Although there has been an extensive search, only three different mutations have been found so far (61, 62, and J. A. Ruizeveld de Winter, unpublished data). All three are in the steroid-binding domain of the receptor.

The first mutation was detected in the androgen receptor of the LNCaP cell line (63,64). Interestingly, the mutation (which causes a threonine-to-alanine substitution at position 868) does not lead to an inactivation of the receptor, as do all mutations causing androgen insensitivity, but increases the affinity of the receptor for progestagens, estrogens, and antiandrogens. Although the tissue from which the LNCaP cell line was derived has not been tested, early passages of the LNCaP cells already contain the mutation, indicating that it could also have been present in the original tumor.

The second mutation found in the androgen receptors of a prostate tumor causes a valine-to-methionine substitution at position 721 (62). No further characterization of the mutated receptor has been reported.

Recently, a third mutation, also leading to a valine-to-methionine substitution, was found at codon 706 in an endocrine-therapy-resistant prostate tumor (61). This mutation does not change the affinity of the receptor for different ligands; however, higher transactivating activity of adrenal androgens and progesterone was observed with the mutant receptor than with the wild-type receptor. These findings suggest that the structure of the steroid-binding domain of the activated mutant receptor is modified, affecting its quality as a transcription factor. It can be speculated from these data that in this tumor nontesticular androgens have contributed to its progressive growth. However, because of the limited number of mutations found so far, it is unlikely that these kinds of mutations are present in the majority of prostate tumors.

GENETIC ALTERATIONS IN PROSTATE CANCER

The pathway leading to malignancy undoubtedly involves both genetic and epigenetic influences. Researchers are attempting to define the molecular events responsible for initiation and progression of prostate cancer, and in the process, to identify urgently needed prognostic markers of aggressiveness of prostate cancer. Unfortunately, the study of the molecular biology of prostate cancer has lagged behind that of most other common solid tumors, and few consistent changes in the sequence, structure, or expression of the prostate cancer cell genome have been identified. Even less is known about how such genetic alterations might be suppressed or even stimulated by the cellular microenvironment (see next section). A summary of some of the efforts to identify consistent genetic alterations in prostate cancer is presented below.

ras Gene Mutations

Genetic aspects of carcinogenesis have received tremendous attention, in large part because of the progress in technology that has enabled the identification and characterization of genetic alterations in cancer. We have applied this technology to the analysis of prostatic DNA derived from both surgical specimens of prostate cancer and established prostate cancer cell lines. One of our first analyses addressed the possible role of mutated *ras* genes in prostate cancer, which had been suggested by a number of previous studies (65–67).

We examined prostate cancers for the presence of *ras* gene mutations using differential oligodeoxynucleotide hybridization of DNA amplified using the polymerase chain reaction (PCR) (68). In this study 24 primary prostate tumors (23 acinar tumors and one ductal tumor) and five prostate cancer cell lines were examined at codons 12, 13, and 61 of the K-*ras*, H-*ras*, and N-*ras* genes. Only two mutations were detected: an A-to-G transition, causing a glutamate-to-arginine substitution, at codon 61 of the H-*ras* gene in a primary prostatic ductal adenocarcinoma, and a G-to-T transversion, causing a glycine-to-valine substitution, at codon 12 of the H-*ras* gene in a prostate cancer cell line (TSU-Pr1) derived from a lymph node metastasis. The presence of these mutations has been confirmed by direct sequencing of the PCR-amplified products.

We have extended this study to include analysis of paraffin-embedded samples, including 12 metastatic prostate cancer lesions, 10 primary ductal adenocarcinomas, and 16 samples of BPH. Of these samples, only one, a metastatic deposit, had a *ras* gene mutation. Two subsequent studies have reported similarly low frequencies of *ras* gene mutations in prostate cancer. These results (all obtained in American patients) suggest that activation of *ras* oncogenes by point mutation is not a common event in either the initiation or the progression of prostatic neoplasia and thus would not seem to have prognostic relevance. However, two reports have suggested that *ras* gene mutations *do* occur at significant frequencies in prostate cancer, both in latent carcinoma and in clinically manifest disease (69,70). Both of these studies examined prostate tissue from Japanese men, raising the intriguing possibility that significant differences may exist in the genetic events associated with prostate cancer in American men versus Japanese men. In addition, in the latter study (70), human papillomavirus (HPV) DNA was detected in more than 40% of the prostate cancer samples analyzed. Whether HPV DNA is present in prostate cancer samples from North American men is controversial.

Tumor Suppressor Genes

We are currently examining the role of tumor suppressor gene inactivation in prostate cancer. Cytogenetic analyses of prostate cancer have not revealed consistent chromosomal deletions (reviewed in 71), which might provide information regarding the location of such genes. One study by Atkin and Baker, referred to by Konig et al. (72), however, showed that four of four patients with late-stage prostate carcinomas exhibited chromosome 10q deletions and three of four had chromosome 7q deletions. In addition, deletions of 10q have been observed in several prostate cancer cell lines (71,72).

To pursue these observations at a molecular level, we studied allelic loss in prostate cancer using polymorphic DNA probes for chromosomes 10q and 7q as well as probes for chromosomes containing documented and putative tumor suppressor genes (3q, 9q, 11p, 13q, 17p, and 18q). We examined 28 primary prostate cancer specimens, most of which were localized, stage B lesions measuring less than 2 cm in diameter. Briefly, we found that the majority (61%) of the tumors in the study exhibited allelic loss on at least one of the chromosomes examined (73). The regions most frequently deleted were found on the long arms of chromosomes 10 and 16. Whereas allelic loss on chromosome 10q was predicted from previous karyotype analyses, allelic loss on chromosome 16q was a novel finding. For neither 10q or 16q, however, have any candidate tumor suppressor genes been identified.

In addition to our study, Bergerheim et al. (74) and Kunimi et al. (75) have examined 10 primary tumor specimens, as well as eight metastatic lesions, for allelic loss on every nonacrocentric autosomal chromosomal arm. Again, chromosomes 10 and 16, along with chromosome 8, most frequently exhibited allelic loss in this study. Taken together, these results provide evidence for the specificity of allelic loss on chromosomes 8, 10, and 16 in prostate cancer.

To investigate chromosome 8p loss of heterozygosity (LOH) in more detail, we performed a deletion-mapping study of this chromosome in more than 50 samples of prostate cancer. As did Bergerheim et al. (74), we found frequent allelic loss on chromosome 8p in these samples. In fact, we found a locus at chromosome 8p22 that was deleted in almost 70% of the prostate cancers examined; in one case a homozygous deletion in this region was observed (76), strongly suggesting the presence of a tumor suppressor gene in this region. Interestingly, the short arm of chromosome 8 shows frequent LOH in a variety of common human tumors including colorectal and lung cancer (77,78). Whether the same gene is the target of these LOH events in prostate, colorectal, and lung cancer is unclear, but it is interesting to speculate that a gene that is inactivated in a large fraction of human cancers may reside in this region of the genome.

To confirm and extend the results of our initial study of LOH in prostate cancer, we examined chromosome 16q in 52 more samples of surgically resected prostate cancer (47 primary tumors, 5 metastases) using four different polymorphic probes for this region. Overall, of 39 tumors for which the markers were informative, 14 (36%) showed loss of alleles on chromosome 16q, a slightly higher fraction than our previous quantitation of 16q LOH (30%) in a smaller number of tumors. Interestingly, whereas 11 (32%) of 34 primary tumors had chromosome 16q deletions, 3 (60%) of 5 metastatic tumors showed deletions on this arm, suggesting a possible role for inactivation of a gene (or genes) in this region in tumor metastasis.

E-cadherin/α-catenin

Three more tumor types, breast carcinoma, hepatocellular carcinoma, and Wilms' tumor, have been identified in which allelic loss on chromosome 16q is a frequent event (79–81). It is not clear whether the affected gene on chromosome 16q is the same in each of these tumor types; however, the gene for the epithelial cell adhesion molecule E-cadherin, located on chromosome 16q22.1 (82), has been proposed to be an important modulator of the invasive potential of a variety of cultured human carcinoma cells (83). In light of this finding, we have evaluated E-cadherin as a putative tumor (invasion) suppressor gene in prostate cancer. In primary prostatic tumors, we found a statistically significant correlation between tumor differentiation as designated by Gleason score and E-cadherin protein levels (84). No poorly differentiated or anaplastic tumors (Gleason scores 9 and 10) had normal E-cadherin levels. In addition, in a small number of tumors, there was a correlation between loss of alleles on chromosome 16q and absent or reduced E-cadherin staining. Most, but not all, metastatic lesions of prostate cancer showed abnormal E-cadherin staining. That two metastatic deposits retained normal E-cadherin levels suggests that alternative pathways may exist for prostate cancer cells to become invasive.

In this regard, we have observed that in at least one prostate cancer cell line (PC-3) there is a complete absence of the protein, α-catenin, which is required for normal E-cadherin function (85). The lack of α-catenin expression in PC-3 cells is due to a homozygous deletion with the α-catenin gene (85). We have mapped the α-catenin gene to chromosome 5q21-q22 (86), and preliminary results suggest that LOH in this region occurs in approximately 25% of prostate cancers analyzed. Thus, the E-cadherin-mediated pathway of cell-cell adhesion may become inactivated by LOH of the α-catenin gene.

Microcell Transfer Studies

As a functional test for the presence on various chromosomes of genes that are capable of suppressing tumor growth, we have used the microcell transfer technique to insert a normal copy of a given chromosome into prostate cancer cell lines. We have successfully transferred chromosomes 5, 8, 10, 11, and 17 into a number of prostate cancer cell lines, including TSU-Pr1, DU145, and PC-3. The presence of transferred chromosomes has been confirmed by in situ hybridization with centromeric probes and by Southern and CA-repeat analysis with polymorphic markers, which can discriminate endogenous chromosomes from exogenous chromosomes. The most striking results are observed after reintroduction of a normal copy of chromosome 5 into PC-3 cells (W. B. Isaacs, C. M. Ewing, and R. A. Morton, manuscript submitted). Not only do the cells undergo a dramatic change in morphology in vitro to form tightly packed, more epithelioid colonies, but the tumorigenicity of the cells is essentially abolished when they are placed into nude mice. These alterations in growth characteristics are accompanied by dramatic alterations in the organization of the actin-based cytoskeleton. Importantly, this reversion to a more normal phenotype is correlated with the reexpression of α-catenin and restoration of E-cadherin-mediated cell-cell adhesion.

In vivo growth suppression is seen following microcell transfer of chromosome 10 into TSU-Pr1 cells, but only in clones of cancer cells that have taken up two or more copies of chromosome 10. The growth suppression is not observed in vitro and is not complete in vivo but rather is manifest as an increase in lag time before tumors appear. Experiments with the PC-3 cell line have been more dramatic, in that following microcell transfer and selection of resistant clones in neomycin, two types of colonies appear. The most numerous are small colonies in which cells undergo only a limited number of divisions and subsequently appear to senesce. Less frequently, colonies of rapidly growing cells form; these cells are fully tumorigenic when injected into nude mice. Experiments with chromosome 17 show a definite tumor-suppressing activity, which, curiously, appears to be independent of restoration of p53 function (see below). Thus far, no consistent suppression of tumor formation has been observed upon transfer of either chromosome 8 or chromosome 11 into prostate cancer cells.

The *p53* and *Rb* Genes

In addition to allelic loss on chromosomes 10q and 16q, approximately one-fifth of the prostate tumors analyzed in our initial study had deletions at chromosomes 17p, 18q, and 13q. These regions harbor the tumor suppressor genes *p53*, *DCC*, and *Rb*, respectively. Thus, the possibility exists that inactivation of these genes plays a role in at least a subset of prostate neoplasms.

The importance of *Rb* gene inactivation in prostate cancer has been suggested by the studies of Bookstein et al. (87,88). We examined additional tumors for allelic loss of the *Rb* gene using PCR-based detection of a variable number of tandem repeats (VNTR) in intron 20 of the gene. Of 41 informative tumors, 11 (27%) had allelic loss of one copy of the *Rb* gene (J. D. Brooks, G. S. Bova, and W. B. Isaacs, manuscript submitted), suggesting that inactivation of the *Rb* gene may be important in a significant subset of prostate adenocarcinomas.

With regard to the *p53* gene, we have found that three of five prostate cancer cell lines (TSU-Pr1, PC-3, and DU145) contain mutations in the coding sequence of the *p53* gene and that the growth of these lines can be suppressed by the introduction of a wild-type copy of this gene (89). We analyzed 71 primary prostate cancers for allelic loss of chromosome 17p and the *p53* gene. Of the 56 informative tumors, 11 (20%) had allelic loss. Furthermore, allelic loss was directly correlated with increasing Gleason grade and appeared correlated with disease stage. In agreement with our analysis of prostate cancer cell lines that are derived from metastases, three of five lymph node metastases had loss of one *p53* allele. Thus, on the basis of allelic loss studies, *p53* inactivation appeared to correlate with disease progression. Curiously, however, sequence analysis of the remaining allele in the 11 primary tumors with LOH on chromosome 17p revealed a lack of mutations in exons 5 to 8 of the remaining *p53* allele, suggesting that a different gene may be the target of these LOH events. Obviously, a closer examination of the roles of these genes in prostate cancer is warranted, particularly with respect to progression of this disease.

DNA Methylation

In addition to deletions, mutations, and chromosomal translocations, alterations in DNA methylation can affect gene expression. Hypermethylation of CpG islands (areas of the genome rich in the sequence CpG, associated with the 5′ regulatory regions of genes) has been associated with gene inactivation. We have examined the methylation status of a highly polymorphic locus, D17S5, located on chromosome 17p13.3, that lies within a cluster of NotI restriction sites (GC\GGCCGC), a finding consistent with the presence of a CpG island. Previous studies by Makos et al. (90) have found hypermethylation at this locus in both colon and lung cancer. Furthermore, deletions of the D17S5 locus at 17p13.3 have been identified in some breast cancers when other commonly deleted areas of 17p, including the *p53* gene at 17p13.1, are retained (80). In prostate cancer, by probing NotI-digested DNA with YNZ22.1, we found hypermethylation at D17S5 in 14 of 15 tumors and four of five prostate cancer cell lines (R. A. Morton and W. B. Isaacs, unpublished observation). These findings further suggest that there may be a tumor suppressor gene on chromosome 17p, separate from *p53*, that plays a role in prostate carcinogenesis and is frequently inactivated by hypermethylation. Further studies, such as identification of the affected gene, will be necessary to clarify the role of this methylation process in the development of prostate cancer.

Hereditary Prostate Cancer

Carter et al. demonstrated that familial aggregation of prostate cancer can be best explained by the mendelian inheritance of a rare autosomal dominant allele that predisposes men to develop this disease (91). From the study population examined, it was estimated that the gene responsible for this predisposition accounts for approximately 9% of the total cases of prostate cancer observed and 45% of the cases observed in men under 55 years of age. In collaboration with Drs. Patrick Walsh, Terri Beaty, Roy Levitt, and Deborah Meyers at our institution, we are initiating efforts to map this gene by linkage studies (92). Analysis of candidate regions, including BRCA1, FCC, and chromosome 8p22, is under way. Although prostate cancer poses several somewhat unique obstacles to such an analysis, it is anticipated that the study will provide critical information regarding the genetic mechanisms not only of hereditary prostate cancer but also of the more common sporadic disease.

Summary

To summarize, a number of genetic changes have been documented in prostate cancer, ranging from allelic loss to point mutations to changes in DNA methylation patterns. Thus far, the most consistent changes are allelic loss events, with the majority of tumors examined showing loss of alleles from at least one chromosomal arm. The short arm of chromosome 8, followed by the long arm of chromosome 16, appear to be the most frequent regions of loss, suggesting the presence of novel tumor suppressor genes. Deletions of one copy of the *Rb* and *p53* genes are observed, although less frequently, as are mutations of the *p53* gene, and accumulating evidence suggests the presence of another tumor suppressor gene on chromosome 17p, which is frequently inactivated in prostate cancer. Alterations in the E-cadherin/α-catenin–mediated cell-cell adhesion mechanism appear to be present in almost half of all prostate cancers and may be critical to the acquisition of metastatic potential of aggressive prostate cancers. Finally, altered DNA methylation patterns have been found in the majority of prostate cancers examined, suggesting widespread alterations in methylation-modulated gene expression. The presence of multiple changes in these tumors is consistent with the multistep nature of the transformation process.

STROMAL-EPITHELIAL INTERACTION IN PROSTATE CANCER GROWTH, DIFFERENTIATION, AND METASTASIS

Interaction between epithelial cells and their microenvironment, such as the underlying stromal cells, is another important factor to be considered in prostate cancer biology. During normal embryonic development, mesenchymal-epithelial interaction produces signals that regulate the morphogenesis of the prostate gland (93,94). As a continuum of the developmental processes, stromal-epithelial interaction also plays a critical role in adult life in defining normal and benign prostate growth and prostate cancer growth, differentiation, and metastasis (57).

Although the dynamics of stromal-epithelial interaction have been recognized for nearly a century, it was only recently that mechanisms of cellular interaction began to be elucidated at the molecular level. In this section we will focus on the molecular and cellular basis of stromal-epithelial interaction in the prostate gland. The critical questions that we wish to address are: What are the evidence for and potential significance of host microenvironment interaction with prostate cancer cells? What are the soluble mediators that are responsible for stromal-epithelial interaction in the prostate gland? And what are the consequences of cell-matrix interaction in prostate growth?

Evidence for and Significance of Stromal-Epithelial Interaction in the Prostate Gland

Using cell or tissue recombination techniques (94,95), we examined the reciprocal roles of stroma and epithelium in organogenesis, tumorigenesis, and commitment of cells to the expression of tissue-specific genes. We concluded that epithelium, even under the influence of testicular androgen, is unable to undergo the "programmed" pattern of cell growth and differentiation unless stroma is present. Rather, signaling between stromal and epithelial components, mediated by yet undefined factors in a reciprocal manner, is required to drive the completion of the predetermined developmental program. These experiments imply that critical information encoded in the developmentally regulated genes in the epithelium needs to be "switched on" by cellular interaction with the underlying stroma.

The signaling between stroma and epithelium continues after embryogenesis, during which cell lineage and eventually the structure and function of the prostate are determined. In the adult organ, stroma is required for epithelial growth (57) and androgen responsiveness (96). Aberrant cellular interaction between stroma and epithelium in the adult may conceivably contribute to hyperplastic growth (97,98) and malignant progression (99) of the prostate gland. Four lines of experimental evidence suggest the roles of stromal-epithelial interaction in normal prostate development and in neoplastic prostate progression.

First, fetal mesenchymal tissues can affect either directly or permissively the developmental programs of the responding normal neoplastic epithelium. For example, fetal urogenital sinus mesenchyme (UGM), under the influence of androgen, can direct the differentiation of the responding epithelium (e.g., fetal urogenital sinus epithelium or adult bladder epithelium) toward that of the prostate. Furthermore, in tissue recombinants, we observed that fetal UGM can redirect Dunning rat prostate tumors to undergo a "normal" pattern of glandular differentiation (100); in other studies, fetal UGM either increased (100) or decreased the growth rate (101) of Dunning rat prostate tumors as tissue recombinants and markedly enhanced the secretory functions of the tumor epithelium (100).

Second, physical localization of androgen receptor has been observed to undergo dynamic changes as described above. Androgen receptor resides exclusively in the mesenchymal compartment during the active phase of prostatic morphogenesis but appears predominantly in the epithelial cells in well-differentiated adult tissues (7,102), suggesting the possible switching of androgen regulatory pathways during normal prostate development. However, during neoplastic prostate cancer progression, the roles of androgen receptor are less clearly defined. Androgen receptor could undergo structural modifications and function as a trans-acting factor (with or without ligands) that regulates the expression of downstream genes in the tumor epithelium (61,62,103). Alternatively, androgen receptor dysfunction, as manifested by androgen independence, may be compensated for by active growth factor–growth factor receptor pathways that support uncontrolled tumor growth and metastasis.

A third line of evidence for the role of stromal-epithelial interaction is that organ-specific fibroblasts (found in the stroma) were observed to promote prostate cancer growth in vivo (58). Interaction between these fibroblasts and prostate cancer cells is reciprocal (57), and such interaction may lead to androgen-independent prostate cancer progression (99,104).

Finally, biochemical and molecular analysis of gene expression in fibroblasts revealed that fibroblasts have the capacity to produce wide varieties of cell-type-specific growth factors and extracellular matrices that have been implicated in supporting prostate cancer growth and progression.

From an oncologic point of view, these four observations support the concept that cancer is more than a single-cell disease. Genetic modifications in the epithelial cells are required, but not sufficient, to account for cancer growth and metastasis. Epithelial tumorigenesis and metastasis can be significantly affected by the host microenvironment, for example, by an altered endocrine and growth factor milieu (57,99). Prostate cancer growth and metastasis can be modulated principally by soluble and matrix-associated signal molecules specified by the organ-specific microenvironment.

Stromal-Epithelial Interaction: Soluble Mediators

Soluble protein growth factors and their receptors that regulate normal prostate development and neoplastic progression are traditionally categorized as autocrine, paracrine, and endocrine mediators. Using a number of assay systems, basic fibroblast growth factor (bFGF), keratinocyte growth factor (KGF), hepatocyte growth factor/scatter factor (HGF/SF), transforming growth factors-β (TGF-βs), transforming growth factor-α/epidermal growth factor (TGF-α/EGF), and insulin-like growth factors (IGFs) I and II have emerged as attractive candidates for being the autocrine or paracrine growth factors that mediate stromal-epithelial communication.

Because of the complexities of stromal-epithelial interaction, however, it is generally regarded as impossible that one single growth factor could confer the reciprocal communications between prostatic stromal and epithelial cells. The perfect candidate molecules should be secreted by effector or affector cells in an androgen-dependent manner. They should act on target cells in an autocrine manner or a paracrine manner or both by delivering the candidate molecules in vivo and should stimulate growth of target cells and maintain their differentiation, bypassing the requirement for androgen. Unfortunately, none of the candidate molecules so far studied meet all of these criteria. In this review, we summarize progress made in understanding these growth factors and their possible roles in the regulation of prostate growth and differentiation.

Fibroblast Growth Factors The presence of a mitogenic factor in human prostate extracts was reported by Jacobs and Lawson (105). This factor was identified by Story et al. (106) as homologous to bFGF. The fibroblast growth factors (FGFs) are a growing family of growth-promoting factors that includes four cellular oncogene products: int-2 (FGF-3), hst/Kfgf (FGF-4), FGF-5, and FGF-6.

DNA isolated from primary human breast tumors was found to cause transformation (foci formation) of NIH3T3 cells (107). Transgenic male mice in which int-2 protein was overexpressed under the influence of the MMTV promoter/enhancer developed prostatic hyperplasia in adulthood (108). Recently, however, what was initially thought to be hyperplastic growth of the prostate actually occurred outside of the prostate proper.

hst/Kfgf (or FGF-4) was isolated from a human stomach tumor and was found to be identical to an oncogene isolated and identified in Kaposi's sarcoma (109,110). hst/Kfgf was found to be expressed by mouse embryonic fibroblasts and tumor endothelial cells.

The *FGF-5* gene was identified in a human bladder tumor following transfection of NIH3T3 cells with human bladder tumor DNA (111). FGF-5 was found to be expressed by mouse embryonic fibroblasts and tumors.

The *FGF-6* gene shares 70% structural homology with the *hst* gene and was identified in a mouse cosmid library after screening the library with *hst* gene (112). FGF-6 was found to be present in tumors.

In addition to cellular oncogene products, the bFGF family consists of other cellular growth factors identified in stromal cells as well as tumor epithelium. These include acidic FGF (aFGF; also called FGF-1 or heparin-bound growth factor [HBGF]-1), bFGF (FGF-2 or HBGF-2),

KGF (FGF-7), and androgen-induced growth factor (AIGF; also called FGF-8).

Acidic FGF was isolated from pituitary and brain tissues. It shares 55% amino acid sequence homology with bFGF (113,114). This growth factor appears to be expressed by the mesenchyme of rat Dunning tumors and rat tissues (115); its expression is less significant in human prostate (116). It is a mitogen for mesoderm- and neuroectoderm-derived cells in vitro.

Basic FGF is the dominant form of FGF found in both normal human prostate and prostate tumors (116,117). It stimulates several important biological functions in vivo, including cell proliferation, angiogenesis and wound healing, induction of chemotaxis, bone growth, and neurite extension (118,119). In the LNCaP human prostate tumor model in vivo (58) and Syrian hamster ductal deferent cells in vitro (120), bFGF can replace androgen in the induction of androgen-dependent tumor cell growth.

Despite the impressive array of bFGF activity that has been described in the prostate gland, some aspects of this growth factor's action remain to be elucidated. For example, although bFGF mRNA has been detected in both primary human prostate stromal cells and epithelial cells, bFGF receptors are associated with prostate stromal cells (derived from human BPH tissues), suggesting that bFGF may be an autocrine growth factor for fibroblast rather than epithelial cell growth (121). However, bFGF receptor was also expressed by both rodent (122) and human (L. W. K. Chung, unpublished observation) prostate cancer cell lines; this result suggests that bFGF can serve as both an autocrine and a paracrine growth mediator in prostate tumors. Unlike other members of the FGF family, however, bFGF and aFGF do not contain signal peptides, so these growth factors are synthesized, but not secreted, by cultured prostatic stromal cells (121). This fact raises the possibility that bFGF serves as an intracrine regulator of prostate tumor growth in vivo (123). Alternative splicing of aFGF (124) and bFGF receptor (125) has been reported, which adds further complexities to the study of the roles of these FGFs in the prostate.

KGF, which was initially isolated from the conditioned medium of human embryonic lung fibroblasts, was identified as an epithelial growth stimulator that promotes organogenesis of the mouse seminal vesicles in organ culture (126) and the growth of Dunning rat prostate epithelial cells in tissue culture (127). It remains to be determined whether KGF expression in the developing and normal adult rat prostate gland is regulated by androgen. In addition, whether KGF or its antibody when delivered in vivo exerts stimulatory or inhibitory action, respectively, on prostate growth is unknown.

The novel growth factor AIGF was identified and cloned from an androgen-dependent mouse mammary cell line, SC-3 (128). This growth factor was found to be highly inducible by the addition of a physiologic concentration of testosterone to SC-3 cells, and it can substitute for androgen in stimulating androgen-dependent growth of the SC-3 cells in culture.

Transforming Growth Factors-β The TGF-βs are bifunctional growth regulators that inhibit epithelial growth but stimulate fibroblast growth (129,130). The TGF-βs constitute a family of structure-related growth factors including TGF-β1, β2, and β3; activin; inhibin; müllerian inhibitory substance; bone morphogenic proteins; and decapentaplegic gene complex in *Drosophila*. Elevated steady-state levels of TGF-β (131) and TGF-β receptor (132) were observed during castration-induced programmed cell death in the rat. The possibility of a role of TGF-β1 in the involution of the rat prostate gland was further supported by a study in which prostatic wet weight decreased after TGF-β1 was infused directly to the prostate (50 ng/day) for 7 days (133). Since TGF-β1 did not affect the host levels of serum testosterone nor seminal vesicle wet weights, these results support the notion that TGF-β1 exerts a *local* growth inhibitory effect on the prostate.

The in vitro growth effect of TGF-β is not only TGF-β receptor dependent but also appears to be biphasic. The growth of A × C rat prostate-derived epithelial cells was inhibited by TGF-β at lower ligand concentrations but stimulated by TGF-β at higher ligand concentrations (134). Higher numbers of TGF-β receptors were detected in human prostatic PC-3 and DU145 cells than in LNCaP cell lines; these levels correspond with the growth-inhibitory effect observed by the addition of TGF-β to these cell lines in tissue culture medium (58,135).

The concentration dependency of TGF-β inhibition of prostatic cell growth in vitro may be true for human cells as well. Wilding et al. (135) did not observe inhibition of the growth of LNCaP cells using a TGF-β concentration greater than 40 pM, whereas Gleave et al. (58) observed inhibition of LNCaP cell growth using concentrations of TGF-β between 0.1 and 0.5 pM.

TGF-β's action in vivo is complex; it has been noted to affect extracellular matrix (ECM) biosynthesis and deposition (136), angiogenesis during tumor growth and wound healing (137), cell growth and differentiation, and immune suppression (138,139). In vivo TGF-β has been reported to stimulate prostate tumor growth and metastasis (140) and has been associated with malignant prostatic epithelium transformation (141). These growth-stimulatory effects of TGF-β in vivo are surprising in view of previous reports that TGF-β inhibited normal mammary epithelial proliferation (142), antagonized liver regeneration following partial hepatectomy (143), and blocked lymphoid and myeloid cell proliferation (130). One interpretation of the growth-stimulatory effect of TGF-β in vivo is that its action may be indirect, overriding the overall growth-inhibitory action of TGF-β. For example, although TGF-β is a growth inhibitor, it was noted to stimulate the overall growth of aortic smooth muscle cells via its indirect inductive action of an autocrine growth regulator, platelet-derived growth factor (PDGF-AA); PDGF-AA overrides the inhibitory action of TGF-β and acts positively on aortic smooth muscle growth (144). In prostate cancer, TGF-β may stimulate the production of bFGF (a potent angiogenic and growth promoter) by prostate fibroblasts, enhancing cancer growth in a paracrine manner (145).

TGF-βs are synthesized as latent higher molecular weight precursors. Latent high molecular weight TGF-β complexes have been detected in both prostatic epithelial and stromal cells (146–148). The latent form of TGF-β must be activated by proteolytic cleavage. Killian et al. (149) proposed that prostate-specific antigen (PSA), an abundant serine protease commonly associated with prostate cancer, can activate the latent form of TGF-β. TGF-β has been proposed to be a potent osteoblast growth stimulator during prostate cancer bony metastasis (150); by exerting a growth-inhibiting effect on osteoclasts but a growth-stimulating effects on osteoblasts, TGF-β results in net new bone growth (151–153). Latent TGF-β could also bind to its binding protein, and there is evidence that the levels of immunoreactive latent TGF-β binding protein may be decreased in malignant, but not in benign, prostate tissues (154).

Transforming Growth Factor-α/Epidermal Growth Factor TGF-α/EGF is frequently secreted by normal and tumor prostatic epithelial cells (155,156). TGF-α/EGF often functions as an autocrine or paracrine factor that mediates growth and angiogenic signals during embryonic development and tumor growth (157,158). Using the LNCaP cell line, Hara (159) reported that TGF-α/EGF and testosterone have cooperative growth-promoting action in vitro. Immunohistochemical localization of TGF-α and its receptor yields variable results with respect to heterogeneity and lobular differences of TGF-α and its receptor distribution (160,161). The intensity of TGF-α staining was found to correlate negatively with the differentiation status of human prostate cancer (162). This result leads to the possibility that as prostate cancer becomes less well differentiated, the TGF-α/EGF receptor (i.e., EGF receptor) may become the major pathway for signal transduction. Consistent with this idea, it has been demonstrated that the human prostate expresses an elevated level of the c-erbB2/*neu* oncogene (163), which shares 55% structural homology with the EGF receptor (164).

Hepatocyte Growth Factor/Scatter Factor HGF/SF was discovered independently by two groups of investigators; one group isolated the growth-promoting factor in support of hepatic regeneration (165), and another group defined the factor from the conditioned medium of fibroblasts that cause epithelial cells to scatter (166). HGF/SF was first synthesized in its precursor form with an M_r of 87 kDa and then processed to form a heterodimer with light peptide chains (M_r 30 kDa) and heavy chains (M_r 57 kDa) linked together by disulfide bonds. This paracrine factor is synthesized and secreted by stromal cells but exerts its biological action on neighboring epithelial cells. It is a potent *mitogen* that can stimulate the growth of hepatocyte (165), prostate (167), melanocyte, and lung endothelial (168) cells in a paracrine manner. During embryonic development, HGF/SF serves as a potent *morphogen* that induces Madin-Darby canine kidney epithelial tubulogenesis (169). HGF/SF is also a *motogen* that increases the motility, migration, and scattering of epithelial cells in culture (170). In addition, when placed in the anterior chamber of the eye of rabbits, HGF/SF was demonstrated to have *angiogenic* activity.

The action of HGF/SF is mediated by its receptor, the c-met oncogene, which encodes a 190-kDa transmembrane tyrosine kinase receptor (171,172). The c-met receptor is also a heterodimer, comprised of 50-kDa (α-chain) and 145-kDa (β-chain) peptides. Immunohistochemical evidence suggests that the c-met oncogene is expressed by normal and tumor epithelial cells (173–175). In human prostate, Pisters et al. (176) demonstrated that c-*met* was not expressed by the luminal epithelial cells of normal and BPH tissues but was expressed abundantly by prostate cancer tissues.

Li et al. (unpublished observation) demonstrated that HGF/SF was expressed by prostate and bone fibroblasts, which exhibit an organ-specific growth-promoting effect on human LNCaP tumor growth in vivo (177). We demonstrated that HGF/SF, bFGF, and nerve growth factor, but not KGF, EGF, TGF-α, TGF-β1, PDGF, IGF, and transferrin, stimulated anchorage-independent growth of prostatic epithelial cells, an activity that was closely linked to prostate-tumor-inducing activity in vivo (177).

Insulin Growth Factors IGF I is a mitogen for prostate cancer cell growth in vitro (178,179). Although IGFs are not produced directly by prostate cancer epithelium, IGF-binding proteins (IGFBPs) are synthesized and secreted by malignant prostate epithelium (180). The sera of patients with metastatic prostate cancer contain markedly elevated levels of IGFBP (181), which presumably can serve as a reservoir for the local release of IGFs following cleavage of IGFBPs by proteolytic enzymes in the prostate (e.g., PSA). Using an osteoblastic metastasis model, Koutsilieris et al.

(182) demonstrated that IGFBP may be the precursor of paracrine stimulator of the growth of both malignant prostate epithelium and bone stroma; IGFBP would be activated initially by a urokinase-type plasminogen activator released by the malignant cancer epithelium.

Cell-Matrix Interaction in the Prostate

It has long been recognized that the behavior of prostate epithelial cells as androgen-responsive target cells is intimately affected by dihydrotestosterone (DHT) and the adhesive yet poorly defined ECM molecules.

Because of the complexities of the ECM biochemistry, the precise nature of cell-matrix interaction in the prostate gland remains unresolved. In 1978, Mosccona and Folkman (183) were the first to recognize the importance of the cell substratum on cell shape and the intrinsic rate of DNA synthesis of the target cells. Gospodarowicz et al. (184) further demonstrated that altering corneal epithelial cells' shape through differential attachment of these cells affected the ultimate responses of the cells to mitogenic hormones. Murphy et al. (185) initiated a study that confirmed that ECM and DHT have complex interactions regulating the growth of, and PSA synthesis and secretion by, LNCaP cells in culture. Fong et al. (186) reached similar conclusions; they have shown that DHT-induced PSA expression by LNCaP cells occurs only if the cells are in close association with ECM.

The roles of ECM in vitro in prostate growth were also clarified by Bulbul et al. (187), who observed that ECM can facilitate the growth of human urologic tumors. (This tumor model has added a new dimension to drug screening programs.) Using that interaction of organ-specific fibroblasts and prostatic epithelial cells in vivo facilitates the growth of tumors (57). Lethally irradiated fibroblasts were also found to stimulate prostate tumor "take" in vivo; this implies the possible role of soluble or matrix-associated molecules in promoting prostate tumor growth in vivo.

In studying the biological effect of reconstituted basement membrane extracted from an EHS tumor (Matrigel) on prostate tumor growth in vivo, Pretlow et al. (188) demonstrated that Matrigel can significantly facilitate the "take" rate of human prostate tumors in vivo. It is conceivable that in this case, Matrigel serves as an ideal substratum in which to anchor local prostate tumor cells and their subsequent growth in vivo. Alternatively, Matrigel could sequester a significant amount of growth regulators that promote tumor growth in vivo. Of course, more complicated reciprocal interaction could take place between stromal and epithelial cells, mediated by soluble growth factors as well as extracellular matrices. Chiquet-Ehrismann et al. (189) demonstrated that both tenacin secreted by the stroma and TGF-β produced by a human mammary tumor epithelial cell line, MCF-7, are required for the growth and development of human mammary tumor epithelium. Reichmann et al. (190) demonstrated that mouse mammary epithelial cells can acquire competence to respond to lactogenic hormone by synthesizing and secreting β-casein as evidenced by the deposition of laminin at the cell interphases, provided that epithelial cells can be differentiated functionally together with fibroblasts. Thus, it can be stated that functional stromal-epithelial interaction occurs only if appropriate growth factors and matrix-associated molecules are involved as mediators.

Ample studies of excellent quality have addressed the structural requirement of ECM in supporting tumor growth and differentiation (191,192). However, the biological responses of tumor cells to ECM have largely not been characterized beyond quantitative analysis of growth and invasive properties. Moreover, many of the previous studies deal with adaptive and "reversible" changes (193,194) often observed following cellular interaction with ECM. In 1994, Freeman and his associates (195) demonstrated for the first time that ECM could impose significant constraints on cultured rat prostate epithelial cells, causing an apparent transdifferentiation of the cells to express constitutively more malignant properties and altered gene expression. The conclusion of that study concurred with our recent observation that selected ECM subtypes facilitated "clonal selection" or actually promoted a permanent change of a cloned rat prostatic epithelial cell line to expression of tumorigenic phenotypes. It is possible that cell-matrix interaction set forward a coordinated change in the synthesis and deposition of ECM, cell adhesion molecules, or intermediate filaments (e.g., a decrease in steady-state levels of expression of E-cadherin and an increase in expression of vimentin) by the tumor cells and increased their aggressive behaviors. The elucidation of the molecular mechanism of cell-matrix interaction using the cell-cell recombination model system described above will shed light on the influence of epigenetic mechanisms on the genetic constituents of the cancer cells. These observations imply that genetic changes of the tumor epithelium may be affected by cell-matrix interactions, which could be inheritable but are modulated by the local tumor microenvironment.

Summary

There is increasing evidence to suggest that stromal-epithelial interaction in the prostate gland is important not only during early organogenesis, cytodifferentiation, and functional differentiation, but also for the maintenance of normal functions. Using the cell-cell recombination model established by our laboratory, we have documented that organ-specific fibroblasts and selected growth factors and extracellular matrices can confer tumorigenic, as well as hormone-responsive, phenotypes to otherwise nontumorigenic human prostatic epithelial cells and bladder epithelial cells, respectively. In this section, we reviewed the potential roles of soluble mediators and extracellular matrices in prostate cancer growth, differentiation, and metastasis. We believe that by understanding the characteristics and the functions of these factors and their receptors, new classes of therapeutic drugs can be developed to interfere with signals mediated by these pathways. Moreover, analysis of the steady-state levels of the relevant factors may offer prognostic values for predicting the rate of tumor growth and progression.

ACKNOWLEDGMENTS

This work was supported in part by the United States Public Health Service (CA55231, DK47596, CA58236) and the Dutch Cancer Society.

REFERENCES

1. Boring CC, Squires TS, Tong T, Montgomery S. Cancer Statistics 1994 CA1994;44:7–26.

2. Coffey DS, The molecular biology, endocrinology, and physiology of prostate and seminal vesicles. In: Walsh PC, Retik AB, Stamey TA, Vaughn ED, Jr, eds. Campbell's Urology, 6th ed. Philadelphia: WB Saunders, 1992:221–251.

3. Cunha GR, Donjacour AA, Cooke PS, et al. The endocrinology and developmental biology of the prostate. Endocr Rev 1987;8:338–362.

4. Sinclair AH, Berta P, Palmer MS, et al. A gene from the human sex-determining region encodes a protein with homology to a conserved DNA-binding motif. Nature 1990;346:240–244.

5. Jost A. Problems of fetal endocrinology: the gonadal and hypophyseal hormones. Recent Prog Horm Res 1953;8:379–418.

6. Thigpen AE, Davis DL, Milatovich A, et al. Molecular genetics of steroid 5alpha-reductase 2 deficiency. J Clin Invest 1992;90:799–809.

7. Shannon JM, Cunha GR. Autoradiographic localization of androgen binding in the developing mouse prostate. Prostate 1983;4:367–373.

8. Husmann DA, McPhaul MJ, Wilson JD. Androgen receptor expression in the developing rat prostate is not altered by castration, flutamide, or suppression of the adrenal axis. Endocrinology 1991;128:1902–1906.

9. Cooke PS, Young P, Cunha GR. Androgen receptor expression in developing male reproductive organs. Endocrinology 1991;128:2867–2873.

10. He WH, Kumar MJ, Tindall DJ. A frame shift mutation in the androgen receptor gene causes complete androgen insensitivity in the testicular feminized mouse. Nucleic Acids Res 1991;19:2373–2378.

11. Charest NJ, Zhou Z-X, Lubahn DB, Olsen KL, Wilson EM, French F.

A frameshift mutation destabilizes androgen receptor messenger RNA in Tfm mice. Mol Endocrinol 1991;5:573–581.

12. Cunha GR, Chung LWK. Stromal-epithelial interactions: I. Induction of prostatic phenotype in urothelium of testicular feminized mice. J Steroid Biochem 1981;14:1317–1321.

13. Donjacour AA, Cunha GR. Assessment of prostatic protein secretion in tissue recombinants made of urogenital sinus mesenchyme and urothelium from normal or androgen insensitive mice. Endocrinology 1993;132:2342–2350.

14. Chodak GW, Kranc DM, Puy LA, Takeda H, Johnson K, Chang C. Nuclear localization of androgen receptor in heterogeneous samples of normal, hyperplastic and neoplastic human prostate. J Urol 1992;147:798–803.

15. Ruizeveld de Winter JA, Trapman J, Vermey M, Mulder E, Zegers ND, Van der Kwast ThH. Androgen receptor expression in human tissues: an immunohistochemical study. J Histochem Cytochem 1991;39:927–936.

16. Prins G, Birch L, Greene G. Androgen receptor localization in different cell types of the adult rat prostate. Endocrinology 1991;129:3187–3199.

17. Husmann D, Wilson C, McPhaul M, Tilley W, Wilson J. Antipeptide antibodies to two distinct regions of the androgen receptor localize the receptor protein to the nuclei of target cells in the rat and human prostate. Endocrinology 1990;126:2359–2368.

18. Evans RM. The steroid and thyroid receptor superfamily. Science 1988;240:889–895.

19. Truss M, Beato M. Steroid hormone receptors: interaction with deoxyribonucleic acid and transcription factors. Endocr Rev 1993;14:459–479.

20. Wahli W, Martinez E. Superfamily of steroid nuclear receptors: positive and negative regulators of gene expression. FASEB J 1991;5:2243–2249.

21. Chang C, Kokontis J, Liao S. Structural analysis of complementary DNA and amino acid sequences of human and rat androgen receptors. Proc Natl Acad Sci USA 1988;85:7211–7215.

22. Faber PW, Kuiper GGJM, Van Rooij HCJ, Van der Korput JAGM, Brinkmann AO, Trapman J. The N-terminal domain of human androgen receptor is encoded by one, large exon. Mol Cell Endocrinol 1989;61:257–262.

23. Tilley WD, Marcelli M, Wilson JKD, McPhaul M. Characterization and expression of a cDNA encoding the human androgen receptor. Proc Natl Acad Sci USA 1989;86:327–331.

24. Trapman J, Klaassen P, Kuiper GGJM, et al. Cloning, structure and expression of a cDNA encoding the human androgen receptor. Biochem Biophys Res Commun 1988;153:241–248.

25. Kuiper GGJM, Faber PW, Van Rooij HCJ, et al. Structural organization of the human androgen receptor gene. J Mol Endocrinol 1989;2:R1–R4.

26. Lubahn DB, Brown TR, Simental JA, et al. Sequence of the intron/exon junctions of the coding region of the human androgen receptor gene and identification of a point mutation in a family with complete androgen insensitivity. Proc Natl Acad Sci USA 1989;86:9534–9538.

27. Sleddens HFBM, Oostra BA, Brinkmann AO, Trapman J. Trinucleotide (GGN) repeat polymorphism in the human androgen receptor gene. Hum Mol

Genet 1993;2:493.

28. Jenster G, Van der Korput JAGM, Van Vroonhoven C, Van der Kwast ThH, Trapman J, Brinkmann AO. Domains of the human androgen receptor involved in steroid binding, transcriptional activation and subcellular localization. Mol Endocrinol 1991;5:1396–1404.

29. Rundlett SE, Wu X-P, Miesveld RL. Functional characterizations of the androgen receptor confirm that the molecular basis for androgen action is transcription regulation. Mol Endocrinol 1990;4:708–714.

30. Simental JA, Sar M, Lane MV, Frebch FS, Wilson EM. Transcriptional activation and nuclear targeting signals of the human androgen receptor. J Biol Chem 1991;266:510–518.

31. Faber PW, King A, Van Rooij HCJ, Brinkmann AO, De Both NJ, Trapman J. The mouse androgen receptor. Biochem J 1991;178:169–178.

32. Jenster G, Trapman J, Brinkmann AO. Nuclear import of the human androgen receptor. Biochem J 1993;293:761–768.

33. Luisi BF, Xu WX, Otwinowski Z, Freedman LP, Yamamoto KR, Sigler PB. Crystallographic analysis of the interaction of the glucocorticoid receptor with DNA. Nature 1991;352:497–505.

34. Lieberman B, Bona BJ, Edwards DP, Nordeen SK. The constitution of a progesterone response element. Mol Endocrinol 1993;7:515–527.

35. Nordeen SK, Suh BJ, Kuhnel B, Hutchison CA. Structural determinants of a glucocorticoid response element. Mol Endocrinol 1990;4:1866–1873.

36. Roche PJ, Hoare SA, Parker MG. A consensus DNA-binding site for the androgen receptor. Mol Endocrinol 1992;6:2229–2235.

37. Tan J, Marschke KB, Ho K-C, Perry ST, Wilson EM, French FS. Response elements of the androgen-regulated C3 gene. J Biol Chem 1992;267:4456–4446.

38. Claessens F, Celis L, Peeters B, Heyns W, Verhoeven G, Rombauts W. Functional characterization of an androgen response element in the first intron of the C3 (1) gene of prostatic binding protein. Biochem Biophys Res Commun 1989;164:833–840.

39. Rennie PS, Bruchovsky N, Leco KJ, et al. Characterization of two cis-acting DNA elements involved in the androgen regulation of the probasin gene. Mol Endocrinol 1993;7:23–36.

40. Adler AJ, Scheller A, Hoffman Y, Robins DM. Multiple components of a complex androgen dependent enhancer. Mol Endocrinol 1991;5:1587–1596.

41. Adler AJ, Danielson M, Robins DM. Androgen-specific gene activation via a consensus glucocorticoid response element is determined by interaction with nonreceptor factors. Proc Natl Acad Sci USA 1992;89:11660–11663.

42. Adler AJ, Scheller A, Robins DM. The stringency and magnitude of androgen-specific gene activation are combinatorial functions of receptor and nonreceptor binding site sequences. Mol Cell Biol 1993;13:6326–6335.

43. Lund SD, Gallagher PM, Wang B, Porter SC, Ganschow RE. Androgen responsiveness of the murine beta-glucuronidase gene is associated with nuclease hypersensitivity, protein binding, and haplotype-specific sequence diversity within intron 9. Mol Cell Biol 1991;11:5426–5434.

44. Riegman PHJ, Vlietstra RJ, Van der Korput JAGM, Romijn JC, Trapman J. Characterization of the prostate-specific antigen gene: a novel human kallikrein-like gene. Biochem Biophys Res Commun 1989;159:95–102.

45. Riegman PHJ, Vlietstra RJ, Van der Korput JAGM, Brinkmann AO, Trapman J. The promoter of the prostate-specific antigen gene contains a functional androgen response element. Mol Endocrinol 1991;5:1921–1930.

46. Riegman PHJ, Vlietstra RJ, Van der Korput JAGM, Romijn JC, Trapman J. Identification and androgen-regulated expression of two major human glandular kallikrein-1 (hGK-1) mRNA species. Mol Cell Endocrinol 1991;76: 181–190.

47. Wolf DA, Schulz P, Fittler F. Transcriptional regulation of prostate kallikrein-like genes by androgen. Mol Endocrinol 1992;6:753–762.

48. Murtha P, Tindall DJ, Young CYF. Androgen induction of a human prostate-specific kallikrein, hKLK2: characterization of an androgen response element in the 5′ promoter region of the gene. Biochemistry 1993;2:6459–6464.

49. Pinsky L, Trifiro M, Kaufman M, et al. Androgen resistance due to mutation of the androgen receptor. Clin Invest Med 1992;15:456–472.

50. Brinkmann AO, Trapman J. Androgen receptor mutants that affect normal growth and development. Cancer Surv 1992;14:95–111.

51. McPhaul JJ, Marcelli M, Zoppi S, Griffin JE, Wilson, JD. Genetic basis of endocrine disease: 4. The spectrum of mutations in the androgen receptor gene that causes androgen resistance. J Clin Endocrinol Metab 1993;76:17–23.

52. Wooster R, Mangion J, Eeles R, et al. A germline mutation in the androgen receptor gene in two brothers with breast cancer and Reifenstein syndrome. Nature Genet 1992;2:132–143.

53. Lobaccaro J-M, Lumbroso S, Belon C, et al. Androgen receptor gene mutation in male breast cancer. Hum Mol Genet 1993;2:1799–1802.

54. Trapman J, Brinkmann AO. Mutations in the androgen receptor. Ann NY Acad Sci 1993;684:85–93.

55. La Spada AR, Wilson EM, Lubahn DB, Harding AE, Fischbeck KH. Androgen receptor gene mutations in X-linked spinal and bulbar muscular atrophy. Nature 1991;352:77–79.

56. Mahtre AN, Trifiro MA, Kaufman M, et al. Reduced transcriptional regulatory competence of the androgen receptor in X-linked spinal and bulbar muscular atrophy. Nature Genet 1993;5:184–188.

57. Chung LWK, Gleave ME, Hsieh JT, Hong S-J, Zhau HE. Reciprocal mesenchymal-epithelial interaction affecting prostate tumour growth and hormonal responsiveness. Cancer Surv 1991;11:91–121.

58. Gleave M, Hsieh J-T, Gao C, von Eschenbach AC, Chung LWK. Acceleration of human prostate cancer growth in vivo by factors produced by prostate and bone fibroblasts. Cancer Res 1991;51:3753–3761.

59. Aronica SM, Katzenellenbogen BS. Stimulation of estrogen receptor mediated transcription and alteration in the phosphorylation state of the rat uterine estrogen receptor by estrogen, cyclic adenosine monophosphate, and insulin-like growth factor-1. Mol Endocrinol 1993;7:743–752.

60. Denner LA, Weigel NL, Maxwell BL, Schrader WT, O'Malley BW.

Regulation of progesterone receptor mediated transcription by phosphorylation. Science 1990;2500:1740–1743.

61. Culig Z, Klocker H, Eberle J, et al. DNA sequence of the androgen receptor in prostatic tumor cell lines and tissue specimens assessed by means of the polymerase chain reaction. Prostate 1993;22:11–22.

62. Newmark JR, Hardy DO, Tonb DC, et al. Androgen receptor gene mutations in human prostate cancer. Proc Natl Acad Sci USA 1992;89:6319–6323.

63. Veldscholte J, Ris-Stalpers C, Kuiper GGJM, et al. A mutation in the ligand binding domain of the androgen receptor of human LNCaP cells affects steroid binding characteristics and response to anti androgens. Biochem Biophys Res Commun 1990;173:534–540.

64. Veldscholte J, Voorhorst-Ogink MM, Bolt-de Vries J, Van Rooij HCJ, Trapman J, Mulder E. Unusual specificity of the androgen receptor in the human prostate tumor cell line LNCaP: high affinity for prostagenic and estrogenic steroids. Biochim Biophys Acta 1990;1052:187–194.

65. Thompson TC, Southgate J, Kitchener G, Land H. Multistage carcinogenesis induced by *ras* and *myc* oncogenes in a reconstituted organ. Cell 1989;56:917–930.

66. Treiger BF, Isaacs JT. Expression of a transfected v-Ha-*ras* oncogene in a Dunning rat prostate adenocarcinoma and the development of high metastatic ability. J Urol 1988;140:1580–1586.

67. Peehl DM, Wehner N, Stamey TA. Activated Ki-*ras* oncogene in human prostatic adenocarcinoma. Prostate 1987;10:281–289.

68. Carter BS, Epstein JI, Isaacs WB. *Ras* gene mutations in human prostate cancer. Cancer Res 1990;50:6830–6832.

69. Konishi N, Enomoto T, Buzard G, Oshima M, Ward JM, Rice JM. K-*ras* activation and *ras* p21 expression in latent prostatic carcinoma in Japanese men. Cancer 1992;69:2293–2299.

70. Anwar K, Nakakuki K, Shiraishi T, Naiki H, Yatani R, Inuzuka M. Presence of *ras* oncogene mutations and human papilloma virus DNA in human prostate carcinoma. Cancer Res 1992;52:5991–5996.

71. Brothman A, Peehl D, Patel A, McNeal J. Frequency and pattern of karyotypic abnormalities in human prostate cancer. Cancer Res 1990;50: 3795–3803.

72. Konig JJ, Hagemeijer A, Smit B, Kamst E, Romijn J, Schroder FH. Cytogenetic characterization of an established xenografted prostatic carcinoma cell line (PC-82). Cancer Genet Cytogenet 1988;34:91–99.

73. Carter BS, Ewing CM, Ward SW, et al. Allelic loss of chromosomes 16q and 10q in human prostate cancer. Proc Natl Acad Sci USA 1990;87:8751–8755.

74. Bergerheim US, Kunimi K, Collins VP, Ekman P. Deletion mapping of chromosomes 8, 10, and 16 in human prostate carcinoma. Genes Chromosom Cancer 1991;3:215–220.

75. Kunimi K, Bergerheim USR, Larsson I-L, Ekman P, Collins VP. Allelotyping of human prostatic adenocarcinoma. Genomics 1991;11:530–536.

76. Bova GS, Carter BS, Bussemakers MJG, et al. Homozygous deletion and

frequent allelic loss of chromosome 8p22 loci in human prostate cancer. Cancer Res 1993;53:3869–3873.

77. Vogelstein B, Fearon ER, Kern SE, et al. Allelotype of colorectal carcinomas. Science 1989;244:207–211.

78. Emi M, Fujiwara Y, Nakajima T, et al. Frequent loss of heterozygosity for loci on chromosome 8p in hepatocellular carcinoma, colorectal cancer, and lung cancer. Cancer Res 1992;52:5368–5372.

79. Tsuda H, Zhang W, Shimasato Y, et al. Allele loss on chromosome 16q associated with progression of human hepatocellular carcinoma. Proc Natl Acad Sci USA 1990;87:6791–6794.

80. Sato T, Tanigami A, Yamakawa K, et al. Allelotype of breast cancer: cumulative allele losses promote tumor progression in primary breast cancer. Cancer Res 1990;50:7184–7189.

81. Maw MA, Grundy PE, Millow LJ, et al. A third Wilms' tumor locus on chromosome 16q. Cancer Res 1992;52:3094–3098.

82. Natt E, Magenis RE, Zimmer J, Mansouri A, Scherer G. Regional assignment of the human loci for uvomorulin and chymotrypsinogen B with the help of two overlapping deletions on the long arm of chromosome 16. Cytogenet Cell Genet 1989;50:145–148.

83. Frixen UH, Behrens J, Sachs M, et al. E-cadherin-mediated cell-cell adhesion prevents invasiveness of human carcinoma cells. J Cell Biol 1991;113:173–185.

84. Umbas R, Schalken JA, Aalders TW, et al. Expression of the cellular adhesion molecule, E-cadherin, is reduced or absent in high grade prostate cancer. Cancer Res 1992;52:5104–5109.

85. Morton RA, Ewing CM, Nagafuchi A, Tsukita S, Isaacs WB. Reduction of E-cadherin levels and deletion of the α catenin gene in human prostate cancer cells. Cancer Res 1993;53:3585–3590.

86. Morton RA, Ewing CM, Wasmuth JJ, et al. Assignment of the human a catenin gene to chromosome 5q21-q22. Genomics 1994;19:188–190

87. Bookstein R, Rio P, Madreperla S, et al. Promoter deletion and loss of retinoblastoma gene expression in human prostate carcinoma. Proc Natl Acad Sci USA 1990;87:7762–7766.

88. Bookstein R, Shew J, Chen P, Scully P, Lee W-H. Suppression of tumorigenicity of human prostate carcinoma cells by replacing a mutated *Rb* gene. Science 1990;247:712–715.

89. Isaacs WB, Carter BS, Ewing CM. Wild-type *p53* suppresses growth of human prostate cancer cells containing mutant *p53* alleles. Cancer Res 1991;51:4716–4720.

90. Makos M, Nelkin BD, Lerman MI, Latif F, Zbar B, Baylin SB. Distinct hypermethylation patterns occur at altered chromosome loci in human lung and colon cancer. Proc Natl Acad Sci USA 1992;89:1929–1933.

91. Carter BS, Beaty TH, Steinberg GD, Childs B, Walsh PC. Mendelian inheritance of familial prostate cancer. Proc Natl Acad Sci USA 1992;89: 3367–3371.

92. Carter BS, Bova GS, Beaty TH, et al. Hereditary prostate cancer: epide-

miology and clinical features. J Urol 1993;150:797–802.

93. Cunha Chung LWK, Shannon JM, Taguchi O, Fuyii H. Hormone-induced morphogenesis and growth: the role of mesenchymal-epithelial interactions. Horm Res 1983;39:559–598.

94. Chung LWK, Anderson NG, Neubauer BL, Cunha GR, Thompson TC, Rocco AK. Tissue interaction in prostate development. In: Murphy GP, Sandberg AA, Karr JP, eds. The prostatic cell: structure and function. New York: Alan R Liss, 1981:177–203.

95. Chung LWK , Chang SM, Bell C, Zhau HE, Ro JY, von Eschenbach AC. Coinoculation of tumorigenic rat prostate mesenchymal cells with nontumorigenic epithelial cells results in the development of carcinosarcoma in syngeneic and athymic animals. Int J Cancer 1989;43:1179–1187.

96. Chung LWK, Hong S-J, Zhau HE, et al. Fibroblast mediated human epithelial tumor growth and hormonal responsiveness in vivo. In: Karr JP, Cofey DS, Smith RG, Tindall DJ, eds. Molecular and cellular biology of prostate cancer. New York: Plenum Press 1991:91–102.

97. McNeal JE. Pathology of benign prostatic hyperplasia. Urol Clin North Am 1990;17:477–486.

98. Chung LWK, Matsuura J, Runner MN. Tissue interactions and prostatic growth: I. Induction of adult mouse prostate hyperplasia by fetal urogenital sinus implants. Biol Reprod 1984;31:155–163.

99. Chung LWK. Implications of stromal-epithelial interaction in human prostate cancer growth, progression and differentiation. Cancer Biol 1993;4:183–192.

100. Chung LWK, Zhau HE, Ro JY. Morphologic and biochemical alterations in rat prostatic tumors induced by fetal urogenital sinus mesenchyme. Prostate 1990;17:165–174.

101. Hayashi N, Cunha GR, Wong YC. Influence of male genital tract mesenchymes on differentiation of Dunning prostatic adenocarcinoma. Cancer Res 1990;50:4747–4754.

102. Schleicher G, Stumpf WE, Drews U, Thiedemann KU, Sar M. Differential distribution of ^{3}H dihydrotestosterone and ^{3}H estradiol nuclear binding sites in mouse male accessory sex organs: an autoradiographic study. Histochemistry 1985;82:453–461.

103. Veldscholte J, Ris-Stalpers C, Kuiper GGJM, et al. A mutation in the liggrud binding domain of the and Roger receptor of human LN cap cells affects steroid binding characteristics and response to anti-androgens. Biochem Biophys 1990;173:534–540.

104. Wu H-C, Hsieh JT, Gleave ME, Brown NM, Pathak S, Chung LWK. Derivation of androgen-independent human LNCaP prostatic cancer cell sublines: role of bone stroma cells. Int J Cancer 1994;57:406–412.

105. Jacobs SC, Lawson RK. Mitogenic factor in human prostate extracts. Urology 1980;16:488–491.

106. Story MT, Sasse J, Jacobs SC, Lawson RL. Prostatic growth factor: purification and structural relationship to basic fibroblast growth factor. Biochemistry 1987;26:3843–3849.

107. Moore R, Casey G, Brookes S, Dixon M, Peters G, Dickson C. Sequence, topography and protein coding potential of mouse int-2: a putative oncogene activated by mouse mammary tumour virus. EMBO J 1986;5:919–921.

108. Muller WJ, Lee FS, Dickson C, Peters G, Pattengale P, Leder P. The int-2 gene product acts as an epithelial growth factor in transgenic mice. EMBO J 1990;9:907–913.

109. Taira M, Yoshida T, Miyagawa K, et al. cDNA sequence of human transforming gene *hst* and identification of the coding sequence required for transforming activity. Proc Natl Acad Sci USA 1987;84:2980–2984.

110. Delli-Bovi P, Curatola AM, Kern FG, Greco A, Ittmann M, Basilico C. An oncogene isolated by transfection of Kaposi's sarcoma DNA encodes a growth factor that is a member of the FGF family. Cell 1987;50:729–737.

111. Zhan X, Bates B, Xu X, Goldfarb M. The human FGF-5 oncogene encodes a novel protein related to fibroblast growth factors. Mol Cell Biol 1988;8:3487–3495.

112. Marics I, Adelaide J, Rayband F, et al. Characterization of the hst-related FGF-6 gene, a new member of the fibroblast growth factor gene family. Oncogene 1988;4:335–340.

113. Gimenez-Gallego G, Rodkey J, Bennet C, et al. Brain-derived acidic fibroblast growth factor: complete amino acid sequences and homologies. Science 1985;230:1385–1391.

114. Esch F, Baird A, Lang N, et al. Primary structure of bovine pituitary basic fibroblast growth factor (FGF) and comparison with the amino-terminal sequence of bovine brain acidic FGF. Proc Natl Acad Sci USA 1985;82:6507–6511.

115. Nishi N, Matuo Y, Kunitomi K. Comparative analysis of growth factors in normal and pathologic human prostates. Prostate 1988;13:39–48.

116. Story MT. Polypeptide modulators of prostatic growth and development. Cancer Surv 1991;11:1–23.

117. Mydlo JH, Bulbul MA, Richon VM, Heston WDW, Fair WR. Heparin-binding growth factor isolated from human prostatic extracts. Prostate 1988;12:343–355.

118. Folkman J. Tumor angiogenesis. Adv Cancer Res 1985;43:175–203.

119. Gospodarowicz D. Fibroblast growth factor and its involvement in developmental processes. Curr Top Dev Biol 1990;24:57–93.

120. Hall JA, Harris MA, Intres R, Harris SE. Acidic fibroblast growth factor gene 5′ non-coding exon and flanking region from hamster DDT1 cells: identification of the promoter region and transcriptional regulation by testosterone and aFGF protein. J Cell Biochem 1993;51:116–127.

121. Sherwood ER, Fong C-Y, Lee C, Kozlowski JM. Basic fibroblast growth factor: a potential mediator of stromal growth in the human prostate. Endocrinology 1992;130:2955–2963.

122. Mansson PE, Adam P, Kan M, McKeehan WL. Heparin-binding growth factor gene expression and receptor characteristics in normal rat prostate and two transplantable rat prostate tumors. Cancer Res 1989;49:2485–2494.

123. Browder T, Dunbar C, Neinhuis A. Private and public autocrine loops

in neoplastic cells. Cancer Cells 1989;1:9–17.

124. Chiu I-M, Wang W-P, Lehtoma K. Alternative splicing generates two forms of mRNA coding for human heparin-binding growth factor 1. Oncogene 1990;5:755–762.

125. Yan G, Fukabori Y, McBride G, Nikolaropolous S, McKeehan WL. Exon switching and activation of stromal and embryonic fibroblast growth factor (FGF)-FGF receptor genes in prostate epithelial cells accompany stromal independence and malignancy. Mol Cell Biol 1993;13:4513–4522.

126. Alarid ET, Rubin JS, Young P, et al. Keratinocyte growth factor functions in epithelial induction during seminal vesicle development. Proc. Natl. Acad Sci. USA 1994;91:1074–1078.

127. Yan G, Fukabori Y, Nikolaropoulos S, Wang F, McKeehan WL. Heparin binding keratinocyte growth factor is a candidate stromal to epithelial cell andromedin. Mol Endocrinol 1992;6:2123–2128.

128. Tanaka A, Miyamoto K, Minamino N, et al. Cloning and characterization of an androgen-induced growth factor essential for the androgen-dependent growth of mouse mammary carcinoma cells. Proc Natl Acad Sci USA 1992;89:8928–8932.

129. Roberts AB, Anzano MA, Wakefield LM, Roche NS, Stern DF, Sporn MB. Type β transforming growth factor: a bifunctional regulator of cellular growth. Proc Natl Acad Sci USA 1985;82:119–123.

130. Massague J. The TGF-β family of growth and differentiation factors. Cell 1987;49:437–438.

131. Kyprianou N, Isaacs JT. Expression of transforming growth factor-β in the rat ventral prostate during castration-induced programmed cell death. Mol Endocrinol 1989;3:1515–1522.

132. Kyprianou N, Isaacs JT. Identification of a cellular receptor for transforming growth factor-β in rat ventral prostate and its negative regulation by androgens. Endocrinology 1988;123:2124–2131.

133. Martikainen P, Kyprianou N, Isaacs JT. Effect of transforming growth factor β1 on the proliferation and death of rat prostatic cells. Endocrinology 1990;127:2963–2968.

134. Shain SA, Lin AL, Koger JD, Karaganis AG. Rat prostate cancer cells contain functional receptors for transforming growth factor-β. Endocrinology 1990;126:818–825.

135. Wilding G, Zugmeier G, Knabbe C, Flanders K, Gelmann E. Differential effects of transforming growth factor β on human prostate cancer cells in vitro. Mol Cell Endocrinol 1989;62:79–87.

136. Bassols A, Massagué J. Transforming growth factor β regulates the expression and structure of extracellular matrix chondroitin/dermatan sulfate proteoglycans. J Biol Chem 1988;263:3039–3045.

137. Yang EY, Moses HL. Transforming growth factor β1-induced changes in cell migration, proliferation and angiogenesis in the chicken chorioallantoic membrane. J Cell Biol 1990;111:731–741.

138. Torre-Amikone G, Beauchamp RD, Koeppen H, et al. A highly immunogenic tumor transfected with murine transforming growth factor β1 cDNA

escapes immune surveillance. Proc Natl Acad Sci USA 1990;87:1486–1490.

139. Fontana A, Freii K, Bodmer S, et al. Transforming growth factor-β inhibits the generation of cytotonic T cells in virus-infected mice. J Immunol 1989;143:3230–3234.

140. Steiner MS, Barrack ER. Transforming growth factor β1 overproduction in prostate cancer: effects of growth in vivo and in vitro. Mol Endocrinol 1992;6:15–25.

141. Truong LD, Kadmon D, McCone BK, Flanders KC, Scardino PT, Thompson TC. Association of transforming growth factor β1 with prostate cancer: an immunohistochemical study. Hum Pathol 1993;24:4–9.

142. Silberstein GB, Daniel CW. Reversible inhibition of mammary gland growth by transforming growth factor-β. Science 1987;237:291–293.

143. Carr BI, Hayashi I, Brayum EL, Moses HL. Inhibition of DNA synthesis in rat hepatocytes by platelet-derived type-beta transforming growth factor. Cancer Res 1986;46:2330–2334.

144. Battegay EJ, Raines EW, Seifert RA, Bowen-Pope DF, Ross R. TGF-β induces biomodal proliferation of connective tissue cells via complex control of an autocrine PDGF loop. Cell 1990;63:515–524.

145. Story MT, Hopp KA, Meier DA, Begun FP, Lawson RK. Influence of transforming growth factor β1 and other growth factors on basic fibroblast growth factor level and proliferation of cultured human prostate-derived fibroblasts. Prostate 1993;22:183–197.

146. Schuurmans AL, Bolt J, Mulder E. Androgens and transforming growth factor β modulate the growth response to epidermal growth factor in human prostatic tumor cells (LNCaP). Mol Cell Endocrinol 1988;60:101–104.

147. Ikeda T, Lioubin MN, Marquardt H. Human transforming growth factor type β-2: production by a prostatic adenocarcinoma cell line, purification and initial characterization. Biochemistry 1987;26:2406–2410.

148. Thompson TC, Truong LD, Timme TL, et al. Transforming growth factor β1 as a biomarker for prostate cancer. J Cell Biochem 1992;16H:54–61.

149. Killian CS, Corral DA, Kawinski E, Constantine RI. Mitogen response of osteoblast cells to prostate-specific antigen suggests an activation of latent TGFβ and a proteolytic modulation of cell adhesion receptors. Biochem Biophys Res Commun 1993;192:940–947.

150. Steiner MS. Role of peptide growth factors in the prostate: a review. Urology 1993;41:99–110.

151. Oreffo RO, Bonewald L, Kukita A, et al. Inhibitory effects of the bone derived growth factors osteoinductive factor and transforming growth factor on isolated osteoclasts. Endocrinology 1990;126:3069–3075.

152. Hock JM, Canalis E, Centrella M. Transforming growth factor β stimulates bone matrix apposition and bone cell replication in cultured fetal rat calvarial. Endocrinology 1990;126:421–426.

153. Joyce ME, Roberts AB, Sporn MB, Bolander ME. Transforming growth factor β and the initiation of chondrogenesis and osteogenesis in the rat femur. J Cell Biol 1990;110:2195–2207.

154. Eklov S, Funa K, Nordgren H, et al. Lack of latent transforming growth factor β binding protein in malignant but not in benign prostatic tissue. Cancer Res 1993;53:3193–3197.

155. Yang Y, Chisholm GD, Habib FK. Epidermal growth factor and transforming growth factor α concentration in BPH and cancer of the prostate: their relationships with tissue androgen levels. Br J Cancer 1993;67:152–155.

156. Connolly JM, Rose DP. Autocrine regulation of DU145 human cancer cell growth by epidermal growth factor-related peptide. Prostate 1991;19:173–180.

157. Derynck R. Transforming growth factor-α. Cell 1988;54:593–595.

158. Snedeker SM, Brown CF, DiAugustine RP. Expression and functional properties of transforming growth factor α and epidermal growth factor during mouse mammary gland ductal morphogenesis. Proc Natl Acad Sci USA 1991;88:276–280.

159. Hara Y. Analysis of anchorage-independent growth of human prostate cancer cell line LNCaP. Nippon Hinyokika Gakkai Zasshi 1992;83:1429–1435.

160. Taylor TB, Ramsdell JS. Transforming growth factor α and its receptor are expressed in the epithelium of the rat prostate gland. Endocrinology 1993;133:1306–1311.

161. Wu HH, Kawamata H, Kawai K, Lee C, Oyasu R. Immunohistochemical localization of epidermal growth factor and transforming growth factor α in the male rat accessory sex organs. J Urol 1993;150:990–993.

162. Harper ME, Goddard L, Glynne-Jones E, Wilson DW. An immunocytochemical analysis of TGFα expression in benign and malignant prostatic tumors. Prostate 1993;23:9–23.

163. Zhau HYE, Wan DS, Zhou J, Miller GJ, von Eschenbach AC. Expression of c-erb B2/neu protooncogene in human prostatic cancer tissues and cell lines. Mol Carcinog 1992;5:320–327.

164. Coussens L, Yang-Feng TL, Liao YC, et al. Tyrosine kinase receptor with extensive homology to EGF receptor shares chromosomal location with neu oncogene. Science 1985;230:1132–1139.

165. Nakamura T, Teramoto H, Ichihara A. Purification and characterization of a growth factor from rat platelets for mature parenchymal hepatocytes in primary cultures. Proc Natl Acad Sci USA 1986;83:6489–6493.

166. Stoker M, Perryman M. An epithelial scatter factor released by embryo fibroblasts. J Cell Sci 1985;77:209–223.

167. Chung LWK. Li W, Gleave ME, et al. Human prostate cancer model: Notes of growth factors and extracellular matrices. J Cell Biochem 1992;16H:99–105.

168. Rubin JS, Chan AML, Bottaro DP, et al. A broad-spectrum human lung fibroblast-derived mitogen is a variant of hepatocyte growth factor. Proc Natl Acad Sci USA 1991;88:415–419.

169. Montesano R, Matsumoto K, Nakamura T, Orci L. Identification of a fibroblast-derived epithelial morphogen as hepatocyte growth factor. Cell 1991;67:901–908.

170. Gheradi E, Stroker M. Hepatocyte growth factor-scatter factor: mito-

gen, motogen and met. Cancer Cells 1991;3:227–232.

171. Giordano S, Di Renzo MF, Narsimhan RP, Cooper CS, Rosa C, Comoglio PM. Biosynthesis of the protein encoded by the c-met proto-oncogene. Oncogene 1989;4:1383–1388.

172. Ferracini R, Longati P, Naldini L, Vigna E, Comoglio PM. Identification of the major autophosphorylation site of the met/hepatocyte growth factor receptor tyrosine kinase. J Biol Chem 1991;266:19558–19564.

173. Prat M, Narsimhan RP, Crepaldi T, Nicotra MR, Natali PG, Comoglio PM. The receptor encoded by the human c-met oncogene is expressed in hepatocytes, epithelial cells and solid tumors. Int J Cancer 1991;49:323–328.

174. Cooper CS. The met oncogene: from detection by transfection to transmembrane receptor for hepatocyte growth factor. Oncogene 1992;7:3–7.

175. Di Renzo MF, Narsimhan RP, Olivero M, et al. Expression of the met/HGF receptor in normal and neoplastic human tissues. Oncogene 1991;6:1997–2003.

176. Pisters LL, Troncoso P, Zhau HE, et al. The role of hepatocyte growth factor/scatter factor and its receptor, the c-met protooncogene in human prostatic tumor growth [abstract]. J Urol 1993;149:482A.

177. Chung LWK, Li W, Gleave ME, et al. Human prostate cancer model: roles of growth factors and extracellular matrices. J Cell Biochem Suppl 1992;16H:99–105.

178. Iwamura M, Sluss PM, Casamento JB, Cockett ATK. Insulin-like growth factor I: action and receptor characterization in human prostate cancer cell lines. Prostate 1993;22:243–252.

179. Pietrzkowski Z, Mulholland G, Gomella L, et al. Inhibition of growth of prostatic cancer cell lines by peptide analogues of insulin-like growth factor 1. Cancer Res 1993;53:1102–1106.

180. Cohen P, Peehl DM, Lamson G, Rosenfeld RG. Insulin-like growth factors (IGFs), IGF receptors, and IGF binding proteins in primary cultures of prostate epithelial cells. J Clin Endocrinol Metab 1991;73:401–407.

181. Kanety H, Madjar Y, Dagan Y, et al. Serum insulin-like growth factor-binding protein-2 (IGFBP-2) is increased and IGFBP-3 is decreased in patients with prostate cancer: correlation with serum prostate-specific antigen. J Clin Endocrinol Metab 1993;77:229–233.

182. Koutsilieris M, Frenette G, Lazure C, Lehoux J-G, Govindan MV, Polychronakos C. Urokinase-type plasminogen activator: a paracrine factor regulating the bioavailability of IGFs in PA-III cell-induced osteoblastic metastases. Anticancer Res 1993;13:481–486.

183. Folkman J, Mosccona A. Role of cell shape in growth control. Nature 1978;273:345–349.

184. Gospodarowicz D. Fibroblast growth factor. Mol Cell Endocrinol 1986;46:187–204.

185. Murphy BC, Pienta KJ, Coffey DS. Effects of extracellular matrix components and dihydrotestosterone on the structure and function of human prostate cancer cells. Prostate 1992;20:29–41.

186. Fong C-J, Sherwood ER, Braun EJ, Berg LA, Lee C, Kozlowski JM.

Regulation of prostatic carcinoma cell proliferation and secretory activity by extracellular matrix and stromal secretions. Prostate 1992;21:121–131.

187. Bulbul MA, Pavelic K, Slocum HK, et al. Growth of human urologic tumors on extracellular matrix. J Urol 1986;136:512–516.

188. Pretlow TG, Delmoro CM, Dilley GG, Spadafora CG, Pretlow TP. Transplantation of human prostatic carcinoma into nude mice in Matrigel. Cancer Res 1991;51:3814–3817.

189. Chiquet-Ehrismann R, Kalla P, Pearson CA. Participation of tenascin and transforming growth factor-β in reciprocal epithelial-mesenchymal interactions of MCF7 cells and fibroblasts. Cancer Res 1989;49:4322–4325.

190. Reichmann E, Ball R, Groner B, Friis RR. New mammary epithelial and fibroblastic cell clones in coculture form structures competent to differentiate functionally. J Cell Biol 1989;108:1127–1138.

191. Lin CQ, Bissell MJ. Multi-faceted regulation of cell differentiation by extracellular matrix. FASEB J 1993;7:737–743.

192. Kleinman HI, McGarvey ML, Hassell JR, et al. Basement membrane complexes with biological activity. Biochemistry 1986;25:312–318.

193. Rubin AL, Yao A, Rubin H. Relation of spontaneous transformation in cell culture to adaptive growth and clonal heterogeneity. Proc Natl Acad Sci USA 1990;87:482–486.

194. Getzenberg RH, Pienta KJ, Coffey DS. The tissue matrix: cell dynamics and hormone action. Endocr Rev. 1990;11:399–417.

195. Freeman MR, Bagli DJ, Lamb CC, et al. Culture of prostatic cell line in basement membrane gels results in an enhancement of malignant properties and constitutive alterations in gene expression. J Cell Physiol 1994;158:325–336.

CHAPTER 13

Molecular Genetic and Biological Aspects of Primary Brain Tumors

Peter A. Steck
Mark A. Pershouse
Hideyuki Saya

Considerable interest has recently been focused on the identification and characterization of specific genes involved in the initiation and progression of various human cancers. These studies are based on experimental evidence in molecular genetics and the current hypothesis that tumor formation involves the creation and accumulation of multiple genetic alterations. The accumulation of damage to genetic material has been proposed to be responsible for the generation and degree of the malignant characteristics of neoplastic cells by numerous investigators (1–3). A model by Fearon and Vogelstein (4) has assigned specific molecular alterations to the various steps in the cascade of carcinogenesis involved in colorectal carcinoma. The major molecular alterations presently implicated in the oncogenic processes include damage to genetic material leading to the activation of oncogenes or loss of function of tumor suppressor genes. In addition, epigenetic events contribute significantly to changes in the phenotypic characteristics of neoplastic cells. For each specific cancer, different sets of oncogenes and tumor suppressor genes appear to be involved in oncogenesis. Paradigms are currently being developed and refined to account for a number of consistent molecular, cytogenetic, and epigenetic alterations observed to occur in each type of cancer. Extensive research efforts are also directed at defining the functional significance of the known alterations. As the previous chapters of this monograph have detailed the molecular mechanisms and results of genetic alterations, we will confine our discussion to those alterations that have been implicated in the oncogenesis of primary brain tumors, particularly glioma, the most common neoplastic disease of the central nervous system (CNS).

Cytogenetic, molecular, and biochemical studies have clearly demonstrated that primary brain tumors are generated and progress through the involvement of multiple, but relatively consistent genetic and epigenetic alterations. These changes include the activation of protooncogenes and the functional loss of putative tumor suppressor genes, as well as increased expression of potent growth regulatory factors or their receptors. The activation of oncogenes and the loss of function of tumor suppressor genes can occur by a variety of molecular mechanisms, including "simple" mutations in the gene structure, major chromosomal translocations, gene amplifications, or deletions to affect gene structure, function, and expression. Epigenetic changes, such as methylation of DNA, or the aberrant switching on or off of embryonic or differentiation pathways may also alter gene expression (5). Both genetic and epigenetic alterations affect the expression or function of the gene product. The apparent difference between oncogenes and tumor suppressor genes is that the latter, when functioning properly, mediate the growth and differentiation of cells in a negative fashion, whereas activated oncogenes stimulate the proliferative and invasive properties of the cells. The oncogenes and tumor suppressor genes, however, may also directly affect the tumor cell's relation with its environment, such as inducing angiogenesis or immune responses.

ANATOMIC AND CLINICAL MANIFESTATIONS OF BRAIN TUMORS

To appreciate more fully the significance of the molecular alterations associated with cancers of the CNS, a brief overview of the anatomic considerations, clinical consequences, and cellular origins of the various CNS tumors would be informative. Malignancies of the CNS are among the most feared cancers, not only for their severe and debilitating neuropathological complications, but also for their potentially psychologically devastating consequences for the patient. The major symptoms and signs of brain tumors include headache, seizures, paralysis, and impaired cognition. A combination of surgery, chemotherapy, and radiation is the principal therapeutic regimen useful in CNS neoplasms. However, all must be administered with an added concern of maintaining significant neurologic function of the afflicted person.

In 1993 the incidence of primary brain tumors was approximately 18,500 newly diagnosed cases for the adult population (6). Unfortunately, a little over two-thirds of these afflicted persons will succumb to the disease within 2 years. Tumors involving the CNS also represent the most common solid neoplasia among the pediatric population, with an incidence of about 2000 per year. These tumors are the second leading cause of cancer deaths in children younger than 15 years of age. The most common intracranial tumors in adults arise from cells of the glial lineage, with an approximate frequency of 45% for glioblastoma, 30% for astrocytoma, 12% for ependymomas and oligodendrogliomas, while meningiomas arising from the arachnoid covering sheath (outer membrane) of the brain account for about 20% of CNS tumors. Medulloblastomas (primitive neuroectodermal tumors, PNETs), thought to arise from precursor neuroectodermal cells, and neuroblastomas, arising from neuroblasts, are more common in the pediatric population, with frequencies of 20% and 4%, respectively, and are rare in the adult population. A variety of other primary neoplasms occur with relatively rare frequency in the different anatomic structures of the CNS, including schwannomas and meningiomas in the spinal cord, adenomas in the pituitary, and pineocytomas in the pineal gland. Some of these less common forms of tumors, however, have been the focus of a number of studies based upon their increased occurrence in certain families with a genetic predisposition to certain afflictions (see discussion of neurofibromatosis genes 1 and 2 below).

The brain also has several anatomic features not associated with other organs. The brain is very sensitive to alterations in its local environment, especially any unregulated cell growth. This is in part due to its confinement within the skull. In addition, the brain lacks a lymphatic system, which would aid in the removal of fluids that accompany cellular proliferation. The blood-brain barrier (BBB) also tightly regulates the exchange of many materials with the brain. The BBB represents a physiologic and anatomic structure formed by tight junctions between the brain endothelial cells, an integral basement membrane, and adjacent astrocytes. The BBB appears to be partially breached in neoplastic areas, although the extent of integrity in regard to possible therapeutic consequences is still controversial. In general, the BBB greatly influences the passage of materials into the brain, limiting the therapeutic effectiveness of certain agents. The CNS also represents a relatively "immunologically privileged" site, although the isolation is relative rather than absolute. As a result, the impact of normal immune surveillance and the possibility of successful immunotherapy may be quite limited.

GENETIC ALTERATIONS IN PRIMARY BRAIN TUMORS

Consistent genetic alterations including chromosomal deletions, translocations, or amplifications have been assumed to represent areas of the genome that may contain an oncogene or tumor suppressor gene. Furthermore, the normal cellular function of protooncogenes and tumor suppressor genes appears to involve components of cellular signal transduction pathways controlling normal cellular growth and differentiation. The abnormal growth stimulatory signaling of oncogene protein products and the loss of inhibitory signals mediated by tumor suppressor gene products combine to generate the neoplastic state. Studies have implicated several genes and their protein products in the oncogenesis of primary brain tumors. These include epidermal growth factor receptor and one of its ligands, transforming growth factor-α, N-myc, gli, p53, neurofibromatosis genes 1 and 2, the putative meningioma tumor suppressor gene (MN1), and putative tumor suppressor genes on chromosomes 4, 9, 10, 11, 13, 14, 17 (non p53), 19, and 22. Similar to colon carcinoma, the grade or degree of histologic malignancy of astrocytic tumors correlates with the accumulation of genetic damage. Low-grade astrocytomas have few identified alterations, whereas the intermediate anaplastic astrocytomas and high-grade glioblastomas have an increasing frequency and number of genetic alterations (Fig. 13.1). From a variety of studies described below, the molecular progression involved in astrocytic tumors can be illustrated to include various genetic events as shown in Figure 13.2. Several points should be emphasized with this "molecular progression" of gliomas. The observed genetic and epigenetic alterations are associated with the different grades of tumor and do not necessarily imply a temporal relation. For instance, the majority of gliomas are first diagnosed as glioblastomas, so determination of the exact order that the alterations occurred may not be possible. In only a few cases has the increase in genetic alterations been shown with an increase in histologic grade in the same patient (7). And not all alterations are found associated with each tumor, suggesting the possibility of alternative pathways to glial cell oncogenesis.

Activation of Growth Stimulatory Pathways

The investigation of the activation of protooncogenes in primary brain tumors has primarily focused on the increased expression of several growth stimulatory signal transduction factors. Certain growth factors and their receptors have been shown to be activated or overexpressed in gliomas and meningiomas. The most prominent among them is the observed gene amplification and overexpression of epidermal growth factor (EGF) receptor in about 40% of gliomas (8,9). Aberrant forms of EGF receptor have been isolated and characterized and appear to be relatively common in gliomas, with approximately 40% of the cases examined with EGF receptor

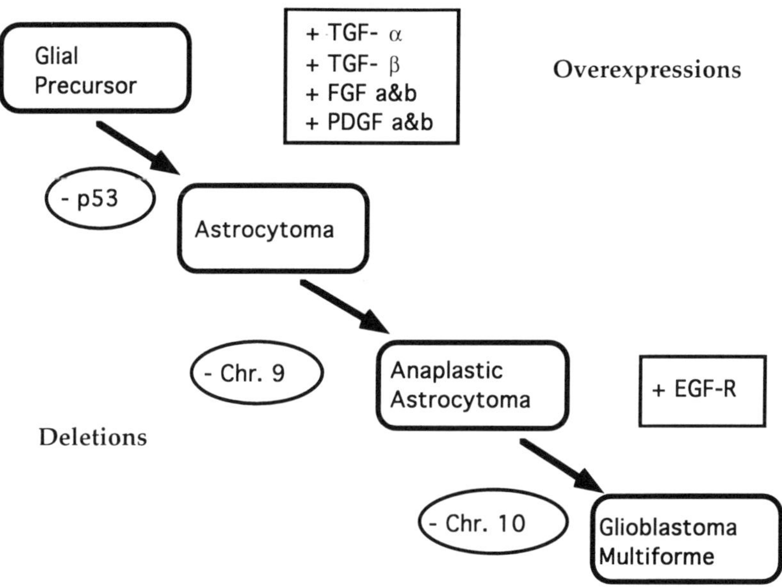

Fig. 13.1: Illustration of known genetic alterations during malignant progression of astrocytic tumors (gliomas). EGF-R, epidermal growth factor receptor.

gene amplification revealing structural alterations (10,11). The most common structural alteration is deletion of a major segment of the external domain, including the ligand-binding domain of the receptor. The altered receptor thus represents a potential tumor-specific antigenic determinant that is being pursued with therapeutic implications (12). Amplification or altered structure of EGF receptor occurs almost exclusively in glioblastoma multiformes; however, only minor biological consequences have been associated with these alterations. Overexpression of EGF receptor has been implicated in increased invasive properties of the tumor cells, more rapid revascularization after partial resection, and altered responsiveness of the tumor cells to biological agents (13,14). Platelet-derived growth factor receptor was also shown to be amplified and overexpressed in a few gliomas (15).

Yamaguchi et al. (16) have demonstrated that astrocytomas exhibit changes in fibroblast growth factor receptor (FGFR) expression during the

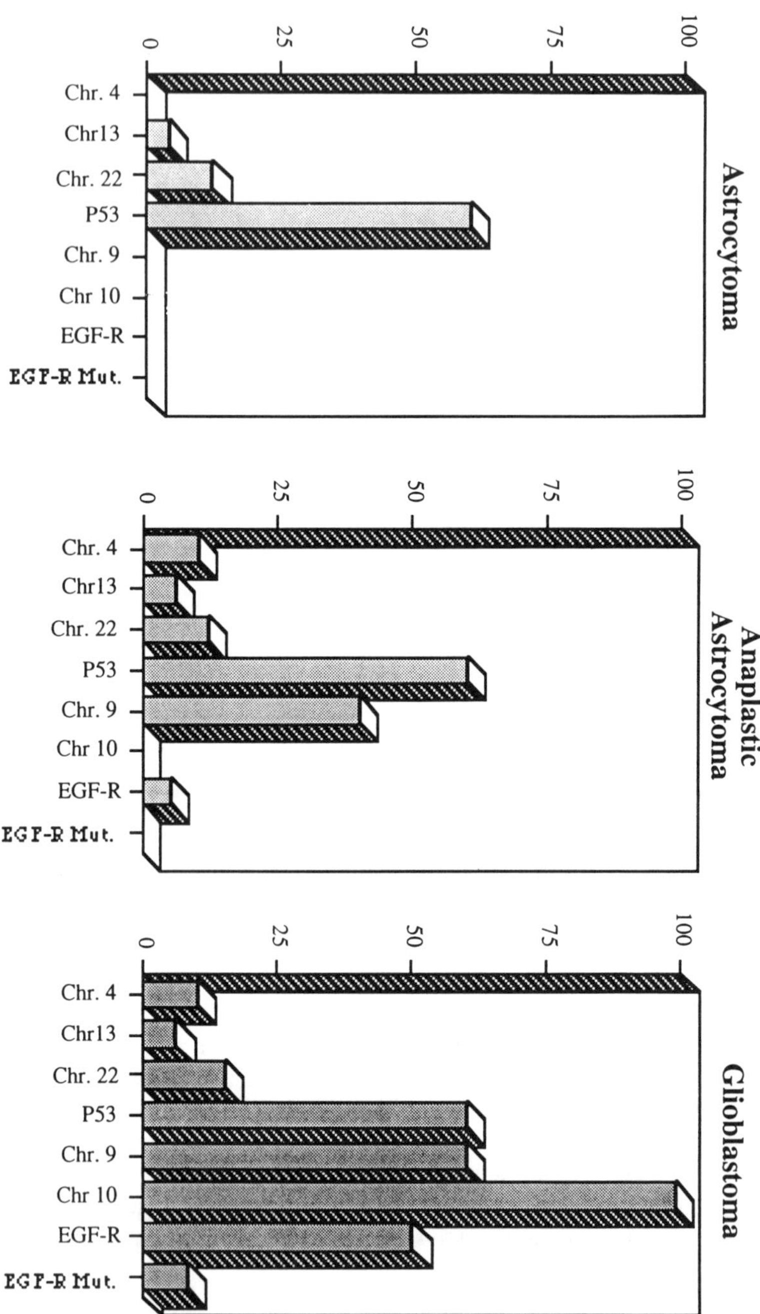
Frequency of Alteration
0
25
50
75
100
Chr. 4
Chr13
Chr. 22
P53
Chr. 9
Chr 10
EGF-R
EGF-R Mut.
Astrocytoma
Anaplastic Astrocytoma
Glioblastoma

progression from a benign to a malignant phenotype. FGFR1 (flg) expression was abundant in glioblastomas, but was low in normal white matter and low-grade astrocytomas. On the contrary, FGFR2 (bek) expression was abundant in normal white matter and benign astrocytomas, but was very low in glioblastomas. Furthermore, glioblastomas expressed an alternatively spliced form of FGFR1 containing two immunoglobulin-like disulfide loops (FGFR1β), whereas normal brain expressed a form of the receptor containing three immunoglobulin-like disulfide loops (FGFR1α). The altered external structures of the FGFRs has been associated with the differential binding of the FGFR ligands (17).

The precise biological significance of the amplification, overexpression, and alteration of these growth factor receptors has yet to be determined, although their association with higher-grade gliomas implies an important association with the malignant progression of the tumor cells.

Overexpression of various growth factors has also been detected in gliomas. These include transforming growth factor-α and -β, acidic and basic fibroblast growth factors, vascular endothelial growth factor, and platelet-derived growth factor a and b chains. The majority of the growth factors appear to have increased expression in all grades of gliomas compared with normal brain, with some factors exhibiting an increased expression in higher grades of tumors (18–20). The growth factors undoubtedly play important roles in tumor cell survival and proliferation, but also are important for proliferation of adjacent normal cells. For example, extensive proliferation of endothelial cells is a common feature of gliomas. Fibroblast growth factors, transforming growth factors-α and -β, and vascular endothelial growth factor have been demonstrated to be potent participants both in autocrine stimulation of tumor cells and in angiogenesis (21). Folkman and Klagsbrun have clearly shown that vigorous angiogenesis is necessary for tumor growth past a diameter of 1 to 2 mm (21). In this regard, vascular endothelial growth factor has been observed to be secreted by glioma cells, but receptors for the factor have been observed only on adjacent endothelial cells. Growth factors may also serve other roles in tumor development, as illustrated by the identification of transforming growth factor-β as an immunosupressive agent in brain tumors (22). These observations demonstrate the multifunctional roles of growth factors and their receptors in oncogenesis.

←

Fig. 13.2: Frequency of known genetic alterations in astrocytic lineage neoplasms.

TUMOR SUPPRESSOR GENES IN PRIMARY BRAIN TUMORS

The p53 Gene

Allelic deletions of chromosome 17p were revealed to be the most frequent chromosomal aberration associated with all malignancy grades of astrocytoma (23–25). The attainment of the somatic homozygosity for loci on chromosome 17p is apparently operative in a wide variety of cancers in addition to astrocytoma. These data have emphasized the importance of identifying genes on 17p whose loss or inactivation could indirectly promote the neoplastic processes. Accumulation of recent results of numerous molecular studies have revealed that the p53 gene maps to region p13 of chromosome 17 and is the most commonly altered gene in a wide range of human malignancies, of both inherited and sporadic forms. The p53 gene was initially considered an oncogene, but several studies have demonstrated that the originally isolated protein was derived from a mutated gene and that the wild-type p53 gene product has tumor-suppressing activity (26–29). Frequent loss of one copy of the p53 gene with accompanying mutation of the remaining allele has been found in a wide variety of transformed cells, cell lines, and tumors of various organ systems, strongly implicating p53 as a tumor suppressor gene (30,31). Strong evidence that p53 is a suppressor gene was demonstrated by replacing the wild-type p53 gene and showing that the expression of p53 had antiproliferative effects in cancer cells including glioblastoma, colorectal carcinoma, osteosarcoma, breast cancer, and gastric carcinoma (32–36). Furthermore, transgenic mice carrying a homozygous mutant p53 allele were observed to develop multiple types of cancer in a short time (37). These observations reveal that the wild-type p53 gene is an important barrier to the progression of neoplastic processes, and that inactivation of the wild-type p53 function may allow cells to develop a more tumorigenic phenotype. Since the wild-type p53 was shown to have ability to activate transcription from specific sequences, the genes induced by p53 have been suggested to mediate its biological role as a tumor suppressor. In fact, several recent reports revealed that the wild-type p53 transcriptionally activates the expression of the *WAF1/CIP1* gene, whose product binds to cyclin complexes and inhibits the function of cyclin-dependent kinases (38,39). The wild-type p53 may control cell-cycle arrest following DNA damage by regulating the expression of *WAF1/CIP1*. Therefore, mutant p53 would permit unregulated cellular proliferation, thus permitting further genetic damage in other critical genes, and result in the genesis of tumors.

Alterations of p53 have been detected in all grades of human malignant astrocytomas, suggesting that the p53 gene alteration is an early event in astrocytoma oncogenesis (23–25). Alterations of the p53 gene are detectable in small numbers of malignant cells in the lower-grade astrocytomas and more frequently in glioblastomas, the highest grade of astrocytoma (40). This observation strongly suggests clonal expansion of the transformed astrocytes carrying the p53 alteration during malignant tumor progression.

Most normal cells express p53 mRNA, but the wild-type p53 protein is not detectable due to its short half-life (41). By contrast, the mutant p53 found in tumor cells is stabilized to an unusually increased degree, thus permitting its detection by immunohistochemical methods. Therefore, the detectable p53 protein in cancer cells is generally considered to be a mutant form or that conformationally changed by abnormal protein binding. Bruner and coworkers (42,43) examined the expression of p53 protein in brain tumors by immunohistochemical analysis, and the overall percentage of astrocytic gliomas with p53-positive immunostaining is 64% in this study. The immunocytochemical cellular labeling was always nuclear and clearly delineates positive and negative cells. Only tumor astrocytes are stained, while vessels, normal brain, and reactive cells are negative, suggesting this may be a useful neoplastic cell marker in positive cases. The results of the p53 immunostaining were also analyzed in relation to tumor grade, primary diagnosis or recurrence, patient age at diagnosis, and bromodeoxyuridine (BUdR) tumor proliferation index. There was no significant correlation with any of these variables. One report suggested an association between p53 gene or protein alterations in brain tumors and age at diagnosis (44), and that study also found a better prognosis in the younger group with more frequent p53 mutation. However, because younger patients with primary malignant astrocytomas usually show a longer survival, the independent predictive significance of p53 gene mutation is still in question.

Li-Fraumeni syndrome (LFS) has been defined as an inherited cancer family syndrome with a proband with sarcoma (45). Heritable germline mutations of the p53 gene were described in patients with LFS and members of their families (46,47). Our recent study has demonstrated that germline mutations of the p53 gene are significantly more frequently identified in patients with multifocal glioma, glioma with another primary malignancy, and glioma associated with a family history of cancer (48). In particular, combinations of these risk factors increased the percentage of germline p53 gene mutations to approximately 43% if two factors were present, or 67% if all three factors were present. All these data strengthen the implication of p53 alterations in the genesis of subsets of malignant astrocytomas.

Involvement of Neurofibromatosis 1 Gene

Von Recklinghausen neurofibromatosis, which is also referred to as neurofibromatosis type 1 (NF1), is one of the most common autosomal dominantly inherited disorders. Patients with NF1 usually display diverse clinical features. Two consistent abnormalities are neurofibromas and cutaneous pigmentations called café au lait spots; less common are mental retardation, renal hypertension, and an increased risk of malignant tumors in the nervous system. In 1990, the gene that is responsible for the genesis of neurofibromatosis type 1, the NF1 gene, was identified on chromosome 17q (49,50). Because mutations and deletions of the NF1 gene were occasionally detected in non-NF sporadic neuroectodermal tumors including neuroblastomas and melanomas (51,52), the NF1 gene is proposed to be a tumor suppressor gene. The NF1 gene encodes a 240-kDa protein (designated as neurofibromin), and has a 360-residue region that shows significant homology to the catalytic domains of both mammalian guanosine triphosphatase-activating proteins (GAP) and yeast IRA proteins (53). The GAP-related domain of the NF1 gene (*NF1*-GRD) product, like GAP and IRA proteins, was shown to stimulate Ras GTPase and consequently inactivate the Ras protein (53–55). This evidence suggests that the NF1 gene product has an important role in the signal transduction pathway by interacting with ras or ras-like proteins, and is thereby related to growth and differentiation.

The *NF1* gene was found to express several alternative spliced transcripts. In particular, the *NF1*-GRD region has two different types of splicing variants (56,57). One (type II) transcript carries a 63-base insert that encodes a region of 21 amino acids in the center of the *NF1*-GRD region, and the other (type I) transcript does not have the insert. Our observation using a number of cell lines and tissue samples revealed that the type I transcript was predominantly expressed in undifferentiated neuronal cells, whereas the type II transcript predominated in differentiated cells (56). In addition, the expression pattern of the two types of *NF1*-GRD transcripts immediately changed, from type I to type II, in a neuroblastoma cell when the neuronal differentiation was induced by retinoic acid treatment. We also examined the differential expression of type I and type II transcripts of *NF1*-GRD in clinical samples of supratentorial malignant brain tumors and demonstrated that malignant astrocytomas had higher ratios of type I to type II than adjacent brain (58). These data suggest that alternative splicing of *NF1*-GRD may play a significant role in neuroectodermal tissue differentiation, and that abnormal regulation of splicing might influence the genesis of neuroectodermal tumors, including malignant brain tumors.

Neurofibromatosis Type 2

Neurofibromatosis type 2 (NF2) is also an autosomal dominantly inherited disease, but the incidence of this disorder (one in 35,000) is much lower than that of NF1. NF1, as described above, predisposes mainly to differentiation abnormalities in neuronal tissues, and the incidence of brain tumor in NF1 patients is approximately 10%. However, the most consistent disorder in NF2 patients is development of intracranial benign tumors, such as bilateral vestibular schwannomas (95%) and meningiomas (45%). The NF2 gene has been identified on chromosome 22q (59,60). Alterations of the NF2 gene were detected not only in NF2 patients but also in patients with non-NF2 meningiomas and schwannomas. Therefore, the NF2 gene is considered to be a tumor suppressor gene involved in these neoplasms. The NF2 gene encodes a 595 amino acid protein (designated as merlin), which has striking homology with the band 4.1 superfamily proteins including moesin, ezrin, and radixin. As the band 4.1 superfamily proteins are known to link cell membrane proteins with cytoskeletal proteins, it can be speculated that merlin also forms a cell membrane–cytoskeleton associated protein complex, which may play a role in cell-cell and cell-matrix contact signaling. Accordingly, alterations in the NF2 gene could impair the cell-cell contact growth inhibition signals and result in generation of benign tumors. Identification of the merlin-related signaling pathway may elucidate one of the potential mechanisms for developing certain intracranial benign tumors, such as schwannoma and meningioma.

CONSISTENT CHROMOSOMAL ABNORMALITIES

Evidence to suggest the presence of unidentified tumor suppressor genes is derived from several independent sources including cytogenetic studies of chromosomal deletions or translocations; molecular evidence of chromosome-specific allelic loss as assessed by restriction fragment polymorphism studies; and somatic cell hybridization studies in which the introduction of a specific chromosome into tumor cells reverts the tumorigenic phenotype of the cells (1,4,61). The most frequent chromosomal deletions observed in astrocytic tumors include losses on chromosomes 9, 10, and 17, suggesting the presence of tumor suppressor genes on these chromosomes (62,63). The genomic alteration associated with chromosome 17 to a large extent represents mutations in the p53 locus (described above), although a distal locus has also been implicated as in breast cancer and possibly in some astrocytomas. The losses on the short arm of chromosome 9 are found mainly in anaplastic astrocytomas and glioblastomas, suggesting that this deletion may be an intermediate event in malignant progression. Alternatively, deletions on chromosome 10 are almost exclusively associated with the high-grade glioblastomas (Figs. 13.1 and 13.2). Specific genes within these regions of loss have not been identified, although they are the objects of intensive investigations. Less frequent allelic losses (less than 30%) have been observed in astrocytic tumors on chromosomes 4, 11, 13, and 22 (63–65). Historically, alterations to chromosome 22 observed in meningiomas were the first consistent chromosomal alteration observed in solid tumors.

Homozygous deletions on 9p have been observed to occur very near or at the interferon-α and -β gene locus (66–69). Similar deletions at 9p21 have been reported to occur in several other human cancers (67). However, replacement of the interferon protein products fails to revert the phenotype of glioma cells, suggesting the putative suppressor gene maps near this region and is not the interferon genes themselves (68). These recent reports have shown that the interferon genes are lost in approximately 60% of the gliomas examined, but the area of consistent loss was toward the centromeric end of the interferon gene locus. Although these results suggest that the interferon genes probably do not represent the tumor suppressor genes involved in the majority of these cancers, the findings do suggest that other biologically significant genes besides tumor suppressor genes themselves may be lost or altered during malignant progression. The altered expression or activity of these other gene products may influence the biological behavior of the tumor cells. Alternatively, the proximity to the interferons may suggest an interferon-related transcriptional control gene is involved in the suppression.

Deletions associated with chromosome 10 are the most prevalent genetic alterations in glioblastomas. Several studies indicated that slightly more than two-thirds of the tumors examined had lost an entire copy of chromosome 10, while about one-fourth of the tumors examined had partial losses on chromosome 10, and less than 10% had no observable losses (63,64,70). These deletions on chromosome 10 are therefore strongly implicated in the acquisition by astrocytic cells of the malignant phenotype observed in glioblastomas. In addition, the amplification of EGF receptor was shown to occur almost exclusively in tumors that had deletions in chromosome 10 (71). The extensive losses also suggest the possibility of the presence of more than one tumor suppressor gene on chromosome 10. The loss of an entire copy of the chromosome has made localization of the putative suppressor gene by molecular methods difficult because no region of consistent loss can be identified (i.e., the losses are too extensive to establish a starting point for the search). Deletions on the long arm, particularly associated with 10q24, have also been reported for prostate, renal, and endometrial cancers (72–74).

One strategy to determine whether a relevant tumor suppressor gene is located on an implicated chromosome is to transfer a normal copy of the suspected chromosome into appropriate tumor cells and then examine the tumorigenicity of the parental and hybrid cells (61). A loss of the tumorigenic ability of the hybrid cells compared with the parental cells would strongly suggest the presence of some form of tumor suppressor gene. The chromosome is transferred by a procedure that is similar to whole cell fusion, except microcells are formed from the donor cells, containing an individual chromosome or several chromosomes enveloped in a small piece of the cell membrane. The microcells are then fused to recipient cells using standard techniques of membrane fusion. Recipient cells containing the chromosome of interest are then selected for by using an endogenous selectable gene on the transferred chromosome or by tagging the chromosome with a selectable gene, such as the gene conferring neomycin resistance, prior to transfer (61,75–78). Consequently, only hybrid cells containing the transferred chromosome are selected and characterized. This technology has several advantages in that a suspected genome region or chromosome can be shown to be functionally important in the behavior of the tumor cells. In addition, the chromosome can be fragmented so only specific regions are transferred into the recipient cells, allowing for localization of the suppressor gene (76).

Using these procedures, we generated hybrid glioma cells containing an introduced normal chromosome 10. The hybrid cells containing the chromosome 10 exhibited a dramatic reversion of the tumorigenic phenotype in vitro and in vivo. The hybrid cells containing chromosome 10 were unable to grow in soft agarose, a classic assay for transformation (78). The cloned cells also showed a decrease in their saturation density and an altered morphology to more closely resemble normal astrocytic cells. However, the exponential growth rate was not altered. Most significantly, the cell clones were unable to form tumors in nude mice. To ensure that the effect of chromosomal transfer was due to chromosome 10 sequences, chromosome 2 was also inserted into glioma cells, and no significant alterations between the parental cells and hybrids containing chromosome 2 were observed. These results strongly imply that chromosome 10 harbors a tumor suppressor gene which is intimately involved in the malignant progression of gliomas.

Using a similar strategy we have obtained strong evidence for the presence of two tumor suppressor genes on chromosome 10 (P. A. Steck, A. Hadi, W. K. A. Yung, H. C. Cheong, M. A. Pershouse, manuscript in press). Subclones were generated that contain only certain fragments of the reinserted chromosome 10. Clones containing an intact chromosome 10 or the short arm and part of the long arm (10q24 to 10q26) exhibited a "fully" suppressed phenotype with an inability to grow in soft agarose and in nude mice. However, other clones that contained only the short arm and centromeric regions of chromosome 10 regained their ability to grow in soft agarose, but still were unable to form tumors in nude mice. These results implicate two separate regions of chromosome 10 that may contain tumor suppressor genes, one on the long arm (10q24 to 10q26) and one on the short arm or near the centromere of chromosome 10. The possible presence of two suppressor regions would explain, at least in part, the frequent loss of an entire copy of chromosome 10 in glioblastomas.

Although the genes on chromosome 10 have not been identified, their biological activities differ from that observed with p53 gene product as previously described above. The chromosome 10 genes appear to mediate the cell's response to its neighbors or environment, as illustrated by the hybrid cells' responsiveness to density-dependent inhibition of cell growth (lower saturation densities) and lack of growth in soft agarose. In contrast, the expression of a normal protein product of the p53 gene results in a significantly decreased growth of the tumor cells, not observed in the chromosome 10 hybrids. Therefore, both gene products mediate loss of tumorigenicity but appear to do so by different mechanisms. This observation has several implications. First, as observed in other model systems, the correction of one genetic effect was sufficient to mediate the tumorigenic phenotype (77), although several defects are necessary for oncogenesis. Furthermore, the observed molecular progression observed in gliomas (Fig. 13.1) is accompanied by independent biological alterations, suggesting a possible and often suspected linkage between the molecular alterations and the biological phenotype of the cells. Consequently, the identification and cataloging of molecular alterations may provide an excellent prognostic indicator of tumor progression and behavior.

Several other chromosomal alterations have been identified that occur at relatively low but apparently nonrandom frequencies in astocytic neoplasms. These include alterations to chromosomes 4, 11, 13, and 22 at frequencies between 15% and 30% of the tumors examined (63,64). The significance of these genetic alterations is generally unknown; however, several different possibilities exist. One possibility is that these less frequent alterations complement the more frequent alterations to achieve a sufficient "accumulation of genetic damage." Alternatively, the genetic alterations may represent involvement of different cell types in the tumor. This may be an important issue in brain tumors as most of the accessory (nonneuronal) cells of the brain are derived from common precursor cells. For instance, the precusor O-2A cell can differentiate into type 2 astrocytes or oligdendrocytes depending on the growth factors present (79). Therefore, the exact origin of some of the astrocytic tumors is currently an important area of research.

Oligodendrogliomas account for about 5% of glial tumors and have a relatively good prognosis with a 5-year survival of approximately 80%. The most common cytogenetic abnormality among the oligodendrogliomas analyzed was deletion of portions of the long arm of chromosome 19 (80). One tumor was subsequently shown to have a homozygous deletion at the polymorphic locus D19S8, suggesting a tumor suppressor gene may reside near this locus. A subset of gliomas present as mixed histologic tumors with both astrocytic and oligodendroglioma characteristics. Consequently, loss of alleles on chromosome 19 is also observed in a small but consistent set of astrocytomas.

Ependymomas arise from cells that line the ventricles of the brain that separate the cerebrospinal fluid from the brain and spinal cord. Ependymomas also have a relatively good prognosis, with a 50% survival peroid of about 5 years. These tumors predominantly occur in children and young adults (<40 years). Although only a relatively few tumors have been studied, combined results from several studies suggest that ependymomas have relatively consistent abnormalities involving chromosomes 11 and 22, suggesting the presence of genes that may be involved in the oncogenesis of these tumors (80,81). Because of the heterogeneity and often mixed histology of brain tumors, these less frequent genetic alterations may prove to be genetic markers of the origin, and possibly of the prognosis, of various CNS tumors.

CONCLUSION

Analysis of molecular genetics, histology, and patient outcomes has strongly indicated an association between specific genetic abnormalities and the malignant phenotype of the neoplastic cells. Although advances have been made in the identification of certain genes involved in the oncogenesis of primary brain tumors, many of the affected genes still need to be cloned and characterized. Once the array of oncogenes and tumor suppressor genes involved in various CNS neoplasms has been characterized, a compilation of the roles and biological significance of the various genes and their protein products still lies ahead.

ACKNOWLEDGMENTS

This work was supported in part by National Institutes of Health grants RO1-CA56041 and PO1-CA55261 and by The Texas Neurofibromatosis Foundation.

REFERENCES

1. Foulds L. The natural history of cancer. J Chronic Dis 1958;8:82–37.

2. Knudson AG. Hereditary cancer, oncogenes, and antioncogenes. Cancer Res 1985;83:1437–1443.

3. Bishop JM. Molecular themes in oncogenesis. Cell 1991;64:235–248.

4. Fearon E, Vogelstein B. A genetic model for colorectal tumorigenesis. Cell 1990;61:759–767.

5. Jones PA, Taylor SM. Cellular differentiation, cytidine analogs and DNA methylation. Cell 1980;32:85–93.

6. Boring CC, Squires TS, Tong T. Cancer statistics, 1993. CA Cancer J Clin 1993;43:7–26.

7. Von Deimling A, Von Ammon K, Schoenfeld D, Wiestler OD, Seizinger BR, Louis DN. Subsets of glioblastoma multiforme defined by molecular genetic analysis. Brain Pathol 1993;3:19–26.

8. Libermann TA, Nusbaum HR, Razon N, et al. Amplification, enhanced expression and possible rearrangement of EGF receptor gene in primary brain tumours of glial orgin. Nature 1985;313:144–147.

9. Wong AJ, Bigner SH, Bigner DD, et al. Increased expression of the epidermal growth factor receptor gene in malignant gliomas is invariably associated with gene amplification. Proc Natl Acad Sci USA 1987;84:6899–6903.

10. Sugawa N, Ekstrand AJ, James CD, Collins VP. Identical splicing of aberrant epidermal growth factor receptor transcripts from amplified rearranged genes in human glioblastomas. Proc Natl Acad Sci USA 1990;87:8602–8606.

11. Wong AJ, Ruppert JM, Bigner SH, et al. Structural alterations of the epidermal growth factor receptor gene in human gliomas. Proc Natl Acad Sci USA 1992;89:2965–2969.

12. Humphrey PA, Wong AJ, Vogelstein B, et al. Anti-synthetic peptide antibody reacting at the fusion junction of deletion-mutant epidermal growth factor receptors in human glioblastomas. Proc Natl Acad Sci USA 1990;87:4207–4211.

13. Lund-Johnasen M, Bjerkvig R, Humphrey P, Bigner SH, Bigner DD, Laerum O-D. Effect of epidermal growth factor receptor on glioma cell growth, migration, and invasion in vitro. Cancer Res 1990;50:8017–8022.

14. Donato NJ, Rosenblum MG, Steck PA. Tumor necrosis factor regulates tyrosine phosphorylation on epidermal growth factor receptors in A431 carcinoma cells: evidence for a distinct mechanism. Cell Growth Differ 1992;3:259–268.

15. Fleming TP, Saxena A, Clark WC, et al. Amplification and/or overexpression of platelet-derived growth factor receptors and epidermal growth factor receptor in human glial tumors. Cancer Res 1992;52:4550–4553.

16. Yamaguchi F, Saya H, Bruner JM, Morrison RS. Differential expression of two fibroblast growth factor receptor genes is associated with malignant progression in human astrocytomas. Proc Natl Acad Sci USA 1994;91:484–488.
17. Cheon H-G, LaRochelle WJ, Bottaro DP, Burgess WH, Aaronson SA. High affinity binding sites for related fibroblast growth factor ligands reside within different receptor immunoglobulin-like domains. Proc Natl Acad Sci USA 1994;91:989–993.
18. Takahashi JA, Mori H, Fukumoto M, et al. Gene expression of fibroblast growth factors in human gliomas and meningiomas: demonstration of cellular source of basic fibroblast growth factor mRNA and peptide in tumor tissue. Proc Natl Acad Sci USA 1990;87:5710–5714.
19. Morrison RS, Gross JL, Herlin WF, et al. Basic fibroblast growth factor–like activity and receptors are expressed in a human glioma cell line. Cancer Res 1990;50:2524–2529.
20. Yung WKA, Zhang X, Steck PA, Hung MC. Differential amplification and rearrangement of TGF-alpha gene in human gliomas. Cancer Commun 1990;2:201–205.
21. Folkman J, Klagsbrun M. Angiogenic factors. Science 1987;235:442–447.
22. Wrann M, Bodmer S, de Martin R, et al. T cell suppressor factor from human glioblastoma cells is a 12.5-kd protein closely related to transforming growth factor-β. EMBO J 1987;6:1633–1636.
23. James D, Carlbom E, Nordenskjold M, Collins VP, Cavenee WK. Mitotic recombination of chromosome 17 in astrocytomas. Proc Natl Acad Sci USA 1989;86:2858–2862.
24. Fults D, Tippets R, Thomas A, Nakamura Y, White R. Loss of heterozygosity for loci on chromosome 17p in human malignant astrocytoma. Cancer Res 1989;49:6572–6577.
25. El-Azouzi M, Chung R, Farmer G, Martuza RL, Black PM. Loss of distinct regions on the short arm of chromosome 17 associated with tumorigenesis of human astrocytomas. Proc Natl Acad Sci USA 1989;86:7186–7190.
26. Eliyahu D, Raz A, Gruss P, et al. Participation of p53 cellular antigen in transformation of normal embryonic cells. Nature 1984;312:646–649.
27. Finlay C, Hinds P, Levine A. The p53 proto-oncogene can act as a suppressor of transformation. Cell 1989;57:1083–1093.
28. Finlay C, Hinds P, Tan T-H, Eliyahu D, Oren M, Levin AJ. Activating mutations for transformation by p53 produce a gene product that forms an hsc 70–p53 complex with an altered half-life. Mol Cell Biol 1988;8:521–539.
29. Lane D, Benchimol S. p53: Oncogene or anti-oncogene? Genes Dev 1990;4:1–8.
30. Baker S, Fearon E, Nigro J, et al. Chromosome 17 deletions and p53 gene mutations in colorectal carcinomas. Science 1989;244:217–221.
31. Nigro J, Baker S, Preisinger A, et al. Mutations of the p53 gene occur in diverse human tumor types. Nature 1989;342:705–708.
32. Mercer WE, Shields MT, Amin M, et al. Negative growth regulation in a glioblastoma cell line that conditionally expresses human wild-type p53. Proc

Natl Acad Sci USA 1990;87:6166–6170.

33. Baker SJ, Markowits S, Fearon ER, Willson JK, Vogelstein B. Suppression of human colorectal carcinoma cell growth by wild-type p53. Science 1990;249:912–915.

34. Chen PL, Chen YM, Bookstein R, Lee WH. Genetic mechanisms of tumor suppression by the p53 gene. Science 1990;250:1576–1580.

35. Casey G, Lo-Hsueh M, Lopez ME, Vogelstein B, Stanbridge EJ. Growth suppression of human breast cancer cells by the introduction of a wild-type p53 gene. Oncogene 1991;6:1791–1797.

36. Kyo E, Yokozaki H, Yanagihara K, et al. Reduced tumorigenicity and cell motility of a gastric carcinoma cell line by introduction of wild-type p53 gene. Int J Oncol 1993;3:265–271.

37. Donehower LA, Harvey M, Slagle BL, et al. Mice deficient for p53 are developmentally normal but susceptible to spontaneous tumours. Nature 1992;356:215–221.

38. El-Deiry WS, Tokino T, Velculescu VE, et al. WAF1, a potential mediator of p53 tumor suppression. Cell 1993;75:817–825.

39. Harper JW, Adami GR, Wei N, et al. The p21 cdk-interacting protein Cip1 is a potent inhibitor of G1 cyclin-dependent kinases. Cell 1993;75:805–816.

40. Sidransky D, Mikkelsen T, Schwechheimer K, Rosenblum ML, Cavanee W, Vogelstein B. Clonal expansion of p53 mutant cells is associated with brain tumor progression. Nature 1992;355:846–849.

41. Oren M. The p53 cellular tumor antigen: gene structure, expression and protein properties. Biochim Biophys Acta 1985;823:67–78.

42. Bruner JM, Moser RP, Saya H. Immunocytochemical detection of p53 in human gliomas. Mod Pathol 1991;4:671–674.

43. Bruner JM, Connelly JH, Saya H. p53 protein immunostaining in routinely processed paraffin-embedded sections. Mod Pathol 1992;6:189–194.

44. Chung R, Whaley J, Kley N, et al. TP53 gene mutations and 17p deletions in human astrocytomas. Genes Chromosom Cancer 1991;3:323–331.

45. Li FP, Fraumeni JF, Mulvihill JJ, et al. A cancer family syndrome in twenty-four kindreds. Cancer Res 1988;48:5358–5362.

46. Srivastava S, Zou ZQ, Pirollo K, et al. Germ-line transmission of a mutated p53 gene in a cancer-prone family with Li-Fraumeni syndrome. Nature 1990;348:747–749.

47. Melkin D, Li FP, Strong LC, et al. Germ line p53 mutations in a familial syndrome of breast cancer, sarcomas, and other neoplasms. Science 1990;250:1233–1238.

48. Kyritsis AP, Bondy ML, Xiao M, et al. Germline p53 gene mutations in subsets of glioma patients. J Natl Cancer Inst 1994;86:344–349.

49. Wallace MR, Marchuk DA, Andersen LB, et al. Type 1 neurofibromatosis gene: identification of a large transcript disrupted in three NF1 patients. Science 1990;249:181–186.

50. Cawthon RM, Weiss R, Xu GF, et al. A major segment of the neurofibromatosis type 1 gene: cDNA sequence, genomic structure, and point mutations. Erratum in Cell 1990;62:193–201.

51. Johnson MR, Look AT, DeClue JE, Valentine MB, Lowy DR. Inactivation of the NF1 gene in human melanoma and neuroblastoma cell lines without impaired regulation of GTP Ras. Proc Natl Acad Sci USA 1993;90:5539–5543.

52. Andersen LB, Fountain JW, Gutmann DH, et al. Mutations in the neurofibromatosis 1 gene in sporadic malignant melanoma cell lines. Nature Genet 1993;3:118–121.

53. Xu GF, O'Connell P, Viskochil D, et al. The neurofibromatosis type 1 gene encodes a protein related to GAP. Cell 1990;62:599–608.

54. Martin GA, Viskochil D, Bollag G, et al. The GAP-related domain of the neurofibromatosis type 1 gene product interacts with ras p21. Cell 1990;63:843–849.

55. Ballester R, Marchuk D, Boguski M, et al. The NF1 locus encodes a protein functionally related to mammalian GAP and yeast IRA proteins. Cell 1990;63:851–859.

56. Nishi T, Lee PSY, Oka K, et al. Differential expression of two types of the neurofibromatosis type 1 (NF1) gene transcripts related to neuronal differentiation. Oncogene 1991;6:1555–1559.

57. Suzuki Y, Suzuki H, Kayama T, Yoshimoto T. Brain tumors predominantly express the neurofibromatosis type 1 gene transcripts containing the 63 base insert in the region coding for GTPase activating protein-related domain. Biochem Biophys Res Commun 1991;181:955–961.

58. Mochizuki H, Nishi T, Bruner JM, Lee PSY, Levin VA, Saya H. Alternative splicing of neurofibromatosis type 1 gene transcript in malignant brain tumors: PCR analysis of frozen section mRNA. Mol Carcinog 1992;6:83–87.

59. Trofatter JA, MacCollin MM, Rutter JL, et al. A novel moesin-, ezrin-, radixin-like gene is a candidate for the neurofibromatosis 2 tumor suppressor. Cell 1993;72:791–800.

60. Rouleau GA, Merel P, Lutchman M, et al. Alteration in a new gene encoding a putative membrane-organizing protein causes neurofibromatosis type 2. Nature 1993;363:515–521.

61. Stanbridge EJ. Human tumor suppressor genes. Annu Rev Genet 1990;24:615–657.

62. Bigner SH, Mark J, Burger PC, et al. Specific chromosomal abnormalities in malignant human gliomas. Cancer Res 1988;48:405–411.

63. James CD, Carlbom E, Dumanski JP, et al. Clonal genomic alterations in glioma malignancy stages. Cancer Res 1988;48:5546–5551.

64. Fults D, Pedone CA, Thomas GA, White R. Alleotype of human malignant astrocytomas. Cancer Res 1990;50:5784–5789.

65. Ransom DT, Ritland SR, Moertel CA, et al. Correlation of cytogenetic analysis and loss of heterozygosity studies in human diffuse astrocytomas and mixed oligo-astrocytomas. Genes Chromosom Cancer 1992;5:357–374.

66. Olopade OI, Jenkins RB, Ransom DT, et al. Molecular analysis of deletions of the short arm of chromosome 9 in human gliomas. Cancer Res 1992;52:2523–2529.

67. Olopade OI, Bohlander SK, Pomykala H, et al. Mapping of the shortest region of overlap of deletions of the short arm of chromosome 9 associated with

human neoplasia. Genomics 1992;14:437–443.

68. Mikyakohi J, Dobler KD, Allalunis-Turner J, et al. Absence of IFNA and IFNB genes from malignant glioma cell lines and lack of correlation with cellular sensitivity to interferons. Cancer Res 1990;50:278–283.

69. James CD, He J, Carlbom E, Nordenskjold M, Cavenee WK, Collins VP. Chromosome 9 deletion mapping reveals interferon α and interferon β-1 gene deletions in human glial tumors. Cancer Res 1991;51:1684–1688.

70. Rasheed BK, Fuller GN, Friedman AH, Bigner DD, Bigner SH. Loss of heterozyosity for 10q loci in human gliomas. Genes Chromosom Cancer chromosome 10 in human glioblastoma multiforme. J Neurosurg 1992;77:295–301.

71. von Deimling A, Louis DN, von Ammon K, et al. Association of epidermal growth factor receptor gene amplification with loss of chromosone 10 in human glioblastoma multifurme. J Neurosurg 1992;77:295–301.

72. Carter BS, Ewing CM, Ward WS, et al. Allelic loss of chromosomes 16q and 10q in human prostate cancer. Proc Natl Acad Sci USA 1990;87:8751–8755.

73. Morita R, Saito S, Ishikawa J, et al. Common regions of deletion on chromosomes 5q, 6q, and 10q in renal cell carcinoma. Cancer Res 1991;51:5817–5820.

74. Simon D, Heyner S, Satyaswaroop PG, Farber M, Noumoff JS. Is chromosome 10 a primary chromosomal abnormality in endometrial adenocarcinoma? Cancer Genet Cytogenet 1990;47:155–162.

75. Fournier REK, Ruddle FH. Microcell-mediated transfer of murine chromosomes into mouse, Chinese hamster, and human somatic cells. Proc Natl Acad Sci USA 1977;74:319–323.

76. Dowdy SF, Fasching CL, Auaujo D, et al. Suppression of tumorgenicity in Wilms tumor by the p15.5-p14 region of chromosome 11. Science 1991;254:293–295.

77. Goyette MC, Cho K, Fasching CL, et al. Progression of colorectal cancer is associated with multiple tumor suppressor gene defects but inhibition of tumorigenicity is accomplished by correction of a single defect via chromosomal transfer. Mol Cell Biol 1992;12:1387–1395.

78. Pershouse MA, Stubblefield E, Hadi A, Killary AM, Yung WKA, Steck PA. Analysis of the functional role of chromosome 10 loss in human glioblastomas. Cancer Res 1993;53:5043–5050.

79. Bogler O, Wren D, Barnett SC, Land H, Noble M. Cooperation between two growth factors promotes extended self-renewal and inhibits differentiation of oligodendrocyte-type2 astrocyte (O-2A) progenitor cells. Proc Natl Acad Sci USA 1990;87:6368–6372.

80. Ransom DT, Ritland SR, Kimmel DW, et al. Cytogenetic and loss of heterozygosity studies in ependymomas, pilocytic astrocytomas, and oligodendrogliomas. Genes Chromosom Cancer 1992;5:348–356.

81. Von Deimling A, Louis DN, Von Ammon K, Peterson I, Wiestler OD, Seizinger BR. Evidence for a tumor suppressor gene on chromosome 19q associated with human astrocytomas, oligodendrogliomas, and mixed gliomas. Cancer Res 1992;52:4277–4279.

Index